VENOMOUS
REPTILES

VENOMOUS REPTILES

OF NORTH AMERICA

CARL H. ERNST

SMITHSONIAN INSTITUTION PRESS
Washington and London

This book is dedicated to Dr. Roger W. Barbour,
teacher, colleague, and friend, under whose
tutelage I learned my profession

Library of Congress Cataloging-in-Publication Data
Ernst, Carl H.
Venomous reptiles of North America / Carl H. Ernst
p. cm.
Includes bibliographical references and index.
ISBN 1-56096-114-8
1. Poisonous snakes—North America. 2. Gila
monster. I. Title.
QL666.O6E77 1992
597.96'0469'0973—dc20   91-3535

British Library Cataloguing-in-Publication Data is
available.

The paper in this publication meets the minimum
requirements of the American National Standard for
Permanence of Paper for Printed Library Materials
Z39.48-1984.

Manufactured in the United States of America.

Front cover: *Crotalus horridus horridus*, yellow phase.
Photo by Roger W. Barbour.

Color illustrations printed in Hong Kong by
Blaze International Productions.

10 9 8 7 6 5 4 3 2
99 98 97 96 95 94 93

NOTE: The location of the specimen shown in Plate
2 is the Salt Lake City Hospital.

Contents

Preface

Venomous reptile populations are decreasing at an alarming rate in the United States and Canada. While some may consider this a blessing, herpetologists and other naturalists see the decrease as further evidence of human failure to maintain an ecological balance between nature and our use and development of the land. If this trend continues, some species may be extirpated, a great loss to the North American system of biological communities. In the near future, wise decisions regarding the welfare of the remaining populations of venomous reptiles will have to be made if they are to survive. Admittedly, they do pose a danger, but in our modern society, this danger is miniscule compared to the annual toll from automobile accidents, for example. To conserve these beasts, we must have adequate knowledge of their life histories in order to identify critical facets of their ecology and behavior. It is the purpose of this book to present the current knowledge of the biology of each species living north of Mexico, and to point out critical gaps in this knowledge worthy of further study.

The literature of venomous reptiles is enormous. It includes hundreds of specialized papers on such topics as morphology, physiology, and biochemistry. My interests, however, lie principally in natural history; I have excluded technical papers that do not have a direct bearing on the life of the reptile in the wild. This is especially true when discussing venoms, on which the literature may become totally confusing to one lacking the necessary biochemical background. My discussions of venoms are general, as are most of the references I cite. Readers deeply interested in this topic are referred to the journal *Toxicon*. In the preparation of this book, I have examined over 1,000 original papers on venomous reptiles and have combined my own observations with pertinent parts of them. With few exceptions, I have listed only those articles and books that have appeared since 1955; older ones can be found listed in Wright and Wright (1962) and Klauber (1972). The cutoff date for listing references in the bibliography was September, 1991.

Skeletal features are sometimes listed among the characteristics of families, genera, and of snakes in general. It is not within the scope of this book to present detailed definitions or descriptions of these characters, and the reader is advised to consult either Romer (1956), Underwood (1967), Dowling (1959, 1975), or Marx and Rabb (1972) for details.

Also included are brief descriptions of the sex cycles of some species. For more detailed information on the stages of either oogenesis or spermatogenesis, the reader should consult a current textbook in general histology or embryology.

A number of persons have contributed in various ways to publication of this book. Ronald J. Crombie, J. Whitfield Gibbons, Steve W. Gotte, Jeffery E. Lovich, Roy W. McDiarmid, Robert P. Reynolds, and George R. Zug gave advice and encouragement. William S. Brown, David Chiszar, David Duvall, William H. Martin, Howard K. Reinert, Stephen M. Secor, and Robert T. Zappalorti supplied data or copies of unpublished manuscripts. Daniel D. Beck, Charles M. Bogert, William S. Brown, David Duvall, Harry W. Greene, William W. Lamar, Roy W. McDiarmid, William W. Palmer, George V. Pickwell, Howard K. Reinert, Robert P. Reynolds, Stephen M. Secor, and Richard A. Seigel reviewed portions of the manuscript and offered valuable suggestions for its improvement. Evelyn M. Ernst, Linda Trimmer, and Addison H. Wynn helped with photography and illustrations. My students Christopher W. Brown, Dale Fuller, Steve W. Gotte, Arndt F. Laemmerzahl, Jeffery E. Lovich, John F. McBreen, Steven W. Sekscienski, and James F. Snyder helped with the field collections and studies, and the following persons supplied specimens or photographs: Roger W. Barbour, Richard D. Bartlett, Ted Borg, Jeff Boyd, Christopher W. Brown, William A. Cox, Dale Fuller, J. Whitfield Gibbons, Ron Goellner,

Steve W. Gotte, David Hild, Jeffrey E. Lovich, Barry Mansell, John R. MacGregor, George V. Pickwell, Earl E. Possardt, John H. Tashjian, and Robert T. Zappalorti. Bonnie Contos, Evelyn Ernst, Cathy Kennedy, and Mary Roper patiently typed the early drafts of the manuscript.

Introduction

Venomous reptiles are an integral part of the natural history and folklore of North America. Yet, if trends continue as in this century, they may totally disappear from our wilds. Populations of the Gila monster, *Heloderma suspectum,* and the rattlesnakes *Crotalus horridus, C. willardi,* and *Sistrurus catenatus* have already been so severely reduced as to be officially listed as "threatened" or "endangered" by several states. Although natural events, such as brush fires, floods, or landslides may kill large numbers of these animals, human interference in some form is the usual cause for their decline.

The general deterioration of the natural environment has eliminated large populations in certain regions, and habitat destruction is probably the major cause of the decline in most venomous species. Swamps, marshes, and bogs have been drained, creating conditions unfit for the aquatic cottonmouth, *Agkistrodon piscivorus,* and the wetland-loving massasauga, *Sistrurus catenatus.* Woodlands have been cleared for additional farmland, or for the construction of highways, housing developments, shopping centers, office complexes, and the like, eliminating the habitat of forest dwellers such as the copperhead, *Agkistrodon contortrix,* and timber rattlesnake, *Crotalus horridus.* Rivers have been dammed and adjacent low-lying

1

woodlands and marshy ground flooded, thus eliminating the flood-plain terrestrial habitats of the copperhead, pygmy rattlesnake, *Sistrurus miliarius,* and possibly, the eastern diamondback rattle-snake, *Crotalus adamanteus.* Western grasslands have been put to the plow or overgrazed, destroying the habitat of the prairie rattle-snake, *Crotalus viridis,* and deserts have been irrigated and culti-vated, making them unsuitable for the Gila monster and several southwestern rattlesnakes.

The automobile has had a detrimental effect, as many venom-ous snakes are killed crossing highways each year. Some drivers seem to take cruel pleasure in running over snakes.

Although there is little direct proof of their harmful effects on venomous reptiles, insecticides and herbicides probably contribute to the decrease in populations. Large quantities of chlorinated hy-drocarbons (ingredients in many pesticides) surely are stored in the body fat in late summer and fall and could well cause the death of many snakes as they use this fat during hibernation.

Overcollecting and wanton killing are certainly factors. The ridgenose rattlesnake, *Crotalus willardi,* the red phase of the eastern pygmy rattlesnake, *Sistrurus miliarius miliarius,* and the Gila mon-ster have suffered severely through collection for the pet trade, and certain populations of the rock rattlesnake, *Crotalus lepidus,* may be seriously depleted in the future. With their relatively slow rate of maturation, our venomous reptiles cannot withstand heavy crop-ping and still maintain their populations. Many local populations of venomous snakes, particularly *Agkistrodon contortrix, Crotalus hor-ridus,* and *C. viridis,* have been purposely eradicated while at com-munal hibernacula, and upon discovery by humans, individual ven-omous snakes are more often killed than let alone. Several states still allow "rattlesnake roundups" which remove hundreds of snakes from the wild each year.

If venomous reptiles are to remain a significant part of our fauna, we must initiate conservation measures. Although we do not yet know enough about their biology to formulate an adequate conservation plan, certain needs are obvious. The waterways and lands harboring important populations should be protected from undue human disturbance and pollution. The trend away from the use of the dangerous residual pesticides must be continued. States must pass and enforce legislation controlling the capture of these creatures in the wild, and definitely must ban snake "roundups."

Equally important, more people must become acquainted with the many fascinating aspects of reptilian biology. Such awareness should make them more interested in the protection of these shy creatures. The creation of such an attitude, not only toward reptiles but also toward all our dwindling wildlife resources, is a major purpose of this book.

If we are to conserve the Gila monster and venomous snakes of the United States and Canada, a more thorough knowledge of their life histories is a necessity. When the following species accounts are examined, it becomes immediately apparent that relatively little is known about most species, particularly concerning their reproductive habits and physiology. While rattlesnake biology, in general, is well known because of the various studies by Gloyd and Klauber, only *Crotalus horridus*, *C. viridis*, and *Sistrurus catenatus* have been adequately studied. The biology of *Agkistrodon contortrix* and *A. piscivorus* has been published in monographs by Fitch (1960b), Burkett (1966), and Gloyd and Conant (1990), but new discoveries are still being made about their life styles. The Gila monster, *Heloderma suspectum*, was chronicled by Bogert and Martín del Campo (1956), Tinkham (1971a,b) and, most recently, Beck (1990). These six reptiles represent only 29% of the 21 venomous species living north of Mexico. Vital information needed to adequately plan management programs is missing. The second major purpose of this book is to summarize the known ecology and behavior of each species. Vital gaps in our knowledge are conspicuous in the accounts owing to the absence of data. It is hoped that herpetologists will take notice of these voids and design research programs to acquire necessary data. I was amazed when writing this book at the amount of knowledge accumulated on venom chemistry, as opposed to that on ecology and behavior. The same amount of effort devoted to these areas of study may give us the necessary information needed to prevent extinction of these animals.

Lizards and snakes belong to the subclass of reptiles known as Lepidosauria, which is characterized by a diapsid skull, one with two temporal openings. Together with the amphisbaenians, they are members of the order Squamata, but are placed into two separate suborders, Lacertilia (lizards) and Serpentes (snakes). In modern squamates the skull has been further modified from the ancestral diapsid form by the loss of the ventral bar closing the lower temporal fossa, thereby leaving only one temporal opening. Approximately 3,000 lizard species and 2,700 snake species live on earth.

Snakes have several skeletal characteristics that separate them from lizards (Romer, 1956), but generally can be told apart by their lack of limbs (and in most cases limb girdles), movable eyelids, or external ear openings.

The first lizards are from the Triassic Period. Snakes apparently evolved later in the Jurassic Period from lizardlike ancestors. Venomous lizards belong to the family Helodermatidae, which dates from the late Eocene of France and the late Paleocene to Recent of North America, and if the species *Paraderma bogerti* is a helodermatid, as contended by Pregill et al. (1986), the family may extend to the late Cretaceous of North America. The immediate ancestors of helodermatids were varanoid lizards.

Venomous North American snakes belong to the families Elapidae (cobras, coral snakes, mambas, sea snakes, and their allies) and Viperidae (vipers and pit vipers). The oldest known elapids are from the early Miocene of France and Spain. The oldest fossil viperids are from the early Miocene of Europe and North America. Pit vipers date back to the Pliocene of North America. The Elapidae and the Viperidae are thought to have evolved from the family Colubridae or its ancestors, but probably from different lines of colubrids.

Several characteristics are important in the identification of the venomous reptiles of the United States and Canada. The scales on the head may be enlarged (Figs. 1–5), and their arrangements and the numbers of each may be used for identification. Also, the body

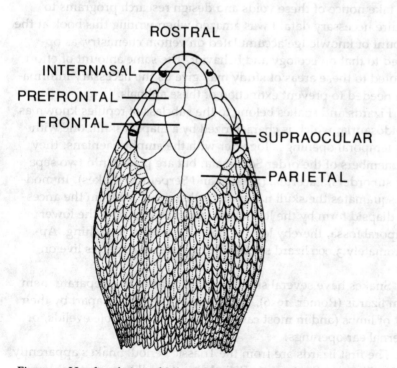

Figure 1. Head scalation of *Micruroides*, *Micrurus*, *Agkistrodon*, and *Sistrurus*, dorsal view (Evelyn M. Ernst).

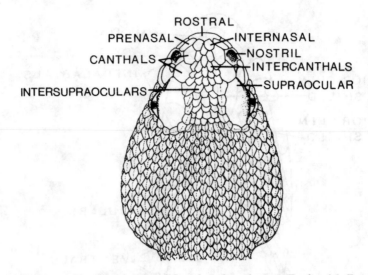

Figure 2. Head scalation of *Crotalus*, dorsal view (Evelyn M. Ernst).

scales may be either keeled (carinate, with a longitudinal ridge) or smooth (lacking a ridge), and the numbers of rows of dorsal scales at various points along the body may be distinctive (Fig. 6, but see Dowling, 1951a, and Kerfoot, 1969). Beneath, the belly scales (ventrals) of snakes are usually transversely elongated and the scales

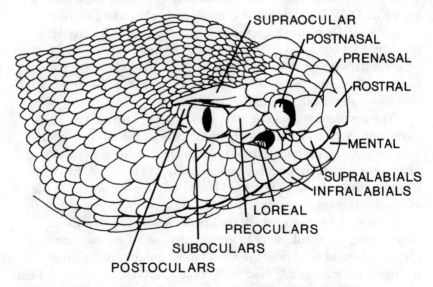

Figure 3. Snake head scalation, lateral view (Evelyn M. Ernst).

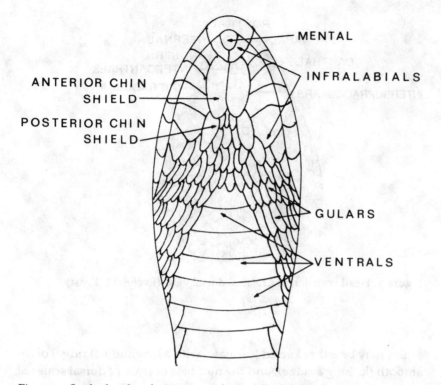

MENTAL

INFRALABIALS

ANTERIOR CHIN
SHIELD

POSTERIOR CHIN
SHIELD

GULARS

VENTRALS

Figure 4. Snake head scalation, ventral view (Evelyn M. Ernst).

underneath the tail (subcaudals) may occur in one or two rows. The
Gila monster has beaded scales completely encircling the midbody
region and tail. The number of ventrals and subcaudals are often
used to separate confusing species. Ventrals are counted from the
most anterior bordered on both sides by the first row of dorsal body
scales and thence posteriorly to the anal plate (Dowling, 1951b). The
anal plate covering the vent is single in some snakes, but subdivided
into two scales in others and the Gila monster.

The copulatory organ of male snakes and lizards is the
hemipenis, which lies in a cavity in the base of the tail that opens
into the cloaca. A retractor muscle is attached to the tip of the
hemipenis, connecting with a series of caudal vertebrae. Everted by
turgidity, the hemipenis is turned inside out to protrude from the
anal vent. It is traversed by a deep groove (sulcus spermaticus) on
the outer surface, along which the sperm flows. The hemipenis
may be single and columnar in appearance, or bifurcate with two
distinct lobes, at least near the tip. The sex organ is quite orna-
mented, and may be adorned with spines, calyces, or hooks, which
serve to anchor it within the female's cloaca. Each taxon has its

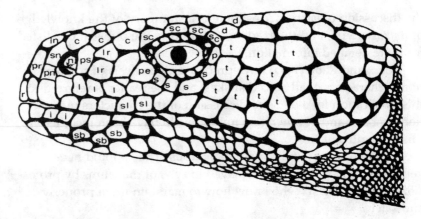

Figure 5. Head scalation of *Heloderma*, lateral view (Evelyn M. Ernst): *c*, canthal; *d*, dorsal; *i*, infralabial; *in*, internasal; *l*, lorilabial; *lr*, loreal; *n*, naris; *p*, postocular; *pe*, preocular; *pn*, prenasal; *pr*, prerostral; *ps*, postnasal; *r*, rostral; *s*, subocular; *sb*, sublabial; *sc*, superciliary; *sl*, supralabial; *sn*, supranasal; *t*, temporal.

own hemipenial type and ornamentation, and the organ is very important in taxonomic analysis. Wright and Wright (1957) and Dowling and Savage (1960) have analyzed and illustrated the variations in hemipenial structure in North American snakes. Branch (1982) has described that of the Gila monster.

Another important taxonomic character is the karyotype, the normal diploid number of chromosomes found in each body cell not producing gametes. This number may vary between genera, and the position of the centromere may vary between species. In

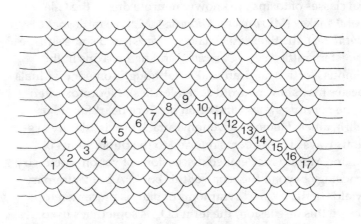

Figure 6. Counting snake scale rows (Evelyn M. Ernst).

the discussion of each species there is a summary of the knowledge of its karyotype, and the reader is referred to the publications listed for more detailed information.

Marx and Rabb (1972) have prepared an analysis of 50 taxonomic characters of advanced snakes, and the reader is referred to this publication for a detailed discussion of the characters mentioned above and of others too numerous to have been included in this brief introduction.

Venomous reptiles are very dangerous. They should never be kept at home, and are best displayed in zoos or museums by professional herpetologists who know how to maintain them properly and safely.

In North America, human envenomation is not as rare as one may believe. Annually, there are about 8,000 cases, by far the greatest majority involving snakes, with about 10–15 fatalities (Brown, 1987). While bites are not uncommon, fatalities are rare, but serious permanent damage may be incurred from a bite. Most of our bites come from the careless handling of these creatures.

Venom is a prey-immobilizing adaptation in snakes, used secondarily as a defensive mechanism. In the Gila monster, however, its primary function may be defensive (see discussion under *Heloderma suspectum*). Venom is not composed of a single substance common to all venomous reptiles; rather, each species has toxic saliva of at least slightly different properties causing different reactions in the prey. The closer the relationship between two venomous reptiles, the more similar are their venom properties and composition. Almost all venoms are composed of approximately 90% protein. These proteins are in the form of enzymes, and various combinations of these enzymes are found in the venom of each individual species. Thus, each venomous reptile usually contains more than one toxic principle, and these combine to produce the overall effect. Two general classes of toxins are known: neurotoxins in the Gila monster, coral snakes (*Micruroides*, *Micrurus*), Mojave rattlesnake (*Crotalus scutulatus*), the tiger rattlesnake (*C. tigris*), and some subspecies of the western rattlesnake (*C. viridis*); and hemotoxins in the other venomous North American snakes (although some individuals of these species may show neurotoxic tendencies). Neurotoxic venoms attack the nervous system and often produce heart failure or breathing difficulties. Hemotoxic venoms attack the circulatory system in various ways. The reader is referred to the publications of Russell, Tu, and Minton for detailed discussions of venom chemistry and pharmacology beyond the scope of this book. A brief summary of venom yields, lethality, and symptoms of bites is presented for each species. In this discussion, the term LD_{50} is sometimes used. This is the median toxicity level of venom, or that concentration of venom which will produce mortality in half of the subject animals injected.

A simple rule to prevent a bite from a venomous reptile is *do not handle it!* As stated previously, most bites occur while the reptile is being held or while one attempts to pick it up. Second, never play with these animals; remain at a sufficient distance (no closer than one snake body length) away from a striking animal. Do not pick up "dead" individuals—they may not yet be dead or they may be playing dead; reflex action may cause the jaws of truly dead individuals to open and close (a fatal envenomation from a decapitated head has been reported by Kitchens et al., 1987). Bites from unseen reptiles may be prevented through common sense and proper dress in the wild. Boots and coarse long trousers should be worn in areas where these creatures occur. Hands and feet should never be put in places that are not first examined, as such places may contain an unpleasant surprise. At night one's path should always be lighted.

If, by accident, you are bitten by a venomous reptile, medical care should be sought at once. As there is now serious doubt about the effectiveness of popular first-aid methods for bites, which may cause more harm than good, prompt medical treatment is the best and safest route. A few generally accepted first-aid measures (adapted from Campbell and Lamar, 1989) are: (1) keep the patient as calm as possible until medical help can be obtained, (2) immobilize the bitten limb, (3) avoid harmful procedures and medicines (tourniquets, cut and suck, cooling the site, ingesting alcohol or aspirin, pouring such things as turpentine onto the bite), (4) if either cardiac or respiratory arrest occur, perform standard cardiopulmonary resuscitation (CPR), and (5) if possible, prior to transport, establish an intravenous line with isotonic fluid (D_5W, normal saline, Ringer's lactate). Antivenin should usually not be injected in the wild, and, if possible, the reptile that has made the bite should be brought along to the medical facility for proper identification before symptoms become too severe.

Once at the medical facility, the bite victim should be observed for at least 24 hours. The bite area should be thoroughly cleaned, and antitetanus serum and a broad-spectrum antibiotic administered. Antivenin is the only proven treatment for envenomation, but should only be administered to persons with symptoms or signs of envenomation. For patients exhibiting hemotoxic effects, the crotalid polivalent antivenin produced by Wyeth International Ltd., Philadelphia, Pa., should be used. As this antivenin also includes antibodies for some neurotoxic rattlesnakes, it may be used for patients bitten by the Mojave rattlesnake (*Crotalus scutulatus*) or any other pit viper, who are exhibiting neurologic effects, particu-

larly respiratory or cardiac arrest. Symptomatic coral snake
(*Micruroides, Micrurus*) bites should be treated with North American
coral snake antivenin, also produced by Wyeth International.

The reader is referred to Campbell and Lamar (1989) for an
excellent summary of snake bite treatment.

Identification of Venomous Reptiles

The accompanying key to the venomous reptiles of the United States and Canada is designed to enable one to identify the animal at hand. The characters most often used are those of the head and body scalation (Figs. 1–6), color, and pattern of adults.

Within a population of reptiles any character will show individual variation; the larger the sample the greater the extremes. In most cases, however, a specimen will have the character as described in the key or at least can be placed within the middle range of measurements. Still, one occasionally encounters a reptile in which a character is quite different or in which a measurement falls outside the given range. For this and other reasons no key is infallible, so after one has arrived at a name by use of the key, the animal should be compared with the photographs and the description in the species account.

Difficulties in keying are most frequent when the reptile is abnormally colored or patterned (or sometimes, a preserved specimen). Then one should follow each alternative down to species and in each case refer to the photographs and text.

If the reader still cannot identify the specimen, it can be taken to the nearest natural history museum, zoo, or university biology department for identification.

KEY TO THE VENOMOUS REPTILES OF NORTH AMERICA

1a. Legs and/or movable eyelids present: a lizard 2
 b. Legs and/or movable eyelids absent: a snake 3
2a. Dorsum covered by separated, beadlike, granular scales; pattern salmon, pink, yellowish red, and black; all teeth grooved .... *Heloderma suspectum*
 b. Dorsum covered by rows of closely touching smooth scales or overlapping rows of rough scales, scales not beadlike; pattern not salmon, pink, yellowish red, and black; no teeth with grooves .... nonvenomous lizards
3a. A pitlike hole present between the nostril and eye 7
 b. No pitlike hole present between the nostril and eye 4
4a. Tail oarlike (flattened from side to side); nostrils valved: .... *Pelamis platurus*
 b. Tail not oarlike; nostrils not valved 5
5a. Body pattern of red, yellow or white, and black bands, **with the red bands separated from the black bands by yellow or white;** face black; a pair of permanently erect fangs at the front of the maxilla 6
 b. Body pattern not of red, yellow or white, and black bands, or if bands are red, yellow or white, and black, the red bands are separated from the yellow or white bands by black bands; face not necessarily black; no permanently erect fangs at the front of the maxilla .... nonvenomous snakes
6a. Most of head black, face black to level of angle of lower jaws; first broad body band behind the light collar red; usually one or two teeth on maxilla behind the fang .... *Micruroides euryxanthus*
 b. Only front of head black, face black to just behind the eyes; first broad body band behind light collar black; no teeth on maxilla behind the fang .... *Micrurus fulvius*
7a. No scaly rattle at tip of tail 8
 b. Scaly rattle present at tip of tail 9
8a. Head yellow-brown to reddish brown; body pattern of brown dumbbell-shaped blotches on a pinkish or grayish to orange- brown ground color; loreal scale present .... *Agkistrodon contortrix*

   b.   Head dark brown, black, or olive-brown; body pattern of dark- brown or black bars on olive or brown ground color, or no pattern present; loreal scale absent .... *A. piscivorous*

  9a.   Dorsal head surface covered with nine enlarged plates 10

   b.   Dorsal head surface covered with small scales or less than nine enlarged plates 11

 10a.   Prefrontal scales in broad contact with loreal scale; preocular scale does not touch the postnasal scale .... *Sistrurus miliarius*

   b.   Prefrontal scale not in contact with loreal scale; preocular scale touches the postnasal scale .... *S. catenatus*

 11a.   Supraocular scales raised or hornlike ..... *Crotalus cerastes*

   b.   Supraocular scales not raised or hornlike 12

 12a.   Tail unicolored black or dark brown, or dark gray with very faint bands 13

   b.   Tail not unicolored black or dark brown; tail bands usually distinct, but may be only confined to anterior portion of tail 14

 13a.   White scales occur within the dark dorsal blotches; usually six or fewer scales in the internasal-prefrontal region .... *C. molossus*

   b.   No white scales within the dark dorsal blotches; usually more than seven scales in the internasal-prefrontal region .... *C. horridus*

 14a.   Tip of rostrum sharply raised; two pale stripes extend backward from the nostril and mental areas to corner of mouth .... *C. willardi*

   b.   Tip of rostrum not sharply raised; no pale stripes on side of face, or two pale stripes extending diagonally backward from orbit toward corner of mouth 15

 15a.   Upper preocular scale subdivided, either horizontally or vertically 16

   b.   Upper preocular scale not subdivided 17

 16a.   No small scales separate the rostral scale from the prenasal .... *C. lepidus*

   b.   A series of small scales separates the rostral scale from the prenasal .... *C. mitchelli*

 17a.   Dorsal body pattern consists of dark crossbands, not blotches .... *C. tigris*

   b.   Dorsal body pattern consists of dark blotches or spots, not crossbands 18

 18a.   Dorsal body pattern consists of a series of paired, small, circular or elliptical, dark spots .... *C. pricei*

   b.   Dorsal body pattern consists of a series of large square or
        diamond-shaped, dark blotches 19
19a.   Dark tail bands decidedly narrower than the light tail
        bands; normally only two rows of intersupraocular scales
        present .... *C. scutulatus*
   b.   Dark tail bands wider than the light tail bands; normally
        more than two rows of intersupraocular scales present
        20
20a.   Tail bands poorly developed or faded; usually more than
        two internasal scales .... *C. viridis*
   b.   Tail bands well developed, not faded; usually only two
        internasal scales 21
21a.   Light tail bands white or cream-colored .... *C. adamanteus*
   b.   Light tail bands gray 22
22a.   Pink, red, or reddish brown; usually no dark speckles in
        dorsal body blotches; anterior infralabials usually trans-
        versely divided .... *C. ruber*
   b.   Gray or brown; dark speckles occur in dorsal body
        blotches; anterior infralabials not transversely
        divided .... *C. atrox*

Helodermatidae: Venomous lizards

Only two venomous lizards exist on earth today, and both are native to North America: the beaded lizard, *Heloderma horridum*, occurs in western Mexico and southern Guatemela, while the Gila monster, *Heloderma suspectum*, is found in the southwestern United States and northern Mexico. The fossil record for the family dates back to the late Cretaceous of North America and the late Eocene of Europe, and the species *Paraderma bogerti* Estes, 1964, is probably the ancestor of both present-day North American helodermatids (Pregill et al., 1986). The family is closely related to the Old World monitor lizard family Varanidae (Pregill et al., 1986).

The venom is neurotoxic, and, contrary to that of snakes, the delivery system is associated with the lower jaw. The dentary bone bears 8–10 long, recurved, pleurodont teeth that are grooved at least on the anterior surface and often on both the anterior and posterior surfaces. The infralabial glands are modified to produce the venom, which flows up the grooves as the lizard chews.

The *Heloderma* are relatively large heavy-bodied lizards, with *H. horridum* usually reaching an adult length of 60–70 cm, but occasionally growing to 1 m. Most adult *H. suspectum* are 25–40 cm in total length. The snout is bluntly rounded, and the lips bulge laterally. An ear hole is present and the eyelids are movable. The fat tail is short and blunt. Scales on the head, body, and limbs are rounded and beadlike with internal osteoderms. Scales at midbody occur in 49–62 transverse rows. On the short limbs, scales may slightly overlap. The toes are heavily clawed. Reproduction is oviparous.

Osteological characters of the family include a skull with a postorbital arch but no supratemporal arch, as the squamosal bone is reduced. The prefrontal and postfrontal bones touch dorsally to the orbit, but do not always prevent the frontal bone from entering the orbit. The lateroventral processes of the paired frontals meet along a suture beneath the olfactory lobe of the brain. The palatine and pterygoidal teeth are reduced, and no parietal foramen is present. The premaxilla is undivided, and the nasal bones are paired. The vomers are narrow and widely separated posteriorly. Both the palatines and pterygoids are also separated.

Heloderma suspectum Cope, 1869
Gila monster
Plates 1, 2

Recognition

The Gila monster is a large (maximum length, 56 cm), stout lizard with beaded yellow, pink, and black scales and a short, fat tail. The body is patterned with irregular dark-brown or black reticulations on a yellow to orange or salmon-pink background. This pattern changes with age, becoming more complex with a pattern consisting of five dark saddlelike bands, one on the neck and four between the front and hind legs, and 4–5 tail bands; however, lizards of populations living in areas of black lava may be darker. The broad head is black from the snout back to the orbits and laterally to at least the corner of the mouth. The chin and neck are black, and the legs and feet are usually predominantly black. The rounded body scales occur in 52–62 rows at midbody. The preanal scales are paired and enlarged. Dorsal head scalation includes a rostral, 2(3) postrostrals, 2 supranasals, 2(3) internasals, 3(1–4) canthals, 3(4–5) superciliaries (a small series of scales positioned dorsally to the orbit where normally a supraocular would occur), 6–8 dorsal scales lying medially between the posterior superciliaries; total scales between the internasals and occiput, 12–13(11–15). On the side of the head are single prenasal, nasal and postnasal scales, 2–4(5) loreals, 4–9 lorilabials (lying between the loreals and the supralabials), 2(1–3) preoculars, 1 postocular, 2–3 suboculars, 5–6(4–7) temporals, 11–12(10–14) supralabials (the first is in broad contact with the rostral), and 13–15(12–16) infralabials. Also present are 3–4 rows of sublabials on the lower jaw below the infralabials, as also are a wedge-shaped mental scale and 4–8 chin shields. The eye is black with a round pupil; the ear opening is a narrow oblique or ovoid slit. The limbs are stout with heavy claws. The hemipenis (from Branch,

1982) is single with an undivided sulcus spermaticus bordered by two folds (proximally it is composed of a single flap). No horns or lateral cups are present. The proximal third of the hemipenis is bare, distally 18–20 fine, soft, transverse flounces encircle the organ, but do not fuse with the folds bordering the sulcus spermaticus. The most distal 6–8 flounces are divided by a bare apical zone that extends as an inverted triangle for a short distance. The dental formula consists of the following teeth: premaxilla, 6–9; maxilla, 7–8; palatine, 0; pterygoid, 0(1–4); dentary, 8–10 (Bogart and Martín del Campo, 1956; Olson et al., 1986).

It is almost impossible to sex *H. suspectum* by external characters. Lowe et al. (1986) reported that the male has a larger and wider head, and that there are subtle differences in the abdominal shape of males and females. Tinkham (1971a) reported that females are slightly larger than males (this is true only of the subspecies *H. s. cinctum*; in *H. s. suspectum*, males are larger), have slightly longer foreclaws, and have more swollen and shorter tails than males. A bulge in the lateral postanal areas on the ventral side just behind the vent is evidence of the hidden hemipenis in males. In the male, the median two scales in each of the first two rows of preanal scales are large and quadrate; these four large median scales are replaced by only two in the female, and they extend into and divide the second anterior row of scales crossing the preanal area. However, these characters are subjective (particularly those regarding the preanal scales, which may be worthless in determining sex), must be used in combination, and seldom lead to a true sex determination. Sexing done by probing during zoo breeding programs has sometimes been successful. Probing is easily done with a long, thin, blunt lubricated rod inserted into the cloaca to-

ward the base of the tail; the rod should slip
into the uneverted hemipenis of males, and
thus penetrate deeper in males (35–50 mm)
than in females (21–31 mm) (Laszlo, 1975; Wagner et al., 1976), but this too does not always
work (Peterson, 1982). During the breeding season, hemipenes can usually be extruded from
adult males in the wild by palpation (Daniel
Beck, pers. comm.). Sex determination of some
individuals may have to be accomplished by
observation of copulation or egg laying.

Karyotype

Diploid chromosomes total 38: 14 macrochromosomes and 24 microchromosomes (Matthey,
1931a,b).

Fossil record

A Pleistocene (Rancholabrean) fossil of *H. suspectum* has been found in Gypsum Cave, Clark
County, Nevada (Brattstrom, 1954b). Yatkola
(1976) and Pregill et al. (1986) have discussed in
depth the fossil history of the family Helodermatidae, tracing it back to the Upper
Cretaceous.

Distribution

The Gila monster ranges from extreme southwestern Utah (Washington County), southern
Nevada (Clark County), and adjacent San
Bernadino County, California, southeastward
through western and southern Arizona to
southwestern New Mexico (Grant and Hildalgo counties), and south in Mexico through
much of Sonora and probably to northwestern
Sinaloa. There is also a record from El Dorado
in west-central Sinaloa that may represent a
disjunct population. The range includes altitudes from approximately sea level to about
1,500 m.

Geographic variation

Bogert and Martín del Campo (1956) recognized two subspecies. *Heloderma suspectum suspectum* Cope, 1869, the reticulated Gila monster, ranges from central (Yavapai County) and
eastern Arizona and southwestern New Mex-

Distribution of *Heloderma suspectum*.

ico southward into Mexico. Adults have the
dark body crossbands obscured by black mottled and blotched pigment that predominates
over the lighter pink or orange ground color
(juveniles have much more sharply defined
crossbands). The light bands on the tail also
contain dark spots or mottling; four to five
dark bands occur on the tail. *H. s. cinctum*
Bogert and Martín del Campo, 1956, the
banded Gila monster, ranges from southwestern Utah, southern Nevada, and adjacent western California, south in the Colorado River
drainage, extending into the Grand Canyon,
and south to Yuma County in southwestern
Arizona. It normally retains most of the juvenile pattern consisting of four well-defined
black saddlelike bands across the body between the fore and hind limbs. Light pink or
orangish pigmentation predominates. Little
dark pigment occurs in the light tail bands,
and there are five black tail bands.

Beck (1985) has reported that specimens of
H. s. cinctum from black basaltic areas in southern Utah contain much more dark pigment than
those from farther south. Funk (1966) reported
a Gila monster that he regarded as an intergrade between the two subspecies from
Yuma County, Arizona, and Daniel Beck has
informed me that intergrades are now commonly found.

Confusing species

No other lizard from the southwestern United States is patterned like *H. suspectum*, or has such beadlike scales.

Habitat

Heloderma suspectum usually lives in arid but not barren areas, particularly those with scattered cacti, shrubs, mesquite, and grasses, but may also be found in pine-oak forests, thorn forests, and irrigated lands. Rocky slopes, arroyos, and canyon bottoms (particularly those with streams) may support relatively dense populations in some parts of Arizona and Sonora.

Pianka (1967) thought *H. suspectum* is restricted to the Sonora Desert because it is a secondary carnivore and almost certainly dependent on the summer rains of this desert, with their concomitant predictable burst of warm-season production and breeding of its prey species. The Gila monster is thus directly dependent upon the length of the growing season and the predictability of warm-season production. Pianka, however, did not consider the Mojave Desert populations of *H. suspectum* that are subject to a more winter-rainfall-dominated regime.

Behavior

The year's activity usually starts in March, when the Gila monster emerges from hibernation, and extends into October, or rarely November. However, some Arizona Gila monsters may bask in January and February (Lowe et al., 1986). In Utah, 64% of the annual activity and 77% of the total distance moved occur between April and early July (Beck, 1990). The maximum days per month that any lizard is surface active is 14. During the active period as much as 97% of the time is spent in shelters and less than 13% of the energy budget is spent on surface activity. In the spring and fall it is mostly diurnal, but the summer heat forces the lizard to become crepuscular or nocturnal from June through September (contrary to Lowe et al., 1986). In Arizona, most are surface active for a total of no more than 190 hours per year, involving a maximum of approximately 115 days; the Gila monster spends about 98% of the year undercover (Lowe et al., 1986).

In Utah, 68% of daily activity occurs between 0830 and 1230 hours (Beck, 1990). Nocturnal activity may also occur in the summer, but *H. suspectum* is not as nocturnal as reported by Smith (1946) and Bogert and Martín del Campo (1956), who stated that it was both diurnal and nocturnal. Beck (1990) observes that reports are biased by the time of day investigators devote to their field studies. In laboratory tests conducted by Lowe et al. (1967), *H. suspectum* showed essentially the same activity pattern regardless of whether subjected to either constant light or constant darkness.

Body temperatures of nondormant animals recorded by Beck (1990) show that in Utah *H. suspectum* spends more than 83% of the year at 25°C or less, and over 50% of the year at or below 20°C. Mean monthly body temperatures (excluding those during activity) range from 12.3°C (December) to 28°C (July). Body temperatures of those in shelters are significantly correlated with surrounding air and soil temperatures. Body temperatures during activity average 29.3°C (17.4–36.8), and activity occurs at air temperatures of 10–34°C and at substrate temperatures of 20.5–32.0°C. Basking occasionally occurs, and one Gila monster observed in late April and early May maintained a mean body temperature of 28.5°C in this way. Bogert and Martín del Campo (1956) reported a mean body temperature of 28.7°C for three captives in a cage with a gradient in heat lamps.

Heloderma suspectum in Arizona may emerge to bask from January to March with body temperatures as low as 12.7°C (Lowe et al., 1986). The mean preferred range in deep body temperature of active Arizona lizards is 22–34°C (mean, 29), and active individuals have been taken with temperatures as low as 16°C and as high as 37°C, but few allow their bodies to heat above 33.8°C without seeking shelter (Lowe et al., 1986).

Based on data from Bogert and Martín del Campo (1956) and information from Warren, Brattstrom (1965) reported a mean activity body temperature of 27.2°C, and minimum and maximum voluntary temperatures of 24.2°C and 33.7°C, respectively. His calculated critical minimum and maximum temperatures were −3.0°C and 48.0°C (although he questioned the latter). It may be noted that Bogert and Martín del Campo (1956) reported that a body temperature of 44.2°C produced potentially lethal paralysis.

Gila monsters have unusually low metabolic rates. The Q_{10} for the metabolic rate between 20 and 25°C is 2.9 (Beck, 1990). A reduced metabolic rate at lower body temperatures provides a significant energy saving during periods of inactivity.

Few are seen above ground in southwestern Utah from late November through February when cold ambient temperatures cause these lizards to seek hibernacula. Typical overwintering sites are animal burrows or rock crevices, preferably on south-facing slopes where the lizard may bask at the entrance on warm winter days. A chamber may be excavated at the end of these holes, and several may hibernate in the same burrow. The same site may be used during estivation. Gila monsters suffer little from the several months of fasting, living instead on fats deposited during the active period (Bogert and Martín del Campo, 1956). Winter body temperatures recorded by Beck (1990) were 11.3–15.2°C, while soil temperatures at 75-, 25- and 5-cm depths were 7.0–15.3°C, 2.0–16.5°C, and −2.5–30.5°C, respectively.

Shelters used by *H. suspectum* in Utah were in rocky areas (59% in loose Navajo sandstone, 41% on basaltic lava slopes or flows). Desert tortoise (*Gopherus agassizi*) burrows and wood-rat (*Neotoma lepida*) mounds were used 8% and 10% of the time, respectively. Sixty-seven percent of the entrances faced east, southwest, or south; the few facing northwest or northeast (2%) were used only in the summer, and all hibernacula faced south. One lizard dug inside its shelter for a total of seven hours during late April and early May, so the amount of digging may be significant (Beck, 1990).

Water availability seems to be a critical problem for *H. suspectum*, as captives do not thrive if not provided free access to water for drinking and soaking. In the field, Gila monsters avoid moisture loss by spending most of their time in burrows, where the relative humidity is nearly always high. Pianka (1967) believed the Gila monster to be restricted to areas of periodic but not infrequent rainfall. Gila monsters are often active after summer rains. It was previously thought that most of their water is obtained from prey, but *H. suspectum* will drink from puddles if available. Also it is now known that its beaded skin is permeable to water, and that it may have quite high rates of water loss or absorption through the skin (Daniel Beck, pers. comm.). During the driest stress periods (July–September) in Utah and other Mojave Desert areas that do not receive significant summer rains, it may remain in its burrow or become more crepuscular or nocturnal. While sleeping or estivating it may even lie on its back (Kauffeld, 1943a).

Home ranges of two males and one female tracked in Utah by Beck (1990) were 66.2, 32.6, and 5.6 ha, respectively. Lizards traveled from only a few meters around their shelters to more than 1 km (mean, 213 m). The rate of travel was about 0.25 km/h, with an average duration of about 51 minutes per bout. The greatest total distances traveled were 1,190 m (15–30 April), 3,555 m (May), 3,150 m (June), and 2,000 m (1–15 July).

Heath (1961) attached trailing devices to nine Gila monsters and followed them for 17 days. When on the move the lizards averaged 215 m per day (almost identical to that noted by Beck, 1990). The distance crawled daily varied from 27 to 457 m, and the most rapid sustained movement was 320 m in three hours. However, movements while foraging are slow with intermittent periods of exploration and inactivity, according to Jones (1983), who also reported increases in the mean daily movement in April and May (mean for both sexes 26.5 m; males 21.7 m, females 32.4 m) and hence higher than in June and July (mean 2.8 m; males 3.1 m, females 2.4 m).

Maximal aerobic speed (MAS) is 0.70 km/h at 25°C and 1.03 km/h at 35°C (John-Alder et al., 1983), but endurance declines as speed is increased toward MAS. Gila monsters are unusual lizards in being unable to engage in brief periods of high-speed burst locomotion, but can sustain slow movement for long periods.

Gila monsters rarely climb above ground, but Cross and Rand (1979) saw two climb 90 cm and 2.5 m up desert willows on separate occasions, and Campbell and Lamar (1989) have observed them as high as 5–7 m in trees.

Agonistic interactions between male *H. suspectum* have been observed in captivity by

Demeter (1986) and in the wild by Lowe et al. (1986) and Beck (1990). According to Lowe et al. (1986) male combat in Arizona is restricted to the period April–July. Beck's (1990) analysis of a three-hour bout between two large males revealed similarities with varanid lizard and pit viper combat. Nine of the 10 major behavioral acts in combat interaction between two captive male *Heloderma suspectum* recorded by Demeter (1986) were observed by Beck who watched two males fight in the field: dorsal straddle, frontal head nudge, lateral head shove, neck arch, tail wrap, lateral head bite, lateral tail thrash, dorsal head pin, and roll. Seven additional major behavioral acts were also identified: head raise (raising of head and stiffening of front limbs by inferior lizard following a neck arch, done in response to head nudge and shove by the other lizard), circling (moving in a semicircular path around another lizard), body twist (while in dorsal straddle, a twisting of body by the inferior lizard and placing its gular region against the neck of the superior lizard so that the bodies entwine; the superior lizard usually responds with a neck arch), lateral rocking (rocking motion from side to side while in dorsal straddle, often resulting in a roll initiated by the inferior lizard; this apparently serves to force the separation of the superior lizard), dorsal body press (turning and pressing the dorsal body surface of the superior lizard against the back of the inferior lizard; usually performed under boulders where the superior lizard pressed the inferior lizard downward by pushing against a boulder with the forelimbs), high stand (standing side by side, each lizard performs a head raise), scoop (pressing the snout under another lizard, scooping it upward).

A typical bout lasts 10(4–15) minutes, but several bouts may be included, extending over three hours (Beck, 1990). Upon approach, the lizards perform head nudge, shove, neck arch, and head raise, often switching roles; circling sometimes precedes these actions. The aggressor then mounts the other male in a dorsal straddle. The lizards repeatedly intertwine tails, untwining them at intervals. The inferior lizard typically responds to a dorsal straddle with a neck arch, while the superior lizard performs a dorsal head pin. The superior lizard may also thrash its tail. Considerable struggling may take place during the next phase.

The inferior lizard often walks with the superior lizard clinging to its back. Lateral rocking sometimes separates the lizards but, if not separated, they usually proceed into a body twist. Finally, the inferior lizard initiates a body twist, and two males remain in this position until one gains the superior position in a dorsal straddle or sometimes they break contact, from the force exerted during twisting.

Apparently, the objective during each bout is to gain and maintain a superior position during a dorsal straddle. The inferior lizard either rolls or twists its body in an effort to break the dorsal straddle and gain the top position.

Male combat suggests that Gila monsters have a definite social system (Beck, 1990). These usually solitary animals are not roaming aimlessly when searching for potential mates or foraging sites. Beck (1990) thought that common shelter use and seasonal movements that bring individuals back to communal areas, establishment of dominance through male-male combat, and scent marking are all elements of a structured social system. During combat, *H. suspectum* may approach the limit of its physical endurance (Beck, 1990), especially while performing dorsal straddles, tail thrashes, and body twists.

Reproduction

The exact size and age of sexual maturity is unknown, but those copulating and ovipositing in captivity have all been over 40 cm in total length (27 cm snout-vent length) (Peterson, 1982).

The sexual cycles have also not been adequately described. Hensley (1950) examined a female preserved in October that contained five large (diameter 35–37 mm) and a series of 25 additional smaller (1.0–7.5 mm) ovarian eggs. As the female was collected in May and had not been allowed to breed, Hensley thought the larger eggs represented those that would normally have been fertilized and laid that year, and that the larger (6.8–7.5 mm) of the smaller groups of eggs presumably would have matured and been laid the next year. Three females collected in April and May by Gates (1956b) contained 40–52 undeveloped eggs, the largest being 5.5 mm long. A female at the Seattle Zoo laid eggs on two successive years (Wagner et al., 1976).

Ortenburger (1924) observed a July copulation in the field, and Lowe et al. (1986) concluded that in Arizona mating takes place principally in May. Captives have mated in January, March to June, September, and December (Gates, 1956b; Shaw, 1964; Wagner et al., 1976; Peterson, 1982). Copulation lasts from 30 minutes to over an hour, but most pairings take 40–60 minutes. The male initiates courtship by moving about with much tongue flicking, apparently seeking the female's scent, and rubbing his cloaca on the ground in a sideways motion (Wagner et al., 1976). When a female is located he lies beside her and rubs his chin against her neck and back while holding her with his hindleg. She may resist and try to bite him while crawling out from underneath, but the male does not bite her. If receptive, she raises her tail and exposes her vent. The male then moves his tail beneath hers, brings their vents together, and inserts his hemipenis. Copulatory movements are slow and convulsive, occurring every 3–4 seconds (the mating posture is illustrated in Gates, 1956b). Mating is a much subdued activity when compared to male combat.

In the wild, clutches of 1–12 eggs (usually 4–6) are laid in July and August (Stebbins, 1985; Lowe et al., 1986). Ditmars (1936) reported 13 eggs as the maximum clutch, but this has never been substantiated. Captive incubation periods have lasted from 117 to 130 days (Shaw, 1964), but average about 125 days (Peterson, 1982). Lowe et al. (1986) reported that in Arizona the eggs overwinter underground and hatch the following May, after a natural incubation period of about 10 months. They probably mean that the young hatch, but remain in the nest cavity overwinter and emerge the next spring. By Lowe et al.'s calculations this would mean an approximately one year period between copulation and hatching of the eggs, which seems much too long in light of the considerably shorter incubation periods recorded in captivity. The eggs are elongate with white, rough, leathery shells: mean length 65.2 mm (55–75), mean diameter 33.3 mm (29–39), and mean weight 38.8 g (31.1–49.6). When first laid, they are shiny and moist, but with slow drying become soft and leathery in 15–20 minutes (Lowe et al., 1986).

Newly hatched young at the Seattle Zoo averaged 158 mm in length (presumably total body length) and 32.5 g in weight (Wagner et al., 1976); young produced at the San Diego Zoo were 150–164 mm in total body length and weighed 28.1–32.9 g (Shaw, 1968); and in Arizona, neonates averaged 165.3 mm in total length and 32.7 g in weight (Lowe et al., 1986). Each had an extremely large amount of yolk in their abdominal cavities. Several days may be required between the pipping of the egg and the final emergence of the neonate. The young are usually lighter in color than adults.

Growth and longevity

Woodin (in Shaw, 1968) reported young brought to the Arizona–Sonora Desert Museum were 183–185 mm long in late spring (5 May–4 June), 172–181 mm on 20 July, and 205 mm on 19 August. Since he never saw newly hatched Gila monsters in the fall, he thought these lengths represented first-year growth since emergence from the nest. Bogert and Martín del Campo (1956) and Tinkham (1971a) reported captive growth rates of 7–10 mm per year for adults. Growth in the wild is also slow in adults. Beck (1990) recorded average growth rates of 4.8 mm per year in lizards initially under 300 mm in snout-vent length, and 2.1 mm per year in larger individuals.

A Gila monster lived 32 years, 2 months in captivity at Ball State University (Cooper and List, 1979).

Food and feeding

A number of foods have been listed for the Gila monster, and there have been at least a few actual observations on natural prey species. Ortenburger (1924) reported one had eaten a lizard (*Cnemidophorus*), Hensley (1949) observed one disgorge cottontail remains (*Sylvilagus*), and Shaw (1948b) and Stahnke (1952) recorded young ground squirrels (*Spermophilus*) from others. Hensley (1950) found what he thought were the remains of desert tortoise eggs (*Gopherus agassizi*) in still another lizard. The only foods recorded by Jones (1983) during a study in foraging ecology were the eggs of ground-nesting Gambel's quail (*Lophortyx*) and mourning doves (*Zenaida*). Bogert and Martín del Campo (1956) also listed young jack rabbits (*Lepus*) as natural prey.

Lowe et al. (1986) thought Gila monsters from Arizona to be carnivore specialists on the newborn young of rodents and lagomorphs, and that these basic foods are opportunistically augmented with birds and lizards and the eggs of birds, lizards, snakes, turtles and tortoises. They also noted that young Gila monsters can consume over 50% of their own body weight at a single feeding, and that adults consume 35%. If an adult can eat 3–4 meals in the spring, it can probably store enough fat to last it until the next spring.

The diet in Utah (based on four direct observations and 20 fecal samples) consisted of young cottontails, 42%; desert tortoise eggs, 29%; young ground squirrels (*Ammospermophilus leucurus*), 16%; mourning dove eggs, 8%; and the carrion of young kangaroo rats (*Dipodomys*), 4% (Beck, 1990). Female desert tortoises sometimes defend nest areas against Gila monsters (Barrett and Humphrey, 1986).

Captives will readily consume eggs or canned dog food mixed with raw eggs. Other foods that have been eaten in captivity include house mice (*Mus*, Hensley, 1949), a dead hamster (*Mesocricetus*, pers. obs.), and a collared lizard (*Crotaphytus*, Stebbins, 1954).

Large quantities of food may be consumed at one time (Lowe et al., 1986; Beck, 1990), probably in response to the difficulties in finding prey at regular intervals in its severe habitat. Prey is rarely if ever envenomated and, in fact, is commonly just overpowered and crushed in the jaws when seized and promptly swallowed. This supports the statements by Bogert and Martín del Campo (1956) and the suggestion by Beck (1990) that the role of the venom delivery system is primarily defensive.

In most cases, prey is detected by olfaction, and foraging Gila monsters frequently flick their tongue. Cooper (1989) tested the use of chemical senses by *H. suspectum* while foraging. The lizard flicked the tongue more often in response to house mouse (*Mus*) odor than to the odors of two control substances, but the number of tongue flicks elicited was no greater for the mouse odor than for the control odors. Nevertheless, the lizard bit in a significantly greater proportion of tests with prey odors than it did with control stimuli.

Nine Gila monsters observed from 1 April to 13 July by Jones (1983) foraged mostly in April and May in areas where eggs of ground-nesting birds were abundant, but egg availability decreased in June and July. Thereafter, the lizards shifted their diets to small mammals and moved to areas where they were fairly plentiful. The Gila monsters rarely backtracked, and consumed an average of 46% (36–83) of the eggs found in each nest. Eggs were either swallowed whole or the contents were partially eaten after the egg shells were broken. No lizard foraged at the same nest twice, and only once did one consume all of the eggs available at a nest. The lizards ate an average of 1.3 eggs per day, foraged on an average of 3.3 eggs per nest, and raided an average of 0.4 nest per day. They spent an average of 11.4 minutes eating eggs, and males foraged on a greater number of nests (0.47) per day than females (0.32). The distance these *H. suspectum* traveled each day was positively correlated with the percentage of eggs eaten per nest in males, and there was also a positive relationship between lizard weight and the number of eggs consumed each day.

In contrast to Jones's (1983) observations, Beck (1990) reported that his Gila monsters spent 5–15 minutes excavating the nests of the birds incubating their eggs or protecting nestlings, and that all of the young or eggs found in a nest were consumed.

Venom and bites

The venom apparatus of *Heloderma suspectum* consists of 8–10 grooved teeth on the dentary (lower jaw), two anterior venom glands, and the ducts leading from the glands to the outer sides at the base of the grooved teeth. Each tooth has an anterior groove, and most contain grooves on both the anterior and posterior surfaces. The three anteriormost teeth are 1–3 mm long; those following are longer, 4–5 mm; the fourth through seventh teeth are longest. The venom is produced in a pair of enlarged inferior labial glands, the glands of Gabe, one on each side of the jaw, which are composed of several (usually 3–4) lobes made up of irregularly columnar cells with large granular nuclei (Tinkham, 1971b). A duct drains each lobe and conducts the venom to the mucous membrane on the outer surface of the gum at the base of the grooved tooth. Russell and Bogert (1981) noted that the venom duct is separated from the tooth pedestal by an outer wall of the cup-shaped dental sac, the mucous membrane fold investing the pedestal. The fold is unique in its

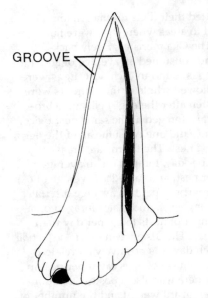

GROOVE

Dentary teeth of *Heloderma suspectum* (Evelyn M. Ernst).

structure and is thought to play an important role in the discharge of the venom and its transfer into grooves on the larger maxillary teeth. The venom is carried from the duct into the grooves on the dentary teeth by capillary action, and in some instances may mix with saliva from the upper jaw glands and perhaps be carried into grooves on the maxillary teeth, as well as those on the premaxilla.

The Gila monster does not strike like a snake, but can lunge quickly forward or turn its head sideways and bite if one gets too close. Once it has sunk its teeth into something it holds on with a bulldog grip (Arrington, 1930) and "chews" in the venom. Bogert and Martín del Campo (1956) reported one clinging to an automobile door handle for 15 minutes. If one keeps his distance from a Gila monster, and does not foolishly reach into crevices or under rocks where it may be hidden, or avoids handling Gila monsters, no bite should occur. These lizards are not aggressive unless provoked, but they are dangerous even though their crude venom apparatus evolved as a protective device. The venom is neurotoxic and electrophoretically similar to that of cobras

(Mebs and Raudonat, 1966). The venom consists mainly of proteins, of which the toxicity lies in those which are nondialyzable (venom chemistry is discussed in detail by Tu and Murdock, 1967; Tu, 1977; and Russell and Bogert, 1981).

Adult Gila monsters may yield 0.75–1.25 ml of venom (Brown and Lowe, 1955; Minton and Minton, 1969). The LD_{50} for a 20-g mouse is 80 ml (subcutaneous, Minton and Minton, 1969), and 1.35 mg per kg of body weight (intracardial) for rats (Patterson, 1967a). The lethal dose for a human adult is probably about half the total yield (0.38–0.63 ml) from an adult lizard, but this quantity is rarely injected in natural bites (Minton and Minton, 1969).

Human envenomation is rarely severe and few deaths have occurred; Woodson (1947) uncritically reported 29 (21%) fatalities in 136 cases, and Grant and Henderson (1957) 7 (29%) fatalities in 24 cases. Case reports are discussed in Englehardt (1914), Brennan (1924), Storer (1931), Duellman (1950), Shannon (1953), Bogert and Martín del Campo (1956), Grant and Henderson (1957), and Russell and Bogert (1981). Even the hatchlings can produce a bite that can be painful for several hours (Lowe et al., 1986). Symptoms are intense pain (in only a few mild cases, no pain), profuse bleeding of wound, severe edema, tenderness of lymph glands, slight ecchymosis about bite, shock, anxiety, general weakness, faintness or dizziness, nausea or vomiting, chills and fever, slight paralysis, swelling of tongue, dyspnea (shortness of breath), dysphonia (slurred speech), increased heart rate, hypotension, cyanosis, increased lymphocytes, decreased neutrophils, and profuse perspiration. Death has occurred in as short a time as 52 minutes (Storer, 1931). Patterson (1967a) reported that in dogs and rats, Gila monster venom causes an immediate reduction in carotid artery blood flow to the brain, and that these animals suffer hypotension, increased heart rates, and breathing difficulties. The venom may also stimulate smooth muscle tissue (Patterson, 1967b), but has little or no effect on blood coagulation (Patterson and Lee, 1969).

There is no commercially available antivenin for venom from *H. suspectum*, but two have been produced on an experimental basis (Russell and Bogert, 1981): (1) Poisonous

Animal Research Laboratory, Arizona State University, and (2) Venom Poisoning Center, Los Angeles County–University of Southern California Medical Center. There has been no urgency in producing a commercial antivenin since bites are rare (and most come from handling captive lizards). Bite victims should be taken to a hospital, their vital signs monitored every 15 minutes, and the diameter of the bitten part, and a point proximal to it, recorded every 15–30 minutes during the first 6–8 hours (Russell and Bogert, 1981). An antitetenus agent (antitoxin/toxoid) should always be given. Symptoms should be treated as they occur. The bitten part should be immobilized in a functional position at, or near, heart level. Do not pack it in ice or try to increase or decrease blood flow in the injured part. The area of the bite should be soaked at least 15 minutes three times a day in 1:20 aluminum acetate solution (Burow's Solution), during which time the fingers (or toes) should be exercised, except on the first day. The wounds should be covered with a light, sterile dressing (Russell and Bogert, 1981).

Predators and defense

Actual observations of predation on Gila monsters are rare. Hensley (1949) mentions an attack by a dog (*Canis*), and Funk (1964a) reported that a black-tailed rattlesnake (*Crotalus molossus*) purportedly ate one. Lowe (in Bogert and Martín del Campo, 1956) saw a badger (*Taxidea*) expel a Gila monster from its burrow, but it did not attack the lizard. Coyotes (*Canis*), foxes (*Vulpes*), eagles (*Aquila, Haliaeetus*), hawks (*Buteo*), and owls (*Bubo*) may occasionally prey on young *H. suspectum* (Bogert and Martín del Campo, 1956; Russell and Bogert, 1981).

Human beings have been the worst enemy of *H. suspectum* in the past, and have often tried to kill these lizards on sight. Tinkham (1971a) reported that some are destroyed by persons who try to run their automobiles over them, and claims that he saw a driver swerve out of his way to hit one. Now, however, habitat destruction plays a larger role. Fortunately, *H. suspectum* is protected in all states, so the pet trade should not be a major factor.

Bogert and Martín del Campo (1956) listed three defensive behaviors by *H. suspectum*. First, it will try to escape by slowly crawling backward and, if possible, seek refuge in an inaccessible burrow or other site. If prevented from fleeing, the Gila monster will try to intimidate its attacker. This involves hissing, open mouth threats, and, if the enemy approaches too close, the lizard lunges. As a last resort, should it have an opportunity, it will bite ferociously. If handled, wild Gila monsters will squirm, try to bite, and defecate.

While most investigators have considered the venom apparatus as primarily a prey-subduing adaptation (Pregill et al., 1986), Bogert and Martín del Campo (1956) as well as Lowe et al. (1986) thought it more likely to have evolved only as a protective mechanism. Since a large part of the lizard's food consists of eggs or nestlings that can be swallowed without previous envenomation, and the rodents it eats are swallowed following the first crunch in the jaws without being envenomated (Beck, 1990), there is no reason to doubt that venom evolved as part of their defensive mechanism.

The body color and pattern may serve two defensive purposes. Against certain-colored soils and vegetation these lizards tend to be cryptic, particularly at twilight or at night. Body color and pattern may also serve as a warning device, supplementing a venom apparatus that evolved primarily as a defensive device capable of causing excessive pain rather than death. Greene (1988) thought the black and yellow pattern of the Mexican *Heloderma horridum* to be primitively cryptic, whereas the black and pink coloration of *H. suspectum* is derived and probably serves to warn.

Populations

Few data have been published on the population structure and dynamics of *H. suspectum*. Beck (1990) marked 27 lizards at a 2-km^2 site in southwestern Utah; 14 (52%) were adults with snout-vent lengths of at least 320 mm, 13 (48%) were over 325 mm, but only 1 (3.7%) immature individual was caught. The total mass of 22 of these Gila monsters was 10.52 kg, mean, 479 g (145–880).

It is very difficult to sex a living Gila monster (see Recognition), so sex ratios of natural populations are unknown. However, Bogert and Martín del Campo (1956) presented data based on 92 preserved *H. suspectum* over 225 mm in snout-vent length. Their total sample included 45 males and 47 females, a 1:1 ratio, but the individual sex ratios of the two subspecies were skewed toward one sex or the other: *cinctum*—16 males to 6 females, 2.7:1, *suspectum*—29 males to 41 females, 1:1.4.

Remarks

Bogert and Martín del Campo (1956) and Russell and Bogert (1981) listed several myths regarding the Gila monster: (1) the tongue is a stinger (it is part of an olfactory mechanism), (2) the breath is venomous (the venom, sometimes mixed with saliva, must be introduced into a puncture to have an effect), (3) the venom may be spat (the venom is not purposely spat at an attacker, but hissing lizards may sometimes acci-

dentally expel droplets of saliva), (4) it has no anus and this is what makes it so mean and venomous (it has a cloaca and an anal vent like other lizards), (5) it jumps at its tormentor (Gila monsters have no jumping ability), (6) it turns on its back so that the venom produced in glands in the lower jaw can better be injected (they seldom, if ever, turn over onto their backs when defending themselves), (7) they will bite out pieces of flesh (the teeth are not adapted for cutting), (8) Gila monsters are not immune to their own venom (they are immune to their own venom; venom is secreted when they bite each other), and (9) Gila monsters are the product of hybridization between rattlesnakes and nonvenomous snakes, especially *Pituophis* (this is ridiculous).

Heloderma suspectum is legally protected in every state in which it lives, and so should never be molested or killed.

Elapidae: Elapid snakes

The approximately 220 species of advanced snakes in this family
have extremely dangerous neurotoxic venom, and include the coral
snakes, cobras, mambas, and sea snakes. The family is well represented in Australia, Southeast Asia, Africa, and South America;
fewer species occur in North America, and only three in the United
States. Elapids are related to vipers, atractaspids, and colubrids,
which are probably ancestral. They differ, however, in having one
or more short, permanently erect, hollow (proteroglyphous) fangs
near the front of the shortened maxillae. These fangs fit into a
pocket on the outside of the mandibular gums (Bogert, 1943). The
venom duct is not attached directly to the fang, but enters a small
cavity in the gum above the entrance lumen of the tooth. Other
shorter teeth may occur behind the fangs on the maxillae, and also
on the pterygoids, palatines, and dentaries. No teeth occur on the
premaxillae. Postfrontal, coronoid, and pelvic bones are absent.
The hyoid is Y or U shaped with two superficially placed, parallel
arms. The body vertebrae have short, recurved hypapophyses.
Only the right lung is present. The spiny hemipenis has a bifurcate
centripetal sulcus spermaticus. Dorsally, the head is covered with
enlarged plates, but no loreal scales are present. The pupils are

round. Body scales are usually smooth and the ventral scutes are well developed only in arboreal and fossorial species. Reproduction is oviparous or ovoviviparous. There are four subfamilies, but only two occur in the United States: the terrestrial Micrurinae with two coral snakes (*Micruroides euryxanthus* and *Micrurus fulvius*) and the marine Hydrophiinae with one sea snake (*Pelamis platurus*).

Micruroides euryxanthus (Kennicott, 1860)
Sonoran coral snake
Plates 3, 4

Recognition

This small (maximum length, 61.5 mm) pretty
snake is red, white (or yellow) and black
banded, most of its face is black posterior to
the level of the angle of the lower jaws, and
the first broad body band behind the light collar is red. The red and white (or yellow) bands
touch, but both the red and black bands are
usually wider than the light bands, which may
be restricted to only a few scale rows in width
in some individuals. The banding extends onto
the venter and follows a red-yellow-black-
yellow-red pattern. This type of coral snake
banding pattern has been termed "tricolor monad" by Savage and Slowinski (1990). Eleven
to 13 black bands cross the body; the tail has
one to two black bands separated by white or
yellow. Black pigment usually does not occur
in the red bands. The dorsal body scales are
smooth and lie in 15 rows at midbody, but in
17 rows on the nape. There are 205–245 ventrals and 19–31 subcaudals; the anal plate is
divided. Dorsal head scalation includes a rostral wider than long, 2 internasals, 2 prefrontals, a frontal, 2 supraoculars, and 2 parietals.
Lateral head scalation includes 2 nasals (the
prenasal touches the first supralabial), a
preocular (in contact with the postnasal), 2
postoculars, 1 + 2 temporals, 7 supralabials,
and 6–7 infralabials. Loreal and subocular
scales are absent. A mental scale on the lower
jaw is separated from the two small chin
shields by a pair of infralabial scales. Posterior
to these are several rows of gular scales before
the ventral scutes. The hemipenis is about
seven subcaudals long, bifurcate with a divided sulcus spermaticus, approaching a capi-
tate condition with the spines expanded in the
distal region, but gradually diminishing in size
toward the apex which bears a small papilla-
like projection (Roze, 1974). A longitudinal naked fold begins at the base and extends almost
parallel to the sulcus spermaticus. A few scattered small spinules may also be present up to
the zone of larger spines. The dental formula
consists of 4 palatine, 0 pterygoidal, and 7
dentary teeth, and the hollow fang on the maxilla is followed, after a diastoma, by 1–2 solid
teeth.

Males average 223 ventrals and 27–28
subcaudals; females average 233 ventrals and
24 subcaudals (Shaw, 1971).

Karyotype and fossil record: Unknown.

Distribution

Micruroides ranges from central Arizona and
southwestern New Mexico southward in Mexico to about Mazatlan, Sinaloa, and east to
western Chihuahua and possibly northwestern
Durango; a hiatus in the range occurs in northern Sinaloa (Campbell and Lamar, 1989). It is
also known from Isla Tiburón, where it occurs
near sea level. In Arizona, this species ranges
to attitudes over 1,900 m.

Geographic variation

Three subspecies are currently recognized
(Roze, 1974), but two are found only in Mexico. The race in the United States is *Micruroides
euryxanthus euryxanthus* (Kennicott, 1860), the
Arizona coral snake, described above.

Confusing species

In coral snakes of the genus *Micrurus* occurring
in northern Mexico or Texas, the first broad
body band after the light neck band is black

Distribution of *Micruroides euryxanthus*.

and there are four chin shields. Harmless red, black, and yellow colubrid snakes from the same area do not have bands crossing the venter; usually their snouts are pale and the red bands touch the black bands.

Habitat

Micruroides has been found in a variety of dry habitats with rocky or gravelly soils. It is most often associated with riparian zones or arroyos, but also inhabits scrub and brushy areas, grasslands, and even cultivated fields. This snake sometimes enters buildings, perhaps to escape the heat. Being a secretive burrower, it is seldom seen above ground, preferring to remain in some subterranean chamber beneath rocks or logs, or within old stumps. The ecological distribution of *Micruroides* is often correlated with that of blind snakes (*Leptotyphlops*), its chief prey, and coral snakes are sometimes found in the channeled burrow systems of these small snakes (Lowe et al., 1986).

Behavior

In 1971, Shaw noted that, despite this species having been known for over 100 years, most details of its ecology and behavior remained unknown. The ensuing 20 years have added

little to our knowledge of this snake. Its secretive nature makes it difficult to research, but an extensive ecological study is needed.

Micruroides probably first becomes seasonally active in March or April and starts to hibernate in October or early November; Gates (1956) found one on 30 November hibernating more than 100 cm below the surface of the ground. Collection records seem to indicate *Micruroides* is most active from July to mid-September. Unfortunately nothing has been published on its thermal requirements.

It is mostly crepuscular or nocturnal in the summer, although sometimes abroad during overcast days or immediately after rains. Spring and fall activity may be more duirnal, possibly during the morning, as in the eastern coral snake, *Micrurus fulvius*.

Reproduction

Reproductive cycles, age and length of maturity, and courtship and mating behavior are unknown. It has been assumed that mating takes place in the spring. Oviposition is correlated to summer rains in July and August (Lowe et al., 1986). The eggs are sometimes laid in rotting wood (Fowlie, 1965). Hatching occurs in the early fall (Shaw, 1971; Lowe et al., 1986). A typical clutch contains two to three eggs (Stebbins, 1985). A 416-mm female dissected 27 July by Funk (1964c) contained two eggs measuring 34.6 × 6.1 mm and 39.3 × 6.2 mm. Both were thin shelled with no apparent embryonic development. Newly hatched young have total body lengths of 190–203 mm (Lowe et al., 1986).

Growth and longevity

No growth data has been reported, and nothing is known of its longevity in the wild. Most *Micruroides* do not adapt well to captivity, and so do not live long, but Shaw (1971) reported that one exceptional snake survived 3 years and 8 months at the San Diego Zoo.

Food and feeding

The leading prey of wild *Micruroides* is the slender blind snake, *Leptotyphlops* (Woodin, 1953; Vitt and Hulse, 1973; Lowe et al., 1986). Lowe et al. (1986) reported that the Sonoran coral snake also eats 11 other species of snakes (ground, *Sonora;* shovel-nosed, *Chionactis;* hook-nosed, *Ficimia;* leaf-nosed, *Phyllorhynchus;* black-headed, *Tantilla;* racers, *Coluber;* and ringneck, *Diadophis*), and four types of lizards (alligator, *Gerrhonotus;* night, *Xantusia;* whiptail, *Cnemidophorus;* and skink, *Eumeces*). Captives, however, have also eaten or tried to eat banded sand snakes (*Chilomeniscus*), night snakes (*Hypsiglena*), and legless lizards (*Anniella*) (Vorhies, 1929; Lowe, 1948b; Woodin, 1953; Gates, 1957; Lindner, 1962; Vitt and Hulse, 1973); but, in feeding tests, *Micruroides* has refused to eat the snakes *Arizona, Phyllorhynchus, Rhinocheilus* and *Thamnophis,* and the lizards *Coleonyx, Sceloporus, Urosaurus, Uta* and *Xantusia* (Vitt and Hulse, 1973). *Micruroides* seems to prefer smooth-scaled reptiles. Shaw (1971) reported his long-lived captive fed regularly on five species of small local lizards and four species of small snakes, but did not name them. The Sonoran coral snake may also eat insect larvae (Fowlie, 1965), but there is no proof of this.

Of 16 preserved museum specimens examined by Vitt and Hulse (1973), only 1 (6%) contained food.

The venom delivery apparatus of *Micruroides* is relatively poor (see below), and reptilian prey is not always immediately incapacitated (Vorhies, 1929; Lowe, 1948b; Woodin, 1953; Lindner, 1962; Vitt and Hulse, 1973). As little time as seven minutes (Lowe, 1948b) or as long as 30–40 minutes (Vorhies, 1929; Lindner, 1962) may elapse before the prey is weak enough to be swallowed, and often the prey puts up a spirited fight during this period.

Feeding behavior generally follows a consistent pattern (Vitt and Hulse, 1973). Coral snakes first respond to prey with increased tongue flicking. The snake then approaches the prey in a series of jerky movements. When close, *Micruroides* touches the prey a number of times with its tongue, then bites it with the mouth opened at approximately 30°. Usually the bite is toward the animal's posterior end. When bitten, the prey usually attempts to crawl away. Then the coral snake uses a chewing motion to move its head toward the head of its victim. In most cases envenomation does not take place (judging from the absence of

tooth puncture marks and behavior of prey after being bitten). The coral snake continues chewing until it reaches the anterior end of the animal's head and then begins swallowing. Swallowing is accomplished by a series of chewing motions separated by irregular short rests. After the prey is swallowed, the snake usually seeks shelter where it can digest its meal at leisure. A feeding observed by Vitt and Hulse (1973) took 22 minutes from the time the coral snake first showed interest (increased tongue flicks) to when it had completely swallowed the prey. The coral snake was 470 mm and the prey, *Leptotyphlops humilis,* was 255 mm. Usually, when *L. humilis* is taken, the initial bite is in the anterior third of the prey's body, whereas in other prey species the bite is normally more posteriorly oriented. Large prey do not exhibit symptoms of envenomation but a few *L. humilis* do.

Behavior is different when prey was refused. When potential prey is located, the frequency of tongue flicks by *Micruroides* increases initially, then stops, and the snake resumes resting. *Micruroides* assumes a defense posture if touched by the potential prey, and all of its movements become jerky. Usually the head is buried under an anterior coil while the tail is raised, exposing the ventral side. Occasionally the cloaca is everted with a popping sound. During the tail display the snake usually attempts to escape by burrowing or crawling under some object. No attempt is made to bite the prey.

Venom and bites

Eight *Micruroides* 215–515 mm long I examined had fangs lengths of 0.1–1.0 mm (average 0.6 mm). Correlation between body length and fang length was positive. Bogert (1943) reported four *Micruroides* 401–456 mm long had fangs 0.7–1.0 mm long.

The venom is neurotoxic. Lowe (1948b) reported the effects on a night lizard (*Xantusia*). The envenomation took place at 2232 hours, and by 2234 slight paralysis of hind limbs was apparent. The hind limbs were completely paralyzed, and the front limbs showed slight paralysis by 2235:30. At 2235:45 the front limbs were nearly completely paralyzed, and the lizard experienced undulatory movements of its entire body. Total loss of righting ability occurred at 2237:30, and at this time the lizard's breathing was deep and rapid. The lizard apparently died at 2238:30, as breathing ceased (time elapsed since bite: 5 min. 15 sec.) At 2239:15 Lowe opened the lizard's body and found the heart beating slowly; normal saline was applied, but at 2239:45 the heart stopped beating (time elapsed since bite: 7 min. 30 sec.).

Its short length, along with a correspondingly small mouth and short fangs, makes it very difficult for *Micruroides* to deliver a penetrating bite to a human. However, human envenomation has occurred. Russell (1967a) reported three cases, all resulting from handling the snake. Common symptoms included immediate, but not severe, pain at the site of the bite, with the pain persisting from 15 minutes to several hours. Nausea, weakness, and drowsiness occurred several hours later. parasthesia (abnormal sensations) also occurred. This was limited to the bitten finger in one case, but spread to the hand and wrist in the other two. Two of the victims experienced no symptoms after 7 to 24 hours of the bite, but symptoms persisted for four days in the third, the one bitten by the largest snake (55 cm). One of the persons bitten was a physician who took detailed notes concerning the bite and its symptoms. He pulled the snake from his finger almost as soon as he was bitten. About three hours later there was parasthesia involving the finger and hand. This was followed by a progressive deterioration of his handwriting ability over a period of about six hours, at the end of which his handwriting looked like that of a five-year-old child. The doctor experienced headache, nausea, and difficulty in focusing his eyes. There was some slight drooping of the upper eyelids. His inability to focus properly resulted in his walking into doors on several occasions. He also thought it possible he experienced some photophobia and increased lacrimation. Weakness and drowsiness occurred, and his memory was vague when he tried to remember the incident. There was no change in heart rate and no difficulty breathing.

From the above, it can be seen that this small snake has very potent venom, and should be considered dangerous, especially to children.

Predators and defense

Few cases of predation on *Micruroides* have been recorded. Owing to its small size, however, it probably is at least occasionally the victim of carnivorous mammals (skunks, badgers, foxes, coyotes, raccoons), kestrels, hawks, owls, roadrunners, and ophiophagous snakes (kingsnakes, whipsnakes), and possibly also tarantulas. It is cannibalized by its own species (Lowe et al., 1986).

This species is rather shy and gentle if not severely disturbed. If startled, it may coil, tuck its head under its body, and elevate its tail, coil it so that the ventral surface faces toward the disturbance and move it about. The moving, banded tail may draw predators away from the tucked-under head in a decoying behavior. Vitt and Hulse (1973) found that three (14%) of the animals they studied had scars on the tail that possibly indicated a predatory attack there.

Another common defensive behavior of this snake is discharging air and fecal matter from the cloaca (Bogert, 1960), producing a moderately loud sound. Air is apparently drawn into the cloaca through the vent and then expelled through the same opening. The sounds last about 0.2 sec and are repeated at 0.3–0.5-sec intervals.

If neither of the above defensive behaviors wards off a predator, and the snake is touched or the predator comes very close, the snake will strike quickly and viciously in a side-sweeping motion, and, if the mouth makes contact, it will bite and continue to chew for some time.

Populations

Micruroides seems rare, but this is probably due to its secretive habits rather than to actual scarcity. It may even prove to be common in some areas.

Remarks

This is the northernmost of the coral snakes in western and tropical America. Owing to this distribution and several characters considered primitive elapid features (such as additional solid teeth behind the fang on the maxilla), it has usually been thought to be near the origins of coral snake evolution (Schmidt, 1928;

Bogert, 1943). Cadle and Sarich (1981) proposed that the New World coral snake genera *Micrurus* and *Micruroides* are more closely related to other elapids, rather than derived from a lineage of South American colubrids. Their conclusions are based on immunological comparisons of serum albumins. They suggested a late Oligocene–early Miocene separation between the New and Old World elapid lines. Murphy (1988) discovered that *Micruroides* and the sea snake *Pelamis platurus* shared five L-

lactate dehydrogenase heterotetramer isozymes, while *Micrurus fulvius* has only two of the isozymes. He proposed that the common ancestor of coral snakes had a five-banded isozyme pattern.

Micruroides euryxanthus may participate in a mimicry complex with other coral snakes and red, yellow, and black banded colubrid snakes. This phenomenon is discussed under *Micrurus fulvius*.

Micrurus fulvius (Linnaeus, 1766)
Harlequin coral snake
Plates 5–7

Recognition

Micrurus fulvius is a banded red, yellow, and black snake (maximum length, 121 cm) on which the red and yellow bands lie beside each other—the tricolor monad type of banding pattern of Savage and Slowinski (1990)—and the snout is black. Black body bands total 13–25, those on the tail 3–7; 12–24 red bands cross the body. The bands continue onto the venter, and there may be some black pigment on the red bands. A bright yellow band occurs on the occiput, separating the black neck band from the parietal scales. The body scales are smooth and occur in 15 rows throughout; the anal plate is divided. Ventrals total 182–232, and subcaudals 26–55. Head scalation usually consists dorsally of a rostral, 2 internasals, 2 prefrontals, a frontal, 2 supraoculars, and 2 parietals; laterally are 2 nasals, a preocular, 2 postoculars, 1 + 1 temporals and 7 supralabials; there are no loreal or subocular scales. On the lower jaw are a mental, which does not touch the chin shields, usually seven pairs of infralabials (the first four pairs touch the anterior chin shields), and two pairs of chin shields (the posterior pair being the larger). The hemipenis has a bifurcated sulcus spermaticus running from the base to nearly the apex of each fork. Each fork tapers gradually toward the apex. The base of the organ is naked after which small spines and scattered spinules cover it up to the bifurcation where large spines begin. The lip of the sulcus is naked for its entire length but is covered on both sides with small spines. Large spines occur before the bifurcation of the organ and gradually diminish in size toward the apex; the area of bifurcation is without spines. Each fork of the hemipenis ends in a spinelike papilla. A large longitudinal naked fold begins almost at the base of the organ and runs approximately parallel to the sulcus, ending shortly before the bifurcation where the large spines begin (Roze and Tilger, 1983). The dental formula consists of the fang on the maxilla, 4–6 palatine teeth, 0 pterygoidal teeth, and 6–7 teeth on the dentary.

Males have 182–217 ventrals and 36–55 pairs of subcaudals; females have 205–232 ventrals and 26–38 pairs of subcaudals (Roze and Tilger, 1983). Also, females attain larger snout-vent lengths, while males have longer tails; tail lengths average 13.8% of snout-vent length in males and 9.3% in females (Jackson and Franz, 1981). Males lack keels on the body scales anterior to and above the anal plate.

Karyotype

The diploid chromosome number of the western subspecies *M. f. tener* is 32: 16 macrochromosomes and 16 microchromosomes (Graham, 1977). All chromosome pairs are homomorphic except pair 6, which is hetermorphic (ZW) in females; males are ZZ.

Fossil record

Pleistocene fossil remains of *M. fulvius* have been found at a Florida Irvingtonian site (Meylan, 1982) and in Rancholabrean deposits in Florida and Texas (Holman, 1981).

Distribution

Micrurus fulvius ranges from southeastern North Carolina southward, mostly on the coastal plain, through Florida, and west through central and southern Georgia, Alabama, and southern Mississippi to southeastern Louisiana. A hiatus occurs in the range in south-central Louisiana, but the species can

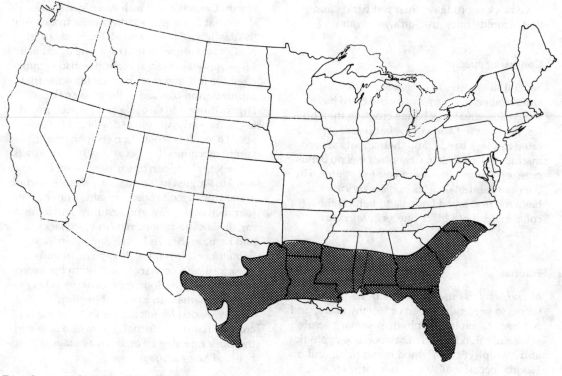

Distribution of *Micrurus fulvius*.

once again be found from central and northern Louisiana and southwestern Arkansas southwestward through southern Texas, and southward in Mexico, through eastern Coahuila, northeastern Nuevo León and Tamaulipas to San Luis Potosi, eastern Guanajuato, Querétaro, and Hidalgo.

Geographic variation

Five subspecies are recognized (Roze and Tilger, 1983), but only two occur in the United States. *Micrurus fulvius fulvius* (Linnaeus, 1766), the eastern harlequin coral snake, occurs east of the Mississippi River from southeastern North Carolina southward through peninsular Florida, and westward to southeastern Louisiana. The red body bands of this subspecies have either no black pigment or only small black spots or black tips on the scales. Often the black pigment concentrates to form a pair

of large dorsal spots. The black neck band does not touch the parietal scales. *Micrurus f. tener* (Baird and Girard, 1853), the Texas harlequin coral snake, ranges from southwestern Arkansas and northern and central Louisiana southwestward through Texas to Coahuila, Nuevo León, and central Tamaulipas. Black pigment in its red bands is widely scattered, and the black neck band touches the parietal scales. Formerly, the name *tenere* was used for this race, but Frost and Collins (1988) have shown that the proper spelling is *tener*. Owing to its apparent allopatry (Conant and Collins, 1991), Collins (1991) has suggested that *M. f. tener* be designated a separate species.

Coral snakes from southern Florida were designated *M. f. barbouri* by Schmidt (1928) on the basis of a few specimens lacking black spots on their red bands. Subsequently, Duellman and Schwartz (1958) reported that many specimens of *M. fulvius* from southern Florida have black spotting on their red bands and that Schmidt's designation was invalid.

Confusing species

Several subspecies of *Lampropeltis triangulum* and the species *Cemophora coccinea* also have red, yellow and black bands crossing their bodies, but their red and yellow bands are separated by black bands, and their snouts are red instead of black. Also, *Cemophora* has no bands crossing its white or cream-colored venter. The Sonoran coral snake, *Micruroides euryxanthus*, has a red first broad body band behind the light collar instead of a black one as in *Micrurus fulvius*.

Habitat

Micrurus fulvius uses a variety of habitats. It seems to prefer dry, open or brushy areas, and has been taken in xerophytic rosemary scrub, seasonally flooded pine flatwoods, xerophytic and mesophytic hardwood hammocks, and, in Florida, occasionally in marshy areas (Neill, 1957; Jackson and Franz, 1981; Dundee and Rossman, 1989), where it spends most of the time buried in the soil, leaf litter, logs, or stumps. It may also hide in gopher tortoise burrows where the two reptiles are sympatric. *Micrurus fulvius* living in Texas prefer either rock-crevice cover or thick plant litter in upland deciduous woodlands where they can hide and also find the semifossorial reptiles upon which they prey.

Behavior

Over most of Florida, the harlequin coral snake is active in every month, but it has a distinct bimodal activity pattern with more activity from March to May and, again, from August to November (Jackson and Franz, 1981); at Long Pine Key in Everglades National Park, however, fewer individuals are seen in August and September (Dalrymple, Steiner et al., 1991). The winter months, December–February, are the period of least activity in Florida. North of Florida *M. fulvius* is forced to hibernate, usually underground.

There is a misconception that coral snakes are nocturnal, but observations on various North, Central and South American species show that these snakes often are active during the daylight (Greene and McDiarmid, 1981). Neill (1957) reported that only one of 121 active *M. fulvius* was taken at night. He also summarized the few literature records for nocturnal captures, and concluded that this snake is largely diurnal in Georgia and Florida, prowling in the early morning shortly after sunrise to about 0900. It is most often seen on bright, sunny mornings, but occasionally also prowls in the late afternoon or early evening. During April to August in Florida, *M. fulvius* is surface active from 0700 to 0900, remains under cover for much of the day, and resumes activity during the late afternoon, 1600 to 1730 (Jackson and Franz, 1981). In March and again in September–November, it appears in mid- or late morning (0900–1000) and, with the exception of a midafternoon quiescent period (1330–1600), remains active most of the day.

In Mexico *M. fulvius* seems to be mostly crepuscular or nocturnal, but is also active in the early morning or on overcast days (Campbell and Lamar, 1989).

Reproduction

Most females mature at a snout-vent length of about 55 cm in 21–27 months (Quinn, 1979b; Jackson and Franz, 1981), but a female *M. f. tener* from northeastern Texas with a 50 cm snout-vent length laid three eggs (Ford et al., 1990).

The smallest male undergoing spermiogenesis found by Quinn (1979b) was 40.2 cm long, snout to vent, and Jackson and Franz (1981) reported that most males 45 cm or longer contain sperm, whereas smaller ones do not. This size is reached in 11–21 months.

In Texas, ovary weights of *M. f. tener* increase from March through April, decline slightly in May, and then more rapidly in June; follicle lengths show the same pattern (Quinn, 1979b). In Florida, vitellogenesis occurs in late winter and early spring (March-May), with follicles reaching preovulatory size by early June (Jackson and Franz, 1981).

In male *M. f. tener* from Texas there is complete regression with spermatogonia and Sertoli cells in May through August, with a peak in June. Early recrudescence with spermatogonial divisions and primary spermatocytes occurs from June through October, with a peak in July. In August and September, males have late recrudescense with secondary spermatocytes and undifferentiated spermatids. Active spermiogenesis with mature sperm in the lumen takes place from August to April. Seminiferous tubule diameter is greatest from November to March, and testes weigh the most from December to February. Sperm resides in the ductus deferens from February to December, and in the epididymis from April to December (Quinn, 1979b). The male sexual cycle in Florida is essentially the same (Jackson and Franz, 1981).

Mating occurs in the spring, and possibly also in the fall. During courtship the male crawls to the female and flicks his tongue several times over her back at midbody. He then raises his head and neck at about a 45° angle and, leaving his neck at that angle, tilts his head down and touches his nose to the female's back. In this position he quickly and smoothly runs his nose along the female's back to about 5 cm behind her head (according to Quinn, 1979b, 60% of the time the male moves along the female's back from rear to front; when he moves from front to back he reverses himself when the area of the female's vent is reached). The male usually does not flick his tongue during this advance, but does align his body over hers. The male's body and tail are dipped at his vent region laterally on the female's side beside her vent. Immediately anterior to the vent the male's body is positioned on the female's back, and immediately posterior to the vent his tail is projected upward at about a 30° angle, but does not touch the female's tail. The female normally rests her tail flat on the ground. Several vent thrusts are then made by the male in an attempt to copulate and his hemipenis may be partially everted. If intromission is at first unsuccessful, the male will move his entire body back and forth on the female's dorsum in about 2.5 cm strokes at a rate of a stroke/sec. This sequence may end in several rapid strokes of the male's snout only. Nose strokes are about 1.25 cm long and occur at a rate of about 2/sec (Quinn, 1979b). An unreceptive female will not gape her vent and will try to crawl away. The male will then pursue her and repeat the sequence (Quinn, 1979b, saw such a courtship sequence occur five times in 40 min).

Although both Zegel (1975) and Quinn (1979b) described the courtship behavior of this species, neither witnessed copulation. However, Vaeth (1984) described a successful mating between a male M. f. tener and a female M. f. fulvius in which the male stopped moving when his head had approached within 2 cm of the female's head and his vent was in apposition to the left side of her vent. He then elevated his tail and tried to wrap it under and around that of the female. She responded by slightly elevating her tail and gaping her cloaca. Intromission was accomplished rapidly with the right lobe of the hemipenis. The elapsed time between introduction and intromission was about 25 minutes. No attempt was made to completely entwine tails nor was there any cloacal rubbing prior to intromission. Shortly after intromission, the male moved the anterior third of his body off the female. After approximately 10 minutes of coital activity, Vaeth observed localized rhythmic contractions in the male's cloacal region which continued intermittently for 12 minutes. In this particular mating, ejaculation was not completely efficient since a pool of seminal fluid accumulated under the two snakes soon after the contractions were observed. Presumably, this seminal fluid was leaking from the exposed part of the sulcus on the basal portion of the hemipenis. There were no other body movements exhibited by either snake during the copulation period. Only an occasional tongue flick was seen, and these may have been in response to Vaeth's movements. After 61 minutes of copulation, the male withdrew his hemipenis and crawled away.

The eggs are laid during May to July (Tryon and McCrystal, 1982, reported a period of 37 days between copulation and oviposition). Two to 13 eggs may be laid, although 4–7 are more common. The eggs are very elongate (20–47 × 6–14 mm, 3–6 g), and are probably usually laid underground or beneath leaf litter; my students and I found four eggs in a hollow depression beneath an old wooden tie on an abandoned railroad embankment in Collier County, Florida.

The young hatch in August and September after an incubation period of about 70–90 days; they are about 177–205 mm in total body length. Hatchlings emerge from slits in the egg shell about 10 mm long; the total time needed for emergence is approximately four hours (Campbell, 1973).

Growth and longevity

Micrurus fulvius doubles its size in less than two years, and by three years may grow to nearly 60 cm snout-vent length (Quinn, 1979b; Jackson and Franz, 1981).

Natural longevity is unknown, but an individual lived almost seven years at the Brookfield Zoo, Chicago (Bowler, 1977), and another is still alive in the Fort Worth Zoo that was given to that institution in 1973 by Harry W. Greene (Greene, pers. comm.).

Food and feeding

Prey lists presented by Jackson and Franz (1981) and Greene (1984) show that *Micrurus fulvius* feeds almost entirely on reptiles: amphisbaenians (*Rhineura floridana*), lizards (*Ophisaurus, Cnemidophorus, Eumeces, Neoseps, Sceloporus, Scincella*), and snakes (*Agkistrodon, Arizona, Coluber, Diadophis, Elaphe, Farancia, Ficimia, Lampropeltis, Leptotyphlops, Micrurus, Opheodrys, Salvadora, Seminatrix, Sonora, Stilosoma, Storeria, Tantilla, Thamnophis, Tropidoclonion, Tropidodipsas, Virginia*). Anurans and rodents are rarely eaten (Greene, 1984).

Although these snakes may feed nearly year round in Florida, feeding activity is most intense in September-November with a lesser peak in April-May.

During foraging, *M. fulvius* usually crawls slowly and pokes or probes with its head beneath leaf litter. These head probes are stereotyped (Greene, 1984). Poking involves repeated forward and lateral head movements, and is accompanied by frequent tongue flick clusters. Apparently both visual and chemical stimuli elicit attacks. Rapid prey movements seem to bring on a quicker attack by the coral snake and sometimes override aversive chemical cues. Approach is slow if the prey is stationary or moving slowly but rapid if the prey is crawling quickly away. Prey is usually seized with a quick forward movement of the anterior or entire body of the coral snake, but may occasionally be seized with a quick sidewise jerk of the head and neck. *Micrurus fulvius* does not have an efficient strike; Greene (1984) thought this perhaps was due to poor vision resulting

from their relatively small eyes. The snake typically holds its prey, sometimes chewing it, until it is immobilized by the venom.

The prey is not usually released before it is swallowed, and is almost always swallowed head first. Pre-ingestion maneuvers suggest that either tactile and/or chemical cues are used to recognize the anterior end of the prey. Greene (1976) showed experimentally that scale overlap on the prey was a cue for locating its head. Swallowing is accomplished, as in other snakes, by alternating jaw movements as the coral snake literally crawls over its prey. Swallowing is almost always followed by much tongue flicking and opening and closing of the mandibles to work the jaws back into position.

Micrurus fulvius is opportunistic, taking whatever suitable prey crosses its path when it is hungry. However, the skinks (*Eumeces, Neoseps, Scincella*) and the smaller snakes it takes may not be cost-effective in the amount of energy gained compared to that needed to find, attack and swallow them (Greene, 1984).

Venom and bites

Fangs of *M. fulvius* (27–88 cm) that I measured were 0.6–2.7 mm in length; adults (65–88 cm) had fangs 1.6–2.7 mm long.

The venom of *M. fulvius* attacks the nervous system, primarily the respiratory center, resulting in difficulty in breathing and death in extreme cases. Pain, sometimes severe, usually occurs at the bite site, and, if the bite is on a limb, may slowly extend up that extremity. The LD_{50} in micrograms for a 20-g mouse is 26 (subcutaneus), and the lethal amount for a human is 4–5 mg (Minton and Minton, 1969). The normal venom yield for *M. fulvius* is 3–5 mg. However, Fix and Minton (1976) reported that 8 of 14 adult *M. fulvius* from which they extracted venom gave dry yields in excess of 6 mg, and that four gave dry yields of 12 mg or more. There was also a distinct positive size correlation ($r = 0.87$), with the longer snakes giving greater venom yields. The enzymatic composition of *M. fulvius* venom is discussed by Aird and da Silva (1991).

Micrurus fulvius can deliver serious, and, in some cases, fatal bites. Fortunately, the percentage of human fatalities (about 20%) is not extremely high, especially if the bite is treated with antivenin (Neill, 1957). Case histories of bites are discussed in Stejneger (1898), Werler and Darling (1950), Niell (1957), Stimson and Engelhardt (1960), and Parrish and Kahn (1967).

As harlequin coral snakes have relatively short fangs, the chance of a human receiving a bite in the wild when not handling one is slim. Nevertheless, Carr (1940) has reported instances of large *M. fulvius* actually striking, but such behavior must be rare, and heavy shoes or boots and thick trousers should be sufficient protection for the hiker. *Micrurus fulvius* is apparently immune to the venom of its own species (Peterson, 1990).

Predators and defense

Jackson and Franz (1981) and Brugger (1989) summarized observations of predation on *M. fulvius*. They mention instances of diurnal predaceous birds, such as the kestrel (*Falco*), red-shouldered and red-tailed hawks (*Buteo*), and loggerhead shrike (*Lanius*), attacking the harlequin coral snake. *Micrurus fulvius* may also eat members of its own species (Jackson and Franz, 1981; Greene, 1984; and Brugger, 1989). Five of nine coral snakes examined by Clark (1949) were taken from the digestive tracks of king snakes (*Lampropeltis getulus*), and Minton (1949) reported that a large bullfrog had eaten a small *M. fulvius*. The imported fire ant (*Solenopsis invicta*) may prey upon the eggs and young of *M. fulvius*, and, at least in Alabama, the population of coral snakes has declined since the introduction of this aggressive insect (Mount, 1981).

Would-be predators of *M. fulvius* do not always fare so well. Brugger (1989) observed an adult male red-tailed hawk (*Buteo jamaicensis*) die with a partially eaten, at least 80 cm, harlequin coral snake in its talons. The bird died of flaccid paralysis typical of the neurotoxic effects of elapid venoms. Six punctures (presumably by fangs) occurred on the left tarsus of the bird and nine more in the skin over the left tarsus gastrocnemius muscle. The muscle was swollen and the tissue surrounding two punctures discolored orange.

When approached, *M. fulvius* will often flatten the posterior portion of the body, tuck its head under an anterior coil, ball up its tail and wave it about, thus drawing the predator's attention away from the head (Greene, 1973), and Gehlbach (1972) has demonstrated that this behavior will deter some potential mammalian predators. When pinned down, however, the snake will strike and chew on the restraining implement. These strikes are usually sidewise, rapid, and often vicious.

Populations

Being a secretive burrower, *M. fulvius* is not readily observable—only 22 (1–2%) of 1,782 snakes collected or observed by Dalrymple, Bernardino et al. (1991) at the Everglades National Park during 1984–1986 were *M. fulvius*—and can live in urban areas without being detected. This gives a false impression of rarity, when these snakes may be quite common. For instance, Beck (in Shaw, 1971) reported that in a 39-month period 1,958 *M. fulvius* were turned in for bounties in Pinellas County, Florida.

The most serious threat to populations of *M. fulvius* is habitat destruction.

Remarks

Immunological assessment of serum albumins by Cadle and Sarich (1981) shows that the American coral snakes (*Micrurus* and *Micruroides*) are close allies with other elapids instead of derivatives of South American colubrids, and that a late Oligocene–early Miocene separation between the New and Old World elapid lineages may have occurred.

The question of whether or not coral snakes and other red, yellow, and black banded snakes form a mimicry complex has been discussed for years. One view is that coral snake patterns are primarily aposematic (Hecht and Marien, 1956; Mertens, 1956b, 1957; Gehlbach, 1972; Pough, 1988), particularly since there is an absence of countershading with the continuation of the dorsal bands onto the venter. Also, the patterns may be cryptic by presenting a visual illusion of disruption as the snake crawls. Can coral snake patterns be aposematic, cryptic, and mimic at the same time? Greene and Pyburn (1973) thought it perfectly feasible.

Grobman (1978) suggested that similar color patterns have arisen independently of natural selection in unrelated sympatric species occupying similar habitats (pseudomimicry), while Wickler (1968) suggested that the dangerously venomous coral snakes are the mimics of mildly venomous colubrid snakes, not the models for these species. In recent summaries of the problem, Greene and McDiarmid (1981) and Pough (1988) concluded that field observations and experimental evidence refute previous objections to the coral snake serving as the Batesian model in the mimicry hypothesis, and that their bright colors serve as warning signals to predators. Smith (1975) showed that the young of some tropical reptile-eating birds instinctively avoid the red-yellow-black banded pattern of coral snakes. Hallwachs and Janzen (in Pough, 1988) reported that in Costa Rica scavengers (mostly birds) that quickly consume road-killed snakes of other species typically leave dead coral snakes alone. Similarly colored and patterned harmless snakes (*Cemophora coccinea, Lampropeltis triangulum, L. pyromelana*) or mildly venomous snakes would also gain protection by their resemblance to the coral snakes. Past objections to the mimicry theory have been based largely on the supposition that once bitten by a coral snake a predator will die from the bite; however, the apparent poor vision and poorly developed venom delivery system of coral snakes precludes that every bite will be fatal. It has also been supposed that coral snakes are nocturnal, and that their bright colors would be meaningless at night,

but reports by Neill (1957), Jackson and Franz (1981) and Greene and McDiarmid (1981) have shown *M. fulvius* to be largely diurnal. Studies of concordant geographic pattern variation by Greene and McDiarmid (1981) strongly suggest that some colubrid species of *Atractus, Erythrolamprus, Lampropeltis,* and *Pliocerus* are involved in mimicry systems with local coral snakes.

Parent birds may respond to the sight of a nearby snake by actions that will reveal the location of their nest. If bright, ringed snake patterns elicit this behavior more than do cryptic patterns, then ringed snakes that eat eggs or nestling birds should have a hunting advantage over cryptic snakes. Smith and Mostrom (1985) examined this theory in field tests with American robins (*Turdus migratorius*) and a red, yellow, and black ringed coral snake model. The model elicited no more response than did a plain brown snake model. So, based on current evidence, the most likely advantage of bright, ringed patterns to snakes is to confer protection against predators, either by camouflage or as warning coloration in mimicry systems.

Coral snakes appear to exhibit near infrared reflectance that renders the snake extremely visible against a leaf litter substrate (Krempels, 1984), and this may advertise a warning to potential avian predators.

Pelamis platurus (Linnaeus, 1766)
Yellow-bellied sea snake
Plates 8, 9

Recognition

This marine snake (maximus length, 113 cm) is easily recognized by its unique oarlike tail which is flattened from side to side, sharply demarcated two-toned body coloration (dark back, yellow belly), and dorsally placed valved nostrils. The dark dorsal body coloring is either black, olive, olive-brown, or brown; the venter is yellow, cream, or pale brown. Juveniles are more brightly colored than adults. Body pigmentation tends to be in the form of broad longitudinal stripes, but some individuals have undulating stripes, bars, or spots. According to Tu (1976) the most common pigment pattern in the eastern Pacific population is a tri-colored one of black(olive)-yellow-brown. The next most common is bicolored black(olive)-yellow. A rare unicolor yellow phase also occurs in less than 1% of the population (4 of 3,077 snakes) (Tu, 1976), but in 3% of the snakes at Golfo de Dulce, Costa Rica (Kropach, 1971b). The laterally compressed tail has a variable pattern of dark and light reticulations or, rarely, large spots. The body scales are smooth, nonoverlapping, and in 44–67 rows at midbody. The 260–465 ventral scutes are divided by a midventral groove, and are almost as small as those on the back; the ventral plate is subdivided. Subcaudals total 39–66. The head is angular, narrowing considerably in front of the eyes. Except in entirely yellow individuals, it is two-toned, dark on top, light beneath. The top of the head is covered with enlarged scales: a rostral, 2 nasals (there are no internasal scales, the enlarged nasals meet dorsally), 2 prefrontals, a frontal, 2 supraoculars, and 2 parietals. Some snakes also have 2 or more small interparietal scales. Lateral head scalation includes 1(2) preocular, 1–2 suboculars, 2(3) postoculars, 2 + 2 + 2–3

+ 3–4 temporals, 7–11 supralabials, and 10–13 infralabials. The chin scales are small and the mental is ungrooved. The tongue is very short, barely capable of protruding from the mouth. The hemipenis is illustrated in Wright and Wright (1957) and Mao and Chen (1980). McDowell (1972) described it as narrow and weakly bilobed, with the sulcus spermaticus forked near the tip, lips of the sulcus prominent and fleshy (but not spiny); the organ with small spines over most of its length, but with papillae at the extreme tip. On the maxilla, the short anterior fang is followed by a diastema separating it from 7(5–6) to 10 solid posterior teeth. Other teeth include 7(5–6) palatines, 22–28 pterygoidals, and 16–18 dentaries.

Three sexually dimorphic characters may help separate adult *Pelamis* (Kropach, 1975): (1) males are shorter (mean snout-vent length, 45.2 cm) than females (mean snout-vent length, 48.1 cm), (2) males have longer tails (mean 12.1 cm) than females (mean 11.4 cm), and (3) male ventrals have longer tubercles than those on the ventrals of females. Unfortunately, none of these is sufficient by itself for sex determination.

Karyotype

The diploid number is 38: 20 macrochromosomes and 18 microchromosomes; pairs 1–2 are metacentric, pair 3 is subtelocentric, and pairs 4–9 have the centromere in a terminal position. Females are ZW, males are ZZ (Gutiérrez and Bolaños, 1980).

Fossil record

No fossils are known, but Hecht et al. (1974) discussed the possible evolution of *Pelamis* in response to marine thermal regimes in the Oli-

Distribution of *Pelamis platurus*.

gocene, and the reader is referred to that paper for more information.

Distribution

Pelamis platurus is the most widely distributed sea snake. Mostly restricted to tropical and subtropical warm oceans, it occurs in the Indian and Pacific oceans from eastern Africa, Madagascar, Arabia, and India, throughout coastal southeastern Asia, Indonesia, Japan, Australia, New Zealand, and the Pacific islands to the western coast of the Americas from Ecuador and the Galápagos Islands north to Baja California and the Gulf of California. It is the only sea snake to have reached the Hawaiian Islands, and waifs have been collected at San Clemente, Orange County, California (Pickwell et al., 1983) and Los Angeles Bay (Shaw, 1961) and seen in the San Diego area (Stebbins, 1985). See Hecht et al. (1974) for a detailed discussion of the distribution of *Pelamis*.

Geographic variation

Subspeciation in *Pelamis* has not been studied seriously; however, differences between populations probably do exist, and such a study would be rewarding.

Habitat

Although usually considered a pelagic species, in the Pacific Ocean off Central America *Pelamis* is most often found within a few kilometers of the coast, and seems to prefer the shallower inshore waters. It normally occurs in waters having temperatures between 22 and 30°C. There it spends most of the time drifting at the surface (rough water will cause it to dive; Tu, 1976), often in large numbers among the flotsam in surface slicks formed at the interface of two currents. Such habitats probably provide good foraging.

Behavior

The natural history presented below is based, as much as possible, on data available concerning the eastern Pacific population of *Pelamis* occurring off the western coast of the Americas. Individuals from this population occasionally reach the California coast.

Pelamis have been collected from 10 February to 14 November in the Galápagos Islands (Reynolds and Pickwell, 1984), but farther north off Panama and Costa Rica they are present in every month, with noticible increases in numbers during the dryer months.

Myers (1945) recorded nocturnal activity between 1930 and 2400 hours at Bahia Honda, Panama. The 22 snakes collected were 220–311 mm long, and were probably juveniles. All made their appearance in the same way by leisurely swimming upward toward the surface. Upon reaching the surface, each snake protruded its head above the water and floated upward until in a nearly horizontal position while breathing in air. They then slowly swam downward toward the bottom, and none seemed interested in feeding. From these observations Myers (1945) suspected that *Pelamis* feeds during the day and spends the nights on the bottom, occasionally rising to the surface to breathe, and Tu (1976) corroborated this. From 14 January to 3 February, he collected 3,077 *Pelamis* between 0700 and 1300 hours, but none from 1500 to 1800 hours. Perhaps surface waters become too warm at that time, or their fish prey is less active. Surfacing depends on the condition of the surface water; fewer surfaced snakes are seen during periods of rough water (Tu, 1976; Rubinoff et al., 1986). Dives to maximum depths of 6.8 m in the dry season and 15.1 m during the wet season have been observed (Rubinoff et al., 1986).

Yellow-bellied sea snakes may remain submerged for long periods of time. Although Dunson (1971) noted that sea snakes are reportedly capable of staying underwater for anywhere from two to eight hours (species not given), Rubinoff et al. (1986) observed that the maximum voluntary submergence time for *Pelamis* was 213 minutes, and of 202 complete dives recorded only 19 exceeded 90 minutes. Diving *Pelamis* may be able to avoid anaerobiosis by having a reduced metabolic rate, an enhanced rate of cutaneous oxygen uptake, or both. The oxygen capacity of the blood of *Pelamis* (as expressed as the average volume of oxygen that can be held as a percent of blood volume) is rather high, 10.2%, but the oxygen capacity of its blood changes with body temperature (Pough and Lillywhite, 1984). It increases when the body temperature is raised from 10 to 20°C, but then drops if the body temperature is raised further to 40°C; however, in nature, the body temperature of *Pelamis* probably varies little during a 24-hour period. Cutaneous breathing is one of the main physiological adaptations of *Pelamis;* it can remove oxygen from water at rates up to 33% of total standard oxygen uptake, and can excrete carbon dioxide into the water at rates up to 94% of standard oxygen consumption (Graham, 1974a). As much as 18 ml of oxygen per gram of body weight can be absorbed through the skin (Heatwole, 1987).

When *Pelamis* dives, it retains enough air in its large lung to keep itself positively buoyant. However, as it swims deeper water pressure squeezes the air in the lung and the snake's overall density becomes greater until it becomes neutral or negatively buoyant at that depth (Graham, 1975). As the oxygen in the lung is used, the resulting carbon dioxide is passed out through the skin rather than diffusing into the lung. Most dives are characterized by a four-phased pattern (Graham, Gee et al., 1987): descent, bounce ascent, gradual ascent, and final ascent. The gradual ascent phase accounts for about 82% of the total underwater period of each dive and may reflect a period when, according to Boyle's Law, the snake compensates for buoyancy lost because of the decline in lung volume. An intracardiac blood shunt assures management of lung oxygen reserves in a manner that augments cutaneous breathing and establishes favorable transcutaneous diffusion gradients that help the removal of built-up nitrogen gases in the blood (to avert the "bends") and the uptake of oxygen (Graham, Gee et al., 1987). Subsurface swimming is slower than that at the surface, and the snake usually assumes a posture in which the tail is elevated and the posterior position of the body is in a nearly vertical position (Graham, Lowell et al., 1987). Undulatory movements when in this posture involve torsional and rolling motions of the body which, through changes in the camber of the tail keel and body, may contribute to the thrust.

Another serious problem confronting *Pelamis* is that of balancing its salt and water content with that of seawater (osmoregulation). While marine mammals have kidneys efficient enough to rid the body of excess sodium ions (Na+), the reptilian kidney is very weak and cannot excrete higher concentrations of Na+ than are found in the blood. The urine of *Pelamis* is always hypoosmotic to the blood plasma. Therefore, additional structures must supplement the kidneys to prevent excess Na+ from building up in the body fluids. Although birds and marine turtles have well-developed salt-secreting glands for this purpose in their nasal passages, no such discrete compact gland has been found in *Pelamis* (Schmidt-Nielsen and Fange, 1958; Taub and Dunson, 1967); however, serial sections of the head prepared by Burns and Pickwell (1972) have confirmed its presence. Well-developed nasal glands are found in some other sea snakes (Burns and Pickwell, 1972). Also, mucoid cell types are absent from the labial glands of *Pelamis,* so these may not aid in salt secretion. *Pelamis* may lack development in the above glands, but it does have a very well developed posterior gland under the tongue (sublingual) which acts as a salt gland, excreting NaCl into the mouth for expulsion (Dunson et al. 1971; Dunson, 1971). This salt gland usually comprises about 0.04% of the total body weight (Dunson and Dunson, 1975), and the fluid excreted has a higher NaCl content than seawater. During periods of salt loading, more Na^+, Cl^-, and K^+ are excreted from the mouth than cloacally (Dunson, 1968). *Pelamis* kept in fresh water for 48 days have shown no decrease in salt-gland Na-K ATPase activity or in gland weight, even though Na^+ concentration dropped markedly. Na-K ATPase has been consistently found in high concentrations in tissues specialized for active ion transport (Dunson and Dunson, 1975). *Pelamis* can survive in fresh water for periods exceeding six months (Dunson and Ehlert, 1971).

The skin of *Pelamis* is permeable to water but not to Na^+ (Dunson and Robinson, 1976); yellow-bellied sea snakes have a very low rate of exchange of Na^+ with seawater. Influx and efflux of Na^+ are balanced, but water is not, and a net loss of water amounting to about 0.4% body weight/day occurs primarily through the skin. So the major osmotic problem of *Pelamis* in seawater is water balance, not salt balance (Dunson and Robinson, 1976; Dunson and Stokes, 1983).

The latitudinal distribution of *Pelamis* in the eastern Pacific Ocean lies between the northern and southern 18°C surface isotherms, and its critical thermal maximum and minimum are 36°C and 11.7°C, respectively (Graham et al., 1971). With rapid cooling this snake will stop feeding at 16–18°C; but it has a high resistance to cold and can withstand 5°C for about an hour. In laboratory tests, it did not acclimate to 17°C (Graham et al., 1971), and thus was not able to survive for long periods in water of that temperature. It generally avoids surfacee temperatures cooler than 19°C (Rubinoff et al., 1986). Body temperatures of *Pelamis* caught in surface waters along the Pacific Coast of Mexico, Costa Rica and Panama were 26.9–31.0°C (Dunson and Ehlert, 1971) and in the Gulf of Panama, 28–29°C (Hecht et al., 1974). These snakes reduce feeding at 26°C, and effectively stop feeding at 23°C (Hecht et al., 1974). Brattstrom (1965) gave the body temperature of one as 24.9° (air temperature 25°C), and Pickwell et al. (1983) found a dying *Pelamis* on a southern California beach after the water temperature had dropped to 16°C. *Pelamis* loses efficient motor control of its swimming and floating at 16°C, but slowly acclimated individuals can survive for at least a week at 33–35°C (Hecht et al., 1974). The optimal temperature range for *Pelamis* seems to be 28–32°.

Body temperature is mostly determined by the surrounding water, but Graham (1974b) reported that its dark dorsal body surface absorbs solar radiation, and consistently elevates the body temperature slightly above that of the water when the snake is at the surface in a calm sea. However, there seems no positive evidence that *Pelamis* moves up or down the water column to thermoregulate, and Graham's laboratory experiments indicate the yellow-bellied sea snake neither seeks nor avoids heat when given thermal choices.

Surface swimming is by sideward undulations aided by the laterally compressed tail which may act as a paddle. When it desires to do so, *P. platurus* can move rapidly through the water. However, such activity is only used for local movements (usually while foraging), and it is doubtful if this snake actively swims for long distances. Instead, long-distance dispersal is probably passive, as the snake is moved about while floating in ocean currents. On occasion they may be blown or drift to the extremes of their range in the eastern Pacific, and this is most likely how they rarely reach southern California.

Although a graceful swimmer, *Pelamis* is poorly adapted for crawling on land. The laterally compressed tail is a hinderance and the lack of elongated ventral scutes prevents gripping the ground for crawling. When washed onto a beach *Pelamis* is almost helpless, and soon dies of heat exhaustion, and it may not be able to breathe normally when out of water (Minton, 1966).

Reproduction

Relatively little is known of the breeding habits of *P. platurus*. Breeding populations apparently occur only where the mean monthly water temperatures are above 20°C (Dunson and Ehlert, 1971). Kropach (1975) expected to find seasonal reproduction in the Gulf of Panama because of ecological seasonality of the region, but instead found young of newborn length in every month. McCoy and Hahn (1979) and Visser (1967) obtained similar results in the Philippines and South Africa, so breeding may take place throughout the year. Males are mature at total body lengths of 500 mm or more, but females must grow to more than 600 mm (Kropach, 1975).

Courtship behavior has not been described, but Dunson (1971) presented a photograph of two individuals "knotted" together in mating activity. All mating occurs in the water, possibly near the surface in the slicks.

Pelamis platurus is ovoviviparous and the young are thought to be born at sea, but Minton (1966) found a 230-mm juvenile in a mangrove swamp, so possibly some females may enter such habitats at the time of birth.

Brood size in the eastern Pacific is two to six young (Kropach, 1975). In South Africa, one to eight young may be born (Visser, 1967; Branch, 1979), but Rose (1950) reported South African *Pelamis* broods of "about ten in number." There seems to be no correlation between the number of young produced and female body size (Visser, 1967). The duration of gestation is unknown, but Kropach (1975) observed gravid females in the laboratory and thought it is at least five months and more likely six or more.

Newborn young are 220–260 mm long (Kropach, 1975) and weigh about 9.5–14.3 g (Dunson and Ehlert, 1971).

Growth and longevity

Growth of young *Pelamis* seems rapid, at least during their first year, but the rate is unknown (Kropach, 1975). Kropach (1975) reported that when males reach 500 mm and mature their growth rate slows, while females maintain their growth rate. Individuals shorter than 300 mm are first-year juveniles; those 300–500 mm long are mostly juveniles and subadults, but may include a few adult males; those 500–600 mm are adult males and subadult females; and snakes over 600 mm are adults.

Pelamis apparently does not do well in captivity. Bowler (1977) reported a female lived only two years, one month and seven days, and Shaw (1962) mentioned another captive that lived over two years and four months.

Food and feeding

Foraging seems mostly to occur only at the surface during the day, and is primarily associated with current slicks. Here *Pelamis* employs a type of ambush behavior to catch its fish prey. It floats motionless among the debris trapped in the slick. Small fish seek out the debris to feed or to hide from other fish predators, and they may congregate among floating materials in large numbers. As the fish swim by, *Pelamis* strikes rapidly sideward with its head and seizes the fish in its mouth. This is usually followed by the snake swimming backward with its prey. The prey is immobilized quickly by the snake's venom, which is chewed into the bite wound, and usually swallowed head-first to avoid impalement by the fish's fin spines. Sensitivity to movement in the water and possibly olfaction seem to be more important in prey detection than vision; Heatwole (1987) reported that finely chopped fish dropped into an aquarium holding these snakes causes a "frenzy" in which an object encountered, even another snake, is bitten. Food is swallowed very quickly. Most snakes collected in slicks contain food, while those caught elsewhere usually do not (Kropach, 1971a).

In the eastern Pacific, fish from at least 19 families and 25 genera are eaten (Klawe, 1964; Kropach, 1975): *Abudefduf, Acanthurus, Anchoviella, Blenniolus, Caranx, Chaetodon, Chloroscombrus, Coryphaena, Engraulis, Fistularia, Hypsoblennius, Kyphosus, Lobotes, Lutjanus, Melanorhinus, Mugil, Mulloidichthys, Peprilus, Polydactylus, Pseudupeneus, Selar, Sphoeroides, Sphyraena, Thunnus,* and *Vomer.* Captives may eat additional species (Shaw, 1962; Klawe, 1964). No preference for food size has been found for any size class of *Pelamis* (Kropach, 1975).

Venom and bites

The venom delivery apparatus of the yellow-bellied sea snake is poorly developed when compared to that of other elapids. More adaptive emphasis seems to have occurred in the development of numerous small teeth to hold prey (see dental formula above) than on well-developed fangs. Indeed, the fang situated at the front of the maxilla is hardly larger than the other teeth on this bone. I examined 25 specimens, 438–755 mm long, that had fangs 0.9–2.8 mm long, and a positive correlation between fang length and total body length was evident.

The venom gland is compartmentalized

and contains cuboidal, columnar, and mucoid cell types, and accessory venom glands may be present that circumscribe the main duct in the suborbital region (Burns and Pickwell, 1972).

The venom is neurotoxic in action and very toxic to fish, which usually die within a minute of being bitten. Tu et al. (1975) isolated a major toxin from eastern Pacific *Pelamis* venom which contained 55 amino acids (low for potent neuro-toxins) with four disulfide linkages which seem essential for toxic action. This toxin (which they named *Pelamis* toxin a) comprised 4.5% of the venom. Two other toxins, b and c, made up 0.95 and 1.6%, respectively. In another study published in 1975, Liu et al. also isolated and partially characterized a toxin from the venom of Taiwanese *Pelamis* which contained 60 amino acids; they named it pelamitoxin a. It is not known if these two toxins are the same. That discovered by Liu et al. was very similar to hydrophitoxin b from the sea snake *Hydrophis* and schistosa 5 toxin from the venom of *Enhydrina*—venoms known to be capable of producing human fatalities.

Thirteen *Pelamis*, 62–83 cm in length, produced dry venom yields of 0.9–5.0 mg, with the greatest yields generally associated with the longest snakes, with an average yield of 2.8 mg (Pickwell et al., 1972). Pickwell et al. (1972) estimated the LD_{25} of *Pelamis* venom for humans to be 3.7 mg for a person of 100 pounds (45 kg), 4.4 mg for one of 120 pounds (54 kg), 5.9 mg for a 160-pound (72 kg) person, and 7.5 mg for a human weighing 200 pounds (91 kg). So, it appears that this snake, with its small mouth and fangs and relatively low venom yield, poses little hazard to most humans. The minimum lethal venom dose to a pigeon is 0.075 mg/kg, and to a mudfish (*Saccobranchus*) 0.25 mg/kg (Rogers, 1903). Other lethal venom dosages include: guinea pig, 1 mg/kg (Nauck, 1929); 20-g mice, 0.5 mg/kg (Barme, 1968); *Rhesus* monkey, 0.16 mg/kg (Pickwell et al., 1972); and dog, 0.05 mg/kg (Pickwell et al., 1972).

Human envenomations that may have occurred have been either very mild or very rare, for there is no literature record of symptoms or fatalities attributed to this snake, except possibly a 1693 report mentioned by Taylor (1953) that the bite is mortal and cannot be treated, and Kinghorn's (1956) statement that one death had definitely been recorded in India

many years ago. Kropach (1972) noted six bites of *Pelamis* where there was no effect at all.

Predators and defense

Records of predatory attacks on *Pelamis* are rare: fish—puffer (*Sphoeroides*; Pickwell et al., 1983), snapper (*Lutjanus*; Weldon, 1988); birds—man-of-war (*Fregata*; Wetmore, 1965), lava gull (*Larus*; Reynolds and Pickwell, 1984); mammals—leopard seal (*Hydrunga*; Heatwole and Finnie, 1980), sea lion(?) (*Zalophus*; Reynolds and Pickwell, 1984). An octopus ate a captive (Branch, 1979). Some of these records may represent carrion eating, rather than predation, as studies by Rubinoff and Kropach (1970) showed that potential predators from the eastern Pacific made no attempts to attack *P. platurus*.

Pelamis is rather mild mannered, and divers have been known to swim among them without being attacked. However, Minton (1966) reported that one tried to bite repeatedly when picked up with a forceps.

Unlike many snakes, *Pelamis* does not discharge scent-gland or other cloacal fluids when first captured, even when molested or restrained (Weldon et al., 1991). It does have a pair of active secretory glands in the base of the tail, but these are reduced in size and may have functions other than defense. Analysis of secretions from these glands indicates cholesterol and palmitic, oleic, stearic, and phenyl acetic acids are present, as well as two proteins, two peptides, and carbohydrates indicating the presence of glycoproteins (Weldon et al. 1991).

The bright contrasting yellow belly may be a warning device to would-be predators, or the contrasting dark back and bright venter may act aposematically so that there is no selection of one particular pattern by a predator.

Pelamis faces a problem not encountered by terrestrial snakes: fouling of the body surface by encrusting barnacles (*Chonchoderma, Lepas, Platylepas*) or sessile marine ectoprocts (*Electra, Membranipora*) (Dean, 1938; Kropach and Soule, 1973; Zann et al., 1975; Reynolds and Pickwell, 1984). Fouling may retard swimming and interfere with courtship or mating behavior. Frequent skin shedding helps remove these unwanted guests, as does also the practice of knotting and tight coiling during which organisms attached to the skin may be scraped off between the coils (Shaw, 1962; Pickwell, 1971). Knotting may also help retard predation.

Populations

Pelamis platurus may sometimes occur in huge aggregations. Belcher (in Smith, 1926) reported having seen thousands swimming on top of the water in the Mindoro and Sulu seas, Tu (1976) collected 3,077 in less than a month off the coast of Costa Rica; and Kropach (1971a) estimated thousands to be drifting in surface slicks off Panama. The sex ratio is usually about 1:1; Kropach (1975) reported 340 males out of a sample of 712 snakes. However, one sample of 73 *Pelamis* from the Philippines collected by McCoy and Hahn (1979) contained only 9 (12%) males. The size-class distribution of this group of snakes was strongly bimodal; 48 (46 females, 2 males) were 343–437 mm

long, while the other 25 (18 females, 7 males) were 476–693 mm long. The frequency of size-class distribution may vary during the year (Kropach, 1975), particularly as newborn young enter the population. Myers (1945) reported that on 1 March at Bahia Honda, Panama, the sea snakes observed were largely juveniles with only a smaller proportion of adults, and Kropach (1975) noted that of 278 *Pelamis* collected in September in Golfo Dulce, Costa Rica, 119 (43%) were neonates.

Remarks

Electrophoretic studies of serum and tissue enzymes have shown *Pelamis* to be most closely related to other hydrophine sea snakes of the genera *Aipysurus, Emydocephalus, Enhydrina, Hydrophis,* and *Thalassophina* (Cadle and Sarich, 1981; Mao et al., 1983; Murphy, 1988). This generally substantiates the sea snake relationships proposed by McDowell (1972) and Voris (1977).

Viperidae: Viperid snakes

The vipers are venomous snakes that evolved from colubrid ancestors that differed from those which eventually gave rise to the elapids. The family consists of about 200 species occurring in Asia, Europe, Africa, and the Americas.

Vipers have evolved an advanced venom-injecting apparatus. Their maxillae have become shortened horizontally while becoming deep vertically and are capable of movement on the prefrontal and ectopterygoid bones. This allows the elongated hollow fangs on the maxillae to be rotated until they lie against the palate when the mouth is closed. During the strike the maxillae move causing the fangs to rotate downward and forward into stabbing positions. Such movable fangs are termed solenoglyphous. In contrast to the elapids, the viperid maxilla does not contain any other teeth but the fang.

Postfrontal, coronoid, and pelvic bones are absent, as also are premaxillary teeth. The prefrontal bone does not contact the nasal bone, and the ectopterygoid is elongated. A scalelike supratemporal bone suspends the quadrate. The hyoid is either Y-shaped or U-shaped with two long superficially placed, parallel arms. All body vertebrae contain elongated hypapophyses. The left lung is absent.

The hemipenis is deeply bilobed or double, with proximal spines, distal calyces, and a bifurcate or semicentrifugal sulcus spermaticus. Head scalation is essentially like that of colubrids, but some have the dorsal surface covered with small scales instead of enlarged plates. Pupils are vertical slits. Body scales are keeled. The secretions produced by the venom glands are predominantly hemotoxic, but several species contain neurotoxic components in their venom. Reproduction is either oviparous or ovoviviparous (ovoviviparous in North American species).

The family comprises three subfamilies, but only one, the Crotalinae, occurs in the Americas. Members of the Crotalinae are called pit vipers because of the small hole that opens between the eye and nostril. The maxilla is hollowed out dorsally to accommodate the pit, and the membrane at the base of the hole is extremely sensitive to infrared radiations, especially those emitted from warm-blooded prey.

The pit organ is usually located in the loreal scale of those crotalids having such a scale, or in the position on the side of the face corresponding to the region of this scale in those that do not. Because of this location, the cavities are sometimes referred to as loreal pits. In rattlesnakes (*Crotalus, Sistrurus*) the opening lies below a line from the nostril to the orbit and slightly closer to the former (Klauber, 1972). Each pit is subdivided into an outer chamber and a smaller inner chamber by a cornified epidermal membrane about 0.025 mm thick. The principal component of the membrane is a single layer of specialized parenchyma cells with osmiophil reticular cytoplasm which lies between two layers of extremely attennuated epidermis (Lynn, 1931; Noble and Schmidt, 1937; Bullock and Fox, 1957). The membrane is innervated by the trigeminal cranial nerve (V), particularly by its ophthalmic and maxillary branches. These pits are similar, but not identical, to the elaborate lip pits present between the labial scales of pythons and some boas, and it is possible the crotalid pits may have evolved from those on the upper jaw of boids (Noble and Schmidt, 1937). Many axons enter the membrane and the innervation is rather dense. The axons lose their myelin, taper to about 1 micron, then expand into flattened palmate structures which bear many branched processes terminating freely over an average area of about 1,500 square microns, overlapping only slightly with adjacent units but leaving virtually no area unsupplied (Bullock and Fox, 1957). *Crotalus* has about 500–1,500 axon endings per mm².

Bullock and Fox (1957) reported the transmission spectrum of the fresh membrane in rattlesnakes (*Crotalus*) shares broad absorption peaks at 3 and 6 microns and about 50% is transmitted in other regions out to 16 microns. The visible light spectrum is at least 50%

transmitted (probably much is lost by reflection); strong absorption takes place at wavelengths shorter than 490 microns. A continual transmission of impulses occurs from the pit membrane to the brain (Bullock and Cowles, 1952). The rate of this continual message is independent of the snake's body temperature but is dependent upon the average radiation from all objects in the receptive field. The membranes are highly sensitive to infrared wavelengths of 15,000–40,000 Å, and any warm or cold object causes a temporary change in the rate of impulse transmission, the response being created to sudden temperature changes. Thus, the pit organ serves to recognize the presence of any object that is warmer or colder than its surroundings. The field is determined to include a cone extending horizontally from 10° across the midline to a point approximately at right angles to the rattlesnake's body from 45° above to 35° below the horizontal. This allows the receptive fields of the two pits to overlap in front of the snake and together they survey a 180° field anterior to them. Sensitivity varies with the wavelength, but is generally greater to infrared emissions in the range of 2–3 microns than to shorter or longer wavelengths. The snakes seem to be able to detect differences in temperature of as little as 0.2°C or less (possibly a temperature change of about 0.003°C in 0.1 second may be detected). However, the pit membrane is not as sensitive as it might be; in fact, Bullock and Fox (1957) reported its sensitivity is not higher than that calculated for human thermal receptors. Still, Bullock and Diecke (1956) have shown it is certainly sensitive enough to detect objects with surface temperatures differing by only 0.1°C, and there is probably no adaptive advantage to having a more sensitive receptor. Interestingly, Chiszar, Duvall et al. (1986) and Dickman et al. (1987) have discovered that certain membranes within the mouths of pit vipers may also be sensitive to thermal stimulation. Recent study has shown that there is a merging of infrared and visual integration in the optic tectum of rattlesnakes (Hartline et al., 1978; Newman and Hartline, 1981). De Cock Buning (1983) thought that, depending on the influence of ecological demands, visual or chemical cues are the main information in the behavioral phases before the strike, but in situations with little input (ie., at night, in rodent's burrows, etc.), hunting behavior is guided mainly by radiation of warm objects.

Pit vipers are believed to have evolved from Old World vipers (Darlington, 1957; Brattstrom, 1964), but how they reached the Americas is unknown. Today, 17 species in three genera (*Agkistrodon, Crotalus, Sistrurus*) occur in the United States and Canada.

Agkistrodon contortrix (Linnaeus, 1766)
Copperhead
Plates 10–14

Recognition

The copperhead is a pinkish to grayish-brown, stout-bodied snake (maximum length, 135 cm) with an orange to copper or rust-red, unpatterned head, and a series of 10–21 brown to reddish-brown, saddle-shaped bands on the body. These dorsal bands are usually broader along the sides of the body and narrower across the dorsum, presenting a dumbbell-like shape, and small dark spots may occur in the light spaces between the bands (see Geographic variation and McDuffie, 1963, for discussions of variation in the banding pattern). The tail lacks a rattle and is yellow, brown, or green. The head pigmentation is usually lighter below the eye than dorsally, and a dark postocular stripe may be present in some, but no light stripes are present. A pair of small dark spots may occur on the parietal scales. The venter is pink, light brown, or cream colored with dark blotches along the sides of the ventrals. As many as 23–25 scale rows may occur anteriorly, but usually only 23(21–25) rows at midbody and 21 before the anal vent. The body scales are keeled and contain apical pits; the anal plate is undivided. Pertinent lateral head scalation includes 2 nasals, a loreal, 2–3 preoculars, 2–3 suboculars, 3–4 postoculars, several rows of temporals, 8(6–10) supralabials, and 10(8–12) infralabials. The eye has an elliptical pupil, and a heat-sensitive pit is located between the nostril and eye. There are 138–157 ventrals and 37–57 subcaudals with no sexual differences. The hemipenis, illustrated by Malnate (in Gloyd and Conant, 1990), is deeply bifurcate with approximately 35 large spines on the basal third of the two lobes. Most of these are straight but some are slightly hooked. The distal two-thirds of each lobe is covered with small flattened papillae,

each ending in a spine, but not arranged in regular rows. No spines occur in the crotch, and the sulcus spermaticus is forked (see Reese, 1947, for the embryological development of the hemipenis). The shortened maxilla contains only the elongated fang. The rest of the dental formula is 5(3–6) palatine teeth, 13–20 pterygoid teeth, and 12–18 teeth on the dentary.

Males have tails 11–19% (mean, 14.2%) of the total body length; those of females are 11–17% (mean, 13.8%) of the total body length.

Karyotype

Diploid chromosomes total 36: 16 macrochromosomes (4 metacentric, 8 submetacentric, 2 subtelocentric; sex determination is ZW in females and ZZ in males), and 20 microchromosomes (Baker et al., 1972; Zimmerman and Kilpatrick, 1973; Cole, in Gloyd and Conant, 1990). The smaller female W chromosome is acrocentric or subtelocentric to telocentric (Zimmerman and Kilpatrick, 1973; Cole, in Gloyd and Conant, 1990).

Fossil record

Fossil remains of the copperhead have been found in deposits from the Miocene (Clarendonian) of Nebraska and upper Pliocene (Blancan) of Kansas, Nebraska and Texas (Holman, 1979; Conant, in Gloyd and Conant, 1990), and from the Pleistocene: Blancan, Kansas; Irvingtonian, Kansas and West Virginia; Rancholabrean, Kansas, Missouri, Pennsylvania, and Texas (Holman, 1981, 1982; Parmley, 1988). Also, a broken vertebra of Rancholabrean age identifiable to the genus *Agkistrodon* but not to species level has been found in Alabama (Holman et al., 1990).

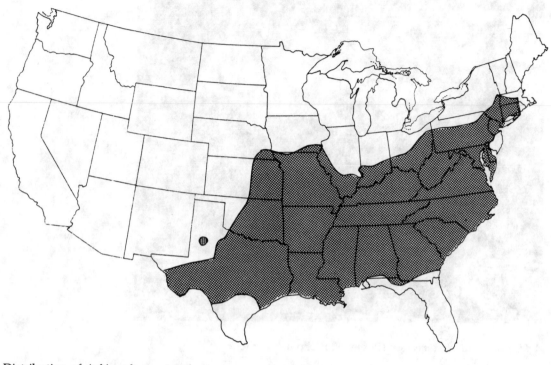

Distribution of *Agkistrodon contortrix*.

Distribution

Agkistrodon contortrix occurs from western Massachusetts and Connecticut and southeastern New York west through the southern two-thirds of Pennsylvania, southern Ohio, Indiana, Illinois, to Missouri and eastern Kansas, and south to Georgia and the panhandle of Florida, Mississippi, Louisiana, eastern and central Texas, and the extreme portions of northern Coahuila and eastern Chihuahua in Mexico. Except in the northeastern part of its range, *A. contortrix* is mainly limited by the southern boundary of the Wisconsinan Glacier (Gloyd and Conant, 1990).

Geographic variation

Five subspecies have been described (Gloyd and Conant, 1943, 1990; Gloyd, 1969; Conant and Collins, 1991). *Agkistrodon contortrix contortrix* (Linnaeus, 1766), the southern copperhead, is pale gray to pinkish in ground color; the crossbands are well marked and very narrow across the back (often they are medially separated), dark lateral spots occur on the venter, and the tip of the tail is pinkish or greenish yellow. It is found from southeastern Virginia southward along the coastal plain to Gadsden and Liberty counties, Florida, and west to eastern Texas and northward in the Mississippi Valley to southern Missouri and southwestern Illinois. *A. c. mokasen* (Palisot de Beauvois, 1799), the northern copperhead, is more reddish brown, has a brighter coppery head and darker, wider crossbands, round dark spots on the side of the venter, and often a greenish tail tip. It ranges from Massachusetts and Connecticut southward on the piedmont and highlands to Georgia, Alabama, and northeastern Mississippi, and west through southern Pennsylvania and the Ohio Valley to Illinois. *A. c. lacticinctus* Gloyd and Conant, 1934, the broad-banded copperhead, can be found from extreme south-central Kansas (Chautauqua and Cowley counties) southward through central Oklahoma and central Texas to Medina and Atascosa counties. The

Agkistrodon contortrix mokasen (Roger W. Barbour, Kentucky).

deep reddish-brown crossbands are not much narrower across the back than along the sides, the tip of the tail is greenish gray, and the venter is cream colored with small reddish-brown or black spots and larger irregularly shaped dark blotches. *Agkistrodon c. pictigaster* Gloyd and Conant, 1943, the Trans-Pecos copperhead, ranges in western Texas from Crockett and Val Verde counties westward through the Big Bend region and Davis Mountains to northern Coahuila and eastern Chihuahua, Mexico. It resembles the broad-banded copperhead in its dorsal pattern, but has a dark reddish-brown to black mottled venter in sharp contrast to cream-colored areas extending onto the belly from the sides, and often the reddish-brown midbody crossbands contain a pair of dark, rounded, lateral spots. *Agkistrodon c. phaeogaster* Gloyd, 1969, the Osage copperhead, resembles the northern copperhead, but has a paler ground color, making the dark dorsal bands more pronounced, no small dark spots between the dark bands, a yellowish-green tail tip, and a grayish, mottled venter. It

is found only in northern and central Missouri, eastern Kansas, and the northeastern corner of Oklahoma; it intergrades with *A. c. lacticinctus* in southern Kansas and Oklahoma (Webb, 1970; Fitch and Collins, 1985).

Broad bands of intergradation occur where the ranges of the various subspecies meet (Gloyd and Conant, 1990).

Confusing species

Most colubrid snakes (*Elaphe, Lampropeltis, Nerodia, Heterodon*) lack a facial pit, have rounded pupils, and may have patterned heads. *Trimorphodon* has elliptical pupils, but also has a patterned head. Cottonmouths, *A. piscivorus*, have a band through the eye and lack a loreal scale, and *Crotalus* and *Sistrurus* have a tail rattle.

Habitat

In the eastern United States, *A. contortrix* inhabits oak-hickory hillsides containing rock crev-

Agkistrodon contortrix mokasen (Roger W. Barbour, Kentucky).

ices and slides in woodlands. Along the southern coastal plains it lives in wet woodlands and along the borders of swamps. It ascends from the coastal plain to elevations of almost 1,200 m. Eastern *A. contortrix* utilize relatively open areas with a higher rock density and less surface vegetation than do sympatric *Crotalus horridus* (Reinert, 1984a). In Texas and northern Mexico, it occurs in dry upland deciduous or coniferous woodlands, or even desert, with sandy soils and rocky outcrops and cliffs. There, it is more commonly found in riparian woodlands near permanent or semipermanent waterways. Over its entire range it occurs at elevations from near sea level to over 1,500 m.

Gravid female *A. c. mokasen* prefer microhabitats that are clearly separated from those occupied by males and nonreproductive females (Reinert, 1984b). Rocky, open, sparsely forested sites with warmer soil temperatures are used until the young are born. Such sites may be close to the hibernaculum or some distance away.

Behavior

Although it is active during daylight in the spring (morning and afternoon) and fall (particularly in the morning), *A. contortrix* becomes crepuscular or nocturnal during the hot summer months and may be quite active in summer after an evening shower. Annually, it is active from April to late October in the northern parts of its range, with most being seen from May to September, but farther south the copperhead may be active from March until early December, or even emerge on warm days during January and February. In Smith County, Texas, it is active from March to November, with peak activity from April to July and again in September and October (Ford et al., 1991).

In the fall they become gregarious and may crawl some distance to a communal hibernaculum, which is sometimes shared with *Crotalus horridus* and smaller numbers of *Coluber constrictor* and *Elaphe obsoleta*. Copperheads usually return to the same area each win-

Agkistrodon contortrix pictigaster (John H. Tashjian, Ft. Worth Zoo).

ter (Fitch, 1960b). Weathered outcrops with crevices extending below the frost line are often used, as also are caves, gravel banks, old stone walls and building foundations, animal burrows, hollow logs and stumps, and sawdust piles. Drda (1968) observed several overwintering in a Missouri cave that were quite active most of the winter; of course, air temperatures in the cave were relatively constant and considerably above freezing. Juveniles were found deeper in the cave than adults. At Mason Neck National Wildlife Refuge, Fairfax County, Virginia, 15–25 copperheads, most about 50 but a few to 88 cm in total length, hibernate each year beneath a broken concrete sidewalk. Formerly this population used the broken foundation of an abandoned house as their overwintering retreat, but moved about 50 m to the sidewalk site when the house was torn down.

Along the coastal plain from Georgia to Virginia many copperheads hibernate singly (Neill, 1948; pers. obs.). In Virginia it is usually the juveniles who hibernate singly, but I have found some adults also alone. Usually hollow logs are used as individual hibernacula.

Fitch (1956) determined the preferred body temperatures of *A. contortrix* from Kansas to be 26–28°C, and the preferred activity ranges probably 23–31°C. Brattstrom (1965) reported the minimum and maximum voluntary body temperatures for 61 individuals to be 17.5 and 34.5°C, but Fitch (1960b) recorded body temperatures as low as 12.4°C for those under rocks in the early spring. The critical thermal minimum and maximum are 4–9°C and 41°C, respectively.

Most basking that I have observed has occurred in either the spring or the fall. Fitch (1960b) found that gravid females basked more often and seemed to prefer warmer body temperatures. Sanders and Jacob (1981) monitored the body temperatures of 20 copperheads by telemetry. Some basking on clear winter days achieved body temperatures of 10°C or higher. They also found that body temperatures recorded in the summer varied among snakes of different body lengths, and that there was a significant negative correlation between snout-vent length and the critical thermal minimum.

Heart rate and breathing rate increase with rising temperatures (Jacob and Carroll, 1982), but temperature has no effect on heart rate or breathing response over the temperature range at which copperheads are normally active.

Three types of movements by copperheads have been identified by Fitch (1960b): movements within the home (or activity) range, abandonment of one home range and occupancy of a second, and migrations to and from hibernacula.

Home ranges of *A. contortrix* in Kansas varied between 3.4 ha for females and 9.8 for males (Fitch, 1958; 1960b). The home range diameter of males averaged 354 m, that of females 210 m. Fitch (1960b) reported movements on the summer range of 1.5–378 m for individuals that remained within their home range, and 442–762 m for those that had apparently shifted home ranges. He also recorded distances of spring or fall captures at hibernacula from points of capture on the summer range of 232–1,183 m. Males traveled longer average distances from hibernacula (656 m) than did females (406 m).

Copperheads in Kansas equipped with radios averaged a mean displacement of 11 m per day (including days not moved). Mean distances moved for only those days when movement occurred were 18 m for a male, 12 m for a nongravid female, and 12 m for one which was gravid (Fitch and Shirer, 1971).

A marked copperhead in the Shenandoah National Park, Virginia, exhibited homing ability when it returned within two days to the place of original capture after having been displaced 0.8 km to a known den site (Martin, in Gloyd and Conant, 1990).

Copperheads may climb into low bushes or trees after prey or to bask. Swanson (1952) observed several young resting on laurel bushes 5 cm or more above ground, Engelhardt (1932) captured one in a fork of a tree 1.2 m above ground, and Johnson (1948) saw one almost 4 m high in a tree. Fitch (1960b) also reported instances of climbing in *A. contortrix*, and I have regularly seen them crawling or basking on the sides of wind-tilted trees at heights of 2–5 m.

Agkistrodon contortrix has no special affinity for water, but does favor damp habitats. It will, however, enter bodies of water voluntarily, and has been observed swimming on numerous occasions (Smith and Sanders, 1952; Fitch, 1960b; Groves, 1977; pers. obs.).

Males sometimes engage in combat dances (Gloyd, 1947; Shaw, 1948a; Fitch, 1960b; Collins, 1974; Mitchell, 1981; Schuett, 1986). After meeting, the male snakes make body contact and then almost immediately rise up facing each other to a height of 30–40% of their body lengths. Much tongue flicking occurs, and the males seem to try to outstare each other. Next, both sway back and forth in unison with heads bent at a sharp angle. Sometimes their elevated bodies are parallel to each other or one snake may actually turn its back toward its opponent. One male then leans over and tries to push the other's head and neck to the ground (topping behavior). The attacked snake responds by entwining its aggressor and tries to pin it by pushing with the anterior body and neck. The pushing match may continue for some time (usually 20–30 minutes, but sometimes over two hours), but eventually one snake, usually the shorter, is pinned to the ground, breaks off contact and crawls away.

Studies by Schuett (1986) indicate that male *A. contortrix* participate in these combat bouts only during the breeding season. Female defense seems to be involved, so Schuett thought mate competition the major function of these encounters. However, this is not certain, as combat will occur in captivity when no female is present. Perhaps male combat is a form of food competition or territoriality involving a food source or critical space.

Reproduction

In Kansas, at an age of three years most, but not all, females are of small adult size (50 cm snout-vent length) and sexually mature (Fitch, 1960b). In northeastern Texas, females with snout-vent lengths of 47–60 cm are capable of reproducing (Ford et al., 1990). The age and size of maturity in the male copperhead is unknown.

At the time of spring emergence in Kansas, the ova are small (1–9 mm) and occur in several size groups, suggesting that they may mature at different times (Fitch, 1960b). In May the ova grow rapidly and ovulation occurs in the latter part of the month. Observations that only about 60% of females breed each season seem to indicate a biennial female reproductive cycle (Fitch, 1960b; McDuffie, 1961), but Vermersch and Kuntz (1986) reported that copperhead females from southern Texas produce young each year.

Numerous matings have been observed by various biologists and these tend to indicate that, although copulation has taken place from April to October, the prime breeding periods are April–May and September–October. Fitch (1960b) found active sperm in the cloaca of a wild-caught female *A. c. phaeogaster* on 19 May. On 24 June a second examination showed still abundant motile sperm, but about 75% of the sperm were dead or very slow in their movements. These data suggest that copperhead sperm can be stored in the cloaca by the female for only a relatively short time. Howarth (1974) reported that sperm storage in female copperheads lasts only about 11 days; however, Schuett and Gillingham (1986) reported overwinter viable sperm storage, and Allen (1955) noted a case of a female *A. c. laticinctus* that gave birth to five young at least a year after separation from any male. Sperm apparently survive much longer in the upper end of the oviducts in vascular tissues specialized as seminal receptacles (Fitch, 1960b). Fitch (1960b) found active sperm in females in April, May, June, and October, but other females examined in April, June, July, August, October and November were negative. Of 59 sexually mature female *A. c. phaeogaster* obtained by Gloyd (1934) in April and May, 21 contained active sperm. He examined the vas deferens of males in April–August and October and found more or less active sperm in all. Schuett (1982) reported that an October copulation resulted in the birth of young on 3 August.

Schuett and Gillingham (1988) analyzed the courtship and mating sequence of *A. contortrix* in detail. Courtship is always initiated and performed by the male, and there are seven distinct male courtship behaviors: (1) touch mounting (contact with the female is made with the snout, and the head and neck are elevated and placed on her back), (2) chin rubbing, (3) dorsal advancing with chin rubbing (simultaneous advancing and chin rubbing while mounted on the female), (4) tail searching (the entire tail is oriented beside that of the female and quivers; while this takes place the tail and vent region are pushed beneath her tail, forming a loop; the quivering stops and her tail and vent region are stroked one to three times; the male's tail remains looped if intromission is not

achieved), (5) no moving while mounted (all motion ceases, with the male's head and neck held close to the female's body), (6) stopping (all body motion ceases except tongue flicking); and (7) dismounting. These behaviors may always take place in sequence. The female's response to male courtship involves 10 behavioral acts: (1) advancing (crawling slowly or rapidly forward), (2) remaining stationary (in an outstretched or coiled position), (3) tail waving (the elevated tip, up to 30°, is moved slowly side-to-side), (4) whipping of elevated tail, (5) tail vibrating (rapid side-to-side movements that may produce a tapping or buzzing sound), (6) body flattening (the body is dorsoventrally compressed), (7) waste elimination from the vent, (8) cloacal gapping (the elevated tail, up to 45°, is arched and the anal plate lowered; intromission is achieved only after this act), (9) head raising (1–3 cm above substrate), and (10) tongue flicking.

During copulation, Schuett and Gillingham (1988) identified eight different male behaviors: (1) intromission, (2) dismounting until only posterior contact is maintained, (3) stationary (no movements, even tongue flicking), (4) cloacal contracting (sporadic contractions of the cloaca and cloacal region), (5) hemipenial enlargement, (6) insemination, (7) backward crawling (moving backward when the female crawls forward), and (8) hemipenial retraction (withdrawal of the hemipenis and its return in a relaxed state within the male's tail). Females react in three ways during mating: (1) remaining stationary (no movements including tongue flicking), (2) advancing (crawling slowly forward pulling the male along), and (3) tail undulating (the elevated tail, > 45°, moving similar to caudal luring movements; Neill, 1960).

The young are ovoviviparous (although Dolley, 1939, reported the embryos to be attached to and developing within the oviduct), and most are born in August or September, but some birthing may take place in July, October and November. The gestation period may range from 105 to about 150 days, and females may aggregate about the time of parturition, particularly near den sites. During birth, the female extrudes the young with a series of muscular contractions of the posterior part of her abdominal region, often while lying in a semicircular coil. The newborn young are enclosed in a membraneous, transparent sac. Emergence from this sac may take as long as 15 minutes. The head is first pushed through the membranes and the young snake then takes a deep breath or a series of deep, open mouthed gasps before continuing its struggles to free itself. Chenowith (1948) observed a parturition in which two young were born in about 10 minutes. The postbirth weights of five females from northeastern Texas were 66.3% (4 young), 67.6% (7), 58.1% (7), 72.7% (10), and 72.2% (7), respectively (Ford et al., 1990).

A single brood may contain 1–21 young (Wright and Wright, 1957; White, 1979; Gloyd and Conant, 1990), but 4–8 are most common. Specific records from published literature and from Fitch's (1960b) Kansas field study indicate a total of 1,068 eggs or young from 203 females with an average brood size of 5.26 young. The longest subspecies *contortrix* and *mokasen* produce the largest broods, up to 18 and 21 young, respectively; while those of *laticinctus* and *phaeogaster* may include up to 11 and 13 young, respectively; and the shortest race *pictigaster* usually produces broods with no more than 3–4 young. Larger females of each subspecies produce the larger broods (Fitch, 1960b; Seigel et al. 1986; Ford et al., 1990; pers. obs.). Brood size may vary from year to year, apparently owing to different environmental conditions (Seigel and Fitch, 1985). Neonate length is also positively correlated to female body length (Ford et al., 1990).

Newborn young are usually 200–250 mm in total length (but may range from 170 to 300 mm), weigh about 4.5–11.0 g, and are patterned like the adults, but paler in ground color. They normally have yellow tails. Functional and replacement fangs are present and the young are venomous (Stadelman, 1929b).

In laboratory tests, Schuett and Gillingham (1986) found that sperm resulting from fall copulations retain their fertilizing ability until spring and that fall and spring ejaculates of different males can overlap prior to spring ovulation producing litters of multiple parentage.

Interspecific mating has occurred in captivity with *A. piscivorus* (Mount and Cecil, 1982) and *Crotalus horridus* (Smith and Page, 1972).

Growth and longevity

Males apparently grow faster than females (Fitch, 1960b). Yearlings in Kansas have snout-vent lengths of 30–40 cm; those two years old, 40–57.5 cm; three years old, 45–62.5 cm; four years old, 53–73 cm; and five years old, 55–75 cm (estimated from Fitch, 1960b). In Indiana, yearlings are 38–43 cm and two-year-olds, 53–59 cm (Minton, 1972).

The longevity record for *A. contortrix* is 29 years, 10 months and 6 days (Bowler, 1977), and several other individuals are known to have survived over 20 years (Gloyd and Conant, 1990).

Food and feeding

A copperhead may consume over twice its body weight in prey each year; Schoener (1977) reported an active individual ate eight meals which totaled 1.25 times its body mass during its annual activity period. Although containing a heat-detecting facial pit, *A. contortrix* is not restricted to warm-blooded prey. It is known to eat millipedes, various insects (dragonflies, beetles, cicadas, mantids, grasshoppers, lepidopteran larvae), salamanders (*Plethodon, Pseudotriton*) chorus frogs (*Pseudacris*), small ranid frogs (*Rana*), narrow-mouthed toads (*Gastrophryne*), small toads (*Bufo*), turtles (*Sternotherus, Terrapene*), lizards (*Cnemidophorus, Crotaphytus, Sceloporus, Eumeces, Scincella, Ophiosaurus*), snakes (*Carphophis, Crotalus, Diadophis, Elaphe, Rhinocheilus, Thamnophis, Tantilla*), small birds, (*Catharus, Dendroica, Petrochelidon, Pipilo, Seiurus, Sturnus, Zonotrichia*), bats (*Tadarida?*), moles (*Parascalops, Scalopus*), shrews (*Sorex, Cryptotis, Blarina*), mice (*Microtus, Peromyscus, Reithrodontomys, Mus, Napaeozapus, Synaptomys, Zapus*), small rats (*Neotoma, Sigmodon*), chipmunks (*Tamias*), young gray squirrels (*Sciurus*), and juvenile rabbits (*Sylvilagus*) (Uhler et al., 1939; Barbour, 1950; Hamilton and Pollack, 1955; Wright and Wright, 1957; Fitch, 1960b, 1982; Herreid, 1961; Murphy, 1964; Garton and Dimmick, 1969; Tennant, 1985; McCrystal and Green, 1986; Gloyd and Conant, 1990; pers. obs.). Carrion may also be consumed (Mitchell, 1977). Smaller individuals eat insects, salamanders and lizards, while adults prey more heavily on mammals. Gravid female copperheads usually do not eat, but when they do, they consume smaller volumes of food than either male or nongravid females (Garton and Dimmick, 1969).

Differences in feeding patterns may also

occur within populations of copperheads. Garton and Dimmick (1969) found that in their study population in Tennessee, males fed mainly on voles (*Microtus*) and caterpillars, nongravid females ate white-footed mice (*Peromyscus*), voles (*Microtus*), and birds (unidentified), and gravid females consumed lizards (*Sceloporus*) and shrews (*Blarina, Cryptotis*). The diet of the gravid females best represented the available species in their selected microhabitat.

The prey most frequently taken by copperheads in Kansas are voles (*Microtus ochrogaster, M. pinetorum*, 24%), cicada (*Tibicen pruinosa*, 15%) and white-footed mice (*Peromyscus* sp., 13%) (Fitch, 1982). *Microtus* sp. and *Peromyscus* sp. compose the greatest estimated biomass (1,051 g/ha), and so are the most readily available prey.

Adults are mostly ambushers, although as they are primarily nocturnal, active hunting behavior may have been missed. Young copperheads actively stalk much of their prey, but Neill (1960) reported that the yellow tail of the newborn may be used to lure small frogs. Large prey is bitten and released to be tracked later when the venom has taken effect; small prey and birds are often retained in the mouth until dead.

Venom and bites

Snakes of the genus *Agkistrodon* have the same typical solenoglyphous fangs as other vipers (Ernst, 1964, 1965, 1982; Kardong, 1979). A series of 214 copperheads 17–110 cm in total body length had fangs 1.1–7.2 mm long (Ernst, 1965, 1982); fang length increased linearly with growth in body and head length.

A newborn of the *Agkistrodon*-complex has fully functional fangs and is capable of injecting venom (Boyer, 1933). These fangs, however, do not remain with the snake throughout its life. They are shed and replaced periodically, an adaptation for replacing broken or loose fangs. A series of 5–7 (in less than 3% of snakes examined) replacement fangs occur in the gums behind and above the functional fang in alternating sockets on the maxillary bone (Ernst, 1982). The replacement fangs lie close together and those distal to the functional fang may be only 0.1–0.3 mm apart. In graduated lengths, they may range from a first reserve fang slightly longer than the functional fang (but never more than 0.2 mm longer) to only a short spike about 0.2 mm long in the last of the series. As with rattlesnakes (Klauber, 1939), the graduated reserve fang series shifts forward to occupy an alternate series of sockets on the maxilla. These sockets are divided by a wall of tissue that separates the developing fangs that will enter the outer socket from those that will enter the inner socket. The socket of the first reserve fang is also separated from that of the functional fang by a membrane. Beside the functional fang is a vacant socket into which the first reserve fang migrates just prior to the shedding of the functional fang.

Such a replacement series is even found in the newborn. The fangs do not develop as a complete unit, but rather from the tip (represented by the most distal replacement fang) upward, and the hollow-tube shape is in evidence from the earliest period of development in which shape can be ascertained. Approximately 19% of the time, one or both fangs were in the process of replacement (Ernst, 1982). The mean replacement rate for the genus *Agkistrodon* is 21.1%; Fitch (1960b) found 19.7% of the *A. contortrix* fangs he examined were being replaced, and reported a 33-day fang shedding cycle.

The venom is highly hemolytic, and mice or rats dissected an hour or two after having been bitten show massive hemorrhaging. The total yield increases with body length (Jones and Burchfield, 1971), and a large copperhead may produce up to 0.29 ml of venom. The usual range of venom yield is 40–70 mg, and probably 100 mg or more are needed to kill an adult human (Minton and Minton, 1969). One bite may discharge 25–75% of the contents of the venom glands (Fitch, 1960b), and juveniles are as toxic as their parents (Minton, 1967). Venoms of the individual subspecies of *A. contortrix* do not differ significantly in their biological activities, and interspecific differences are more pronounced than individual variation in biological activities (Tan and Ponnudurai, 1990). Copperhead venom exhibits moderately high levels of protease, low alkaline, phosphomonoesterase and L-amino acid oxidase levels, high arginine esterhydrolase and hyluronidase activity, but no phosphodiesterase activity (venom chemistry is discussed in detail by Markland, 1988; Markland et al., 1988; Dyr et al., 1989, 1990; Tan and Ponnudurai, 1990; and Sturzebecher et al., 1991).

Symptoms of copperhead bites include pain and swelling, weakness, giddiness, breathing difficulty, hemorrhage, either an increased or a weakened pulse, occasionally shock and hypotension, nausea and vomiting, gangrene, ecchymosis, edema, unconsciousness or stupor, fever, sweating, headache, and intestinal discomfort (Hutchison, 1929; Campbell and Lamar, 1989). *A. contortrix* is responsible for many bites each year, but mortality from these bites is almost nonexistent. Although other fatal cases may have been documented, the only record of a death-producing bite that I could find was reported by Amaral (1927) of a 14-year-old boy bitten on the finger and treated too late with antivenin. Campbell and Lamar (1989) reported the fatality rate from a copperhead bite has been estimated to be 0.01%, and Minton and Minton (1969) reported a similar low rate of 0.3%. Probably only the very young or old need worry about this snake. Case histories of bites are given by Boyer (1933), McCauley (1945), Fitch (1960b), and Diener (1961).

The serum of *A. contortrix* contains components capable of neutralizing enzymes from vemon of its own species (Weinstein et al., 1991).

Predators and defense

Bullfrogs (*Rana*), ophiophagous snakes (*Coluber, Drymarchon, Lampropeltis, Micrurus*), opossums (*Didelphis*), coyotes (*Canis*), cats (*Felis*), horned owls (*Bubo*), and red-tailed hawks (*Buteo*) are natural enemies, and in captivity moles (*Scalopus*) have killed and eaten young copperheads (Fitch, 1960b). A young captive *A. contortrix* has also been seen eating a dead litter mate of similar size (Mitchell, 1977), but it is not known if this species is naturally cannibalistic. Over much of the range, habitat destruction, insecticide poisoning, and the automobile have severely reduced populations.

Copperheads are rather docile creatures. They usually lie motionless in a loose coil or freeze in place if crawling when an intruder is first detected, and this habit, along with their camouflaged pattern, makes them very dangerous. Many bites have resulted from persons unwittingly stepping on, sitting on, or touching unseen snakes. When touched, they often quickly strike, but at other times just remain quiet or try to crawl away. When handled, they spray musk on their handler. Contrary to the folktale, this musk has its own odor and does smell like cucumbers.

Populations

Copperheads can occur in large populations, especially around hibernacula or sites with a high prey density. In July, 1960, Barbour (1962) captured seven adults in less than 15 minutes in an area no larger than 3 × 6 m in Breathitt County, Kentucky, and I have caught 10 in as many minutes at another Kentucky site. Over 34% (105/305) of the snakes collected by Ford et al. (1991) at Sheff's Wood, Smith County, Texas, during a four-year period were copperheads. The total population, based on a 10-year census, in Fitch's (1960b) Kansas study area was approximately 1,664 individuals, a density for the 88 hectares of 18.8/ha. During 30 years of study, 2,681 copperheads were captured at the site (Fitch, 1982). The biomass of *A. contortrix* was estimated to be 0.80 kg/ha on the plot, the adult to juvenile ratio to be 4.26:1, and the adult male to female ratio to be 1.22:1 (Fitch, 1982).

In the populations studied by Fitch (1960b) and Vial et al. (1977), individuals of the smaller and younger size and age classes were more numerous. Snakes older than eight years represented only 5% of the population, while those no older than two years comprised 55%. Vial et al. (1977), in constructing life tables for *A. contortrix* based on data from Fitch (1960b), estimated a skewed sex ratio of 74% males to 26% females at birth, but calculated a greater mortality rate in males and predicted the sex ratio would reach unity by the eighth year. This is interesting since the sex ratios of litters are usually much closer to 1:1; Gloyd (1934) reported that in a total of 69 young in 16 litters, 29 (42%) were females and 40 (58%) were males.

Remarks

Studies of the immunoelectrophoretic patterns of venom within the genus *Agkistrodon* indicate that the American species are closely related to the Japanese species *A. halys* (Tu and Adams, 1968). Further comparisons of venom from the three New World *Agkistrodon* by Jones (1976) have revealed geographical differences among and within the three species, and that differences exist in the venoms of the various subspecies of *A. contortrix*. His work also shows a greater affinity of the neotropical species *A. bilineatus* to *A. contortrix* than to *A. piscivorus*, contrary to former thought (Brattstrom, 1964). However, comparisons of skin kerotin biochemistry by Campbell and Whitmore (1989) suggest that *A. contortrix* is not so closely related to *bilineatus*, and that the three American species of *Agkistrodon* may not be monophyletic. *A. contortrix* and *A. piscivorus* seem most closely related, while *A. bilineatus* is widely divergent. A further complication is the fact that the surface contours of the body scales of *A. contortrix* and *A. bilineatus* are caniculate, while those on *A. piscivorus* are caniculate/cristate (Chiasson et al., 1989). How this relates to the evolution of the three New World species is unclear, as Price (1989) in another study of microdermatoglyphics within the genus *Agkistrodon* thought the American species to comprise closely related taxa, since they have a similar basal scale pattern of 1–2-micron punctuations without any other visible small elements. A neonatal basal pattern consisting of polygonal cells with diameters of approximatedly 25 microns is shared by all three, as is also a similar apical pattern.

Klauber (1956) and Gloyd and Conant (1990) have evaluated the use of the names *Agkistrodon* and *Ancistrodon* for this genus, and have presented convincing legalistic arguments for the validity of the former. Smith and Gloyd (1964) and Gloyd and Conant (1990) discuss the use of the name *mokasen* and conclude that *Cenchris mokeson* Daudin, 1803, is a junior synonym of *Agkistrodon contortrix mokasen* Palisot de Beauvois 1799.

Agkistrodon piscivorus (Lacépède, 1789)
Cottonmouth

Plates 15–17

Recognition

This is a heavy-bodied, large (maximum length, 189.2 cm), olive, dark-brown or black, amphibious pit viper that lacks both a loreal scale and broad dumbbell-shaped transverse dorsal bands. Juveniles are lighter olive or brown, and individual snakes become progressively darker with age until almost or totally black; the 10–17 transverse dorsal bands characteristic of juveniles fade with age until absent in large adults. The venter is tan to gray and heavily patterned with dark blotches. Some have a light-bordered, dark cheek stripe and dark bars on the rostrum. The keeled, doubly pitted body scales occur in 25(21–27) scale rows at midbody, but in only 21 rows near the anal vent; the anal plate is undivided. Dorsally on the head are nine large plates, as in *A. contortrix*; laterally the scalation is similar to that of *A. contortrix*, but no loreal scale is present and there are 2–3 preoculars, 2–4 postoculars, 5 longitudinal rows of temporals, 7–8(6–9) supralabials, and 10–11(8–12) infralabials. Despite the absence of loreal scales, a heat-sensitive pit is situated between the nostril and eye. On the underside of the body are 128–145 ventrals, and 30–56 subcaudals forming an un-

Juvenile *Agkistrodon piscivorus* (left) and *A. contortrix* (right) (Roger W. Barbour, Kentucky).

divided row (Burkett, 1966, reported a female with only 17 subcaudals). The hemipenis is bilobed with a divided sulcus spermaticus. It has recurved spines near the base and calyces near the apex. The dental formula is: maxillary, 1 (the fang); palatine, 5(4–6); pterygoid, 15–16(14–18); and dentary, 17–18(15–20).

Males have 128–145 ventrals, 30–54 subcaudals, and tail lengths 13–19% (mean, 16.4%) of the total body length; females have 129–144 ventrals, 36–56 subcaudals, and tails 11–19% (mean, 15.5%) of the total body length.

Karyotype

As reported for *A. contortrix* (Fischman et al., 1972; Baker et al., 1971; Zimmerman and Kilpatrick, 1973; Cole, in Gloyd and Conant, 1990), but Zimmerman and Kilpatrick (1973) found that the female *A. piscivorus* has a

submetacentric W chromosome while this chromosome is acrocentric in female *A. contortrix*.

Fossil record

Pleistocene fossils of *A. piscivorus* have been found in Rancholabrean deposits in Florida and Texas (Holman, 1981), and a broken Rancholabrean vertebra identifiable only to the genus *Agkistrodon* has been discovered in Alabama (Holman et al., 1990).

Distribution

A. piscivorus ranges from southeastern Virginia south on the Atlantic Coastal Plain to the Florida Keys, west along the Gulf Coastal Plain, and south through the Mississippi Valley from southern Indiana, southern Illinois, and Missouri southwestward to central Texas.

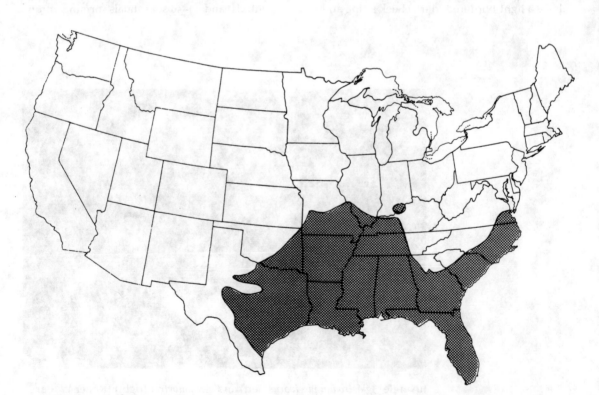

Distribution of *Agkistrodon piscivorus*.

Agkistrodon piscivorus conanti (Barry W. Manzell, Florida).

Geographic variation

There are three subspecies (Gloyd, 1969). *Agkistrodon piscivorus piscivorus* (Lacépède, 1789), the eastern cottonmouth, has no pattern on its light-brown snout, the transverse dorsal bands strongly contrast with the relatively lighter body coloration, and males have 39–51 subcaudals while females have 40–50. This race ranges from southeastern Virginia along the Atlantic Coastal Plain to east-central Alabama. *A. p. leucostoma* (Troost, 1836), the western cottonmouth, has no pattern on its dark-brown to black snout, the transverse dorsal bands blend into the relatively dark body coloration, and males have 30–54 subcaudals, females 36–56. This is the shortest subspecies, reaching a maximum of less than 160 cm, while the other two may exceed 180 cm in total length. It ranges in the Mississippi Valley from southern Indiana (Wilson and Minton, 1983; Forsyth et al., 1985) and southwestern Illinois through southern Missouri to extreme southeastern Kansas, and south to the Gulf Coast from central Texas to Mobile Bay. *A. p. conanti* Gloyd, 1969, the Florida cottonmouth, has the rostrum conspicuously marked with a pair of dark vertical bars,

the transverse body bands are lost in the relatively dark body coloration, and males have 43–54 subcaudals, females 41–49. It is found from southern Georgia and Mobile Bay, Alabama, south through peninsular Florida.

A broad zone of intergradation between the subspecies *piscivorus* and *leucostoma* occurs in Alabama and east-central Mississippi.

Confusing species

Water snakes of the genus *Nerodia* have no facial pits, divided anal plates, round pupils, and two rows of subcaudals. Also, when swimming, *A. piscivorus* inflates its lung resulting in much of its body floating on the surface, while the species of *Nerodia* elevate only the head and neck to the surface. Copperheads, *A. contortrix*, are reddish in color, and have dark dumbbell-shaped bands across the dorsum. Species of *Crotalus* and *Sistrurus* have a tail rattle.

Habitat

This species is the only member of the *Agkistrodon* complex to have taken up an aquatic existence. It lives in almost any type of

Agkistrodon piscivorus leucostoma (Roger W. Barbour, Kentucky).

water body, from brackish coastal marshes to freshwater cypress swamps, bayous, ponds, lakes, rivers, and streams with sand or mud bottoms. It may even occur in the drainage ditches of various southern cities, and I have seen or collected this snake in such unlikely habitats as clear, gravelly or rocky piedmont streams in Alabama, Arkansas, and Missouri. Occupied waterways usually have abundant basking sites (logs, brush piles, mud banks). In Florida, it is found on some offshore keys, and may be common in the rookeries of wading aquatic birds, such as ibis, anhingas, herons, and egrets (Wharton, 1969). Cottonmouths may also live in the lodges of round-tailed muskrats, *Neofiber alleni* (Lee, 1968).

During droughts most cottonmouths remain active and aggregate at remaining waterholes; apparently they can withstand much drying. *Agkistrodon piscivorus* has a low skin water permeability similar to its more terrestrial relative, *A. contortrix* (Dunson and Freda, 1985). In eastern Texas the cottonmouth is restricted to lowland floodplains, and has only a niche overlap value of about 0.68 with *A. contortrix* (Ford et al., 1991).

Behavior

With the advent of warm spring weather, cottonmouths emerge from hibernation and return to their water bodies. This usually occurs in April over most of the range, although they may first appear in March if the daily temperatures rise high enough. In late August and early September the snakes begin a rather leisurely exodus from the water to upland hibernacula, and by October or early November most have disappeared from their aquatic habitats. Exceptions occur, however, and they have been seen in early December in Virginia (Wood, 1954). In southern Florida, cottonmouths are active in every month, but are less so from January to April (Dalrymple, Steiner et al., 1991), and in eastern Texas they seem most active in September and October (Ford et al., 1991).

The cottonmouth is more tolerant of cold than most snakes, and is one of the last to enter hibernation (Neill, 1947, 1948). The winter is spent on shore at some upland site, such as a rock crevice on a hillside with a southern exposure (sometimes with *Elaphe obsoleta* and

Agkistrodon piscivorus leucostoma, juvenile (Roger W. Barbour, Kentucky).

Crotalus horridus, Smith, 1961), hollow logs and stumps in woodlands, under the roots of over-turned trees, palmetto patches, in piles of leaves, or in crayfish, rodent, or gopher tor-toise (*Gopherus*) burrows. Dundee and Burger (1948) found them in a limestone cliff 0.3 km from and 36 m above the nearest body of wa-ter. Often several cottonmouths will use the same hibernaculum. Wharton (1969) discov-ered the cloacal temperatures of cottonmouths in a subsurface den in Florida were 4.2–16.5°C, and within a few degrees of the air tempera-ture, adnd thought the soil temperatures kept the snake's temperature intermediate between that of the soil and air.

The cloacal temperature of 11 active *A. piscivorus* from western Texas recorded by Brattstrom (1965) were 24.6–27.7°C (average, 27.0°C), and Bothner (1973) reported that ac-tive cottonmouths in Georgia had cloacal tem-peratures of 21.0–35.0°C (average, 25.7°C).

During the spring and fall, activity is mostly diurnal, but in summer cottonmouths are predominately nocturnal. During the day they either bask, especially in the morning, re-main undercover, or lie quietly beside logs or among cyprus trees. At the latter sites, they often come in contact with unexpecting hu-mans, making them quite dangerous.

Wharton (1969) reported the home ranges of *A. p. conanti* on Sea Horse Key, Florida, were 0.04–1.22 ha; males had slightly larger activity ranges (average, 0.17 ha) than females (aver-age, 0.14 ha). Some made long movements. A 132-cm male crawled 320 m in 27 months, while a 63.5-cm female moved 380 m in six months. Two other Sea Horse Key females trav-eled 450 and 498 m, respectively, and another left its home range and established a new one on the other side of a mangrove inlet. How-ever, movements of 60–70 m between captures were most common. An *A. p. leucostoma* stud-ied by Tinkle (1959) had a home range of ap-proximately 0.16 ha.

While foraging, *A. piscivorus* characteristi-cally crawls with its head and neck elevated above the ground.

Like many other pit vipers, male *A. piscivorus* sometimes participate in combat dances (Carr and Carr, 1942; Ramsey, 1948; Wagner, 1962; Burkett, 1966; Perry, 1978, Mar-tin, 1984; Blem, 1987). Typically, the combat-ants face each other with the upper 30–40% of their bodies raised vertically, and the rest of their bodies extended out behind them. Their heads are held parallel to the substrate. After sizing each other up for a few seconds to a minute, the males begin swaying back and forth in unison, then suddenly lunge at each other with mouths closed, and entwine a few coils of their anterior bodies. A pushing match then begins to see which male can pin the other's head and neck to the substrate. When one has been topped (pinned), the two snakes usually break off and crawl or swim away. No biting occurs during these bouts. Combat may occur in shallow water or on land.

Blem (1987) observed the development of combat in two litters of *A. piscivorus* born in captivity. The older litter consisted of four cottonmouths (two males) born 5 September 1981, and the younger litter of five (three males) born 7 September 1982. The older snakes averaged 548 mm (514–596) in snout-vent length when combat was first observed. At the same time the second litter averaged only 406 mm (387–474) in snout-vent length. Combat was first observed in the larger males of the first litter on 5 June 1984, and always began shortly after the introduction of food (fish) and was observed at no other time. Minor bouts lasted only 3–6 minutes while prolonged series took 45 minutes (average bout duration, 17 minutes). The second litter showed less intense combat displays during this period, but short bouts of topping behavior occurred between males when food was present. No reproductive behavior was noted during these captive combat bouts. Apparently such behavior only manifests itself at about three years of age, and may be coincident with sexual maturity.

Reproduction

Few data are available regarding the attainment of sexual maturity by female cottonmouths. Burkett (1966) reported that a female *A. p. leucostoma* with a snout-vent length of 45.5 cm was gravid. The smallest mature female *A. p. leucostoma* found in Louisiana by Kofron (1979) was 55.2 cm in snout-vent length, and Penn (1943) dissected another Louisiana female with a total length of 63 cm that contained six embryos. Blem (1981) found that female *A. p. piscivorus* 70.7–79.1 cm in total length are mature in southeastern Virginia. Ford et al. (1990) reported two female *A. p. leucostoma* from north-

eastern Texas with snout-vent lengths of 56 and 58 cm were mature, and Arny (in Wharton, 1966) reported that a female *A. p. leucostoma* was mature at a total length of 59.4 cm. Wharton (1966) thought a length of 80 cm was required for the attainment of sexual maturity in Florida *A. p. conanti* (which reaches a greater total length than the other two subspecies). Conant (1933) reported that a female raised in captivity produced young at an age of 2 years, 10 months.

The ovarian follicles begin to enlarge in August and September. Vitellogenesis begins in February in Louisiana, with ovulation in May (Kofron, 1979). Ovulation also occurs in late May and early June in Florida (Wharton, 1966). Although Burkett (1966) and Wharton (1966) reported a biennial female reproductive cycle, Arny (in Kofron, 1979), Krofron (1979) and Blem (1981, 1982) have presented convincing evidence for an annual female reproductive cycle, and Burkett (1966) has also reported a case of a female producing young in two consecutive years. Larger (older?) females are more likely to be gravid than smaller females in a population (Blem, 1982).

Male *A. p. conanti* over 65 cm are sexually mature (Wharton, 1966). In Alabama, testicular recrudescense occurs in April, spermiogenesis begins in June and peaks in July and August, and spermatogenesis stops in October (Johnson et al., 1982). Sperm is stored overwinter in the vas efferens, epididymides and vas deferens.

Mating apparently can occur anytime during the year. Actual dates on which copulation has been observed in the wild or in captivity include 21 January, 10 and 11 March, 31 August, and 10 and 19 October (Allen and Swindell, 1948; Wright and Wright, 1957; Anderson, 1965; Dundee and Rossman, 1989; Gloyd and Conant, 1990), and Wharton (1966) observed pairing of the sexes in every month but January in Florida. Females possibly store viable sperm for long periods. An adequate description of cottonmouth courtship and mating behavior has not been published.

The ovoviviparous young are normally born in August or September after a gestation period of about 160–170 days (Wright and Wright, 1957; Dundee and Rossman, 1989; Ford et al., 1990). There is a direct correlation between the number and size of the young and the female body length and weight; larger females produce more young of greater body length (Blem, 1981). Litters of from one to 16(20?) young have been reported (Clark, 1949; Wright and Wright, 1957; Gloyd and Conant, 1990), but five to eight are probably most common. Females may congregate before giving birth. Funk (1964b) observed a parturition during which four young were born in about 30 minutes. The neonates were enclosed in fetal membranes which they ruptured in five to eight minutes; however, the young did not emerge from the membrane sac until about an hour later. All four took their first breaths after rupturing their sacs but before leaving them. These first breaths were deep with widely opened mouths, and were taken within one minute of rupturing the membranes.

Newborn young of *A. p. leucostoma* are about 260 mm (228–299) long (Barbour, 1956a; Burkett, 1966; Gloyd and Conant, 1990; Ford et al., 1990), while those of *A. p. piscivorus* also average 260 mm (222–293) in total length (Gloyd and Conant, 1990; pers. obs.). These young are much shorter than the average 335-mm (285–350) young produced by *A. p. conanti* (Wharton, 1966; Gloyd and Conant, 1990; pers. obs.). The masses of the young, amnionic fluid and membranes of two northeastern Texas litters of four and five *A. p. leucostoma* were 22.5% and 34.8% of the females' prebirth weight (Ford et al., 1990). Young *A. p. leucostoma* weigh 11.0–14.9 g at birth. The young are light brown with darker brown transverse bands and yellow tails.

Hybridization with *A. contortrix* has occurred in captivity (Mount and Cecil, 1982).

Growth and longevity

Barbour (1956a) studied *A. p. leucostoma* in western Kentucky and presented the following notes on size classes: 7–8 months, 260–298 mm (having grown about 25 mm since birth); 19–20 months, 312–337 mm (having grown about 45 mm in the year before capture); and 31–32 months, average of 425 mm (a mean length increment of 95 mm over the preceding group); after this growth was too variable to calculate.

The maximum known life span for *A. piscivorus* is 21 years (Gloyd and Conant, 1990).

Food and feeding

From the many types of prey listed in the literature for *A. piscivorus*, one forms the idea that this snake does not care what it eats as long as it moves. This is not entirely correct, as it is also known to consume carrion (Burkett, 1966). Apparently it is an opportunist, taking as prey that animal which is most available and easiest to catch at the time it is hungry.

Known prey are as follows: fish—minnows, shiners (*Cyprinodon, Notemigonus, Notropus*), killifish, mollies, top minnows (*Fundulus, Gambusia, Heterandria, Lucania, Poecilia*), catfish (*Bagre, Ictalurus*), drum (*Aplodinotus, Cynoscion*), gizzard shad (*Dorosoma*), goby (*Gobionellus*), mullet (*Mugil*), freshwater eel (*Anguilla*), bowfin (*Amia*), pirate perch (*Apherdoderus*), mudminnow (*Umbra*), pickerel (*Esox*), sunfish, black bass (*Elassoma, Lepomis, Micropterus*), and crappie (*Pomoxis*); amphibians—salamanders, newts (*Ambystoma, Desmognathus, Eurycea, Notophthalmus, Amphiuma, Siren*), frogs (*Rana*), tree frogs (*Acris, Hyla, Osteopilus*), narrow-mouthed toad (*Gastrophryne*), spadefoot (*Scaphiopus*), and true toad (*Bufo*); reptiles—juvenile alligators (*Alligator*), turtles (*Chelydra, Kinosternon, Sternotherus, Pseudemys, Terrapene, Trachemys, Trionyx*), lizards (*Anolis, Eumeces, Scincella*), and snakes (*Agkistrodon piscivorus, Crotalus, Elaphe, Farancia, Heterodon, Lampropeltis, Masticophis, Nerodia, Regina, Sistrurus, Storeria, Thamnophis*); birds (including eggs, nestlings and fledglings)—*Anhinga*, comorant (*Pahalacrocorax*), sora rail (*Porzana*), pied-billed grebe (*Podilymbus*), glossy ibis (eggs, *Plegadis*), Louisiana heron (eggs, *Hydranassa*), American egret (*Casmerodius*), chicken (eggs, *Gallus*), mourning dove (*Zenaida*), fish crow (*Corvus*), chicadee (*Parus*), seaside sparrow (*Amnospiza*), cardinal (*Cardinalis*), towhee (*Papilo*), and wood thrush (*Hylocichla*); mammals—moles (*Scalopus*), shrews (*Blarina, Cryptotis, Sorex*), bats (species not named), squirrels (*Sciurus*), muskrats (*Ondatra*), mice (*Microtus, Mus, Perognathus, Peromyscus*), rats (*Oryzomys, Rattus, Sigmodon*), and cottontail rabbits (*Sylvilagus*); snails; various insects (beetles, cicadas, damselflies, grasshoppers, lepidopteran larvae); and crayfish (Allen and Swindell, 1948; Yerger, 1953; Barbour, 1956a; Klimstra, 1959; Laughlin, 1959; Wharton, 1969; Collins and Carpenter, 1970; Kofron, 1978; Gloyd and Conant, 1990; pers. obs.) Fish and amphibians are the most frequently eaten prey (Barbour, 1956a; Klimstra, 1959; Burkett, 1966; Kofron, 1978). Marine fishes eaten are probably first captured and then dropped by birds.

Sometimes cottonmouths gorge themselves. In one instance, a captive ate a brown water snake (*Nerodia taxispilota*) about 150 mm longer and 230 g heavier than itself, but could not crawl and died the next day (Allen and Swindell, 1948). Another swallowed nine fish in 85 minutes (Bothner, 1974), and I observed a young cottonmouth in Florida swallow a full-grown cotton rat (*Sigmodon hispidus*).

When searching for prey, *A. piscivorus* usually swims with its head elevated above the water; however, Bothner (1974) saw one which explored a pool with its head under water. They may aggregate at wading-bird rookeries to eat the young which fall from the nests (Wharton, 1969). In contrast to the behavior of gravid females of other pit vipers, those of *A. piscivorus* may continue to feed during gestation (Burkett, 1966).

Odor, sight, and heat radiation (from birds and mammals) are used to detect prey. If in the water when prey is identified, cottonmouths quickly swim to it, seize it, and hold it in their mouths. On land, they either ambush prey or actively pursue it. If the prey is large and struggles, it is usually held in the mouth until the venom immobilizes it. This can sometimes cause problems; a cottonmouth I kept in the laboratory had its tongue and part of its mouth chewed away by a brown rat (*Rattus*) before the rodent died. Kardong (1982b) reported that if more than one mouse is presented to *A. piscivorus* in close succession, the first is released after the bite, but mice subsequently bitten may be held.

The striking and tasting of prey trigger a chemosensory search pattern in *A. piscivorus* causing it to trail released prey (Chiszar, Scudder, Knight et al., 1978; Chiszar, Simonsen et al., 1979; Chiszar, Andren et al., 1982; Chiszar, Radcliffe, Overstreet et al., 1985; Chiszar et al., Chiszar, Radcliffe, Boyd et al., 1986). This involves a sustained, high rate of tongue flicking. Significantly more tongue flicks occur after striking rodent prey than after seeing, smelling, or detecting thermal cues from rodents (Chiszar, Radcliffe, Overstreet et al., 1985).

Young cottonmouths may wave their yellow tails about to lure frogs and other small prey (Wharton, 1960), and Studenroth (1991) reported that an *A. p. conanti* hooked its tail and posterior third of its body around half of the top of a pitfall bucket while it lowered itself into the bucket and apparently searched its contents for prey.

Bothner (1974) observed an apparent dominance of two *A. piscivorus* over a third that arrived late at a feeding pool in Georgia.

Venom and bites

A. piscivorus is a large, sometimes aggressive, dangerous snake. Like other vipers, it has a solenoglyphous venom delivery system with fangs to 11 mm (Ernst, 1964, 1965, 1982; Kardong, 1974). Newborn young have mean fang lengths of 2.7 mm and fully developed venom glands (Ernst, 1982). The fangs are shed periodically and usually replaced on one side at a time (Ernst, 1982). The replacement process takes about five days to complete, and during this time venom may be ejected through either the old or the new fang, or both, depending on the stage of development. During my studies of fangs, several cottonmouths examined had four functional fangs.

Venom yield per bite may be great. Wolff and Githens (1939a,b) reported an average yield of 0.158 g (0.08–0.24), but that a 152-cm cottonmouth yielded 1.09 g (4 ml) of venom during one extraction. Allen and Swindell (1948) also reported that a 150-cm *A. piscivorus* once yielded 2.5 ml of venom. Average dried venom yields are 100–150 mg (Minton and Minton, 1969). The minimum lethal dose for 350-g pigeons is 0.06–0.09 g (Wolff and Githens, 1939a). The LD_{50} in micrograms for a 20-g mouse is 80 (intravenous), 102 (intra-peritoneal), and 516 (subcutaneous) (Minton and Minton, 1969), and for rats is 28–45 mg/kg of body weight (Pollard et al., 1952). The estimated lethal dose for humans is 100–150 mg (Minton and Minton, 1969). Venom chemistry is discussed in Tu (1977).

Cottonmouth venom is very hemolytic, destroying red blood cells and exhibiting strong overall anticoagulant activity (although coagulation at the site of the bite may occur in some cases), and fatalities have occurred (Hutchison, 1929; Allen and Swindell, 1948; Anderson, 1965; Burkett, 1966). Venom from *A. piscivorus* has high hyaluronidase, arginine ester hydrolase, and phospholipase A activities (Tan and Ponnudurai, 1990). Symptoms of cottonmouth bites include swelling and pain at the site, weakness, giddiness, difficulty in breathing, hemorrhage, weakened pulse or heart failure, lowered blood pressure, nausea and vomiting, occasional paralysis, a drop in body temperature, unconsciousness or stupor, and nervousness (Hutchison, 1929; Essex, 1932; Burkett, 1966). Nasty secondary bacterial infections may also occur, such as tetanus or gas gangrene. Allen and Swindell (1948) reported that 50% of the bites result in crippled fingers or toes due to gangrene. Case histories of bites are given by Hulme (1952) and Burkett (1966).

Interspecific differences in certain biological activities of the venom are greater than intraspecific variations of these activities, and can be used for differentiation of the various species of *Agkistrodon* (Tan and Ponnudurai, 1990). Venoms of the subspecies of *A. piscivorus* do not differ significantly in their biological activities.

Predators and defense

Adults have few enemies other than alligators and humans. Since the alligator has been protected, there has been a noticible decrease in cottonmouth populations at certain areas of Florida, but the habitat has also dried and this may have played a role in the decline in numbers (pers. obs.). Juveniles are preyed on by a multitude of animals: large fish—gars (*Lepisosteus*), bullheads (*Ictalurus*), largemouth bass (*Micropterus*); snapping turtles (*Chelydra*); snakes—kingsnakes (*Lampropeltis*), indigo snakes (*Drymarchon*), and even larger cottonmouths; large wading birds—wood storks (*Mycteria*), egrets (*Cosmeroides*), herons (*Ardea*); horned owls (*Bubo*), hawks (*Buteo*), eagles (*Haliaeetus*); raccoons (*Procyon*), otters (*Lutra*), dogs (*Canis*), and cats (*Felis, Lynx*).

The disposition of individual cottonmouths varies greatly. Those I have kept ranged from very timid to extremely aggressive, but all would bite if handled. They can and will bite underwater. When first disturbed they usually try to escape. If this is not possible, they coil and strike, often repeatedly. The well-publicized behavioral trait of gaping open the mouth to show the inner pinkish-white lining does not always occur. In fact, I seldom saw this display while capturing several hundred cottonmouths; when it did occur, the body was often also either thrown into an S-shaped coil or dorsal-ventrally flattened. Schuett (in Gloyd and Conant, 1990) has reported that gaping is present in neonates and is probably inate, and that it is more frequently displayed by cool snakes, 10–18 C. When first grasped or pinned down, cottonmouths thrash about violently, striking at any near object, and may even bite themselves. Because of their commonness, size, unpredictibility, and violent tempers, cottonmouths are among the most dangerous of our snakes, and should be approached with care.

Populations

Cottonmouths may be the most common reptile in certain habitats. At Murphy's Pond, Hickman County, Kentucky, they formerly occurred in densities of over 700 per hectare (Barbour, 1956a), and I have seen as many as eight basking on one fallen cyprus tree in Florida. The late Percy Viosea once collected 114 cottonmouths in one day on Delacroix Island, St. Bernard Parish, Louisiana (Dundee and Rossman, 1989). At Lone Pine Key, Everglades National Park, *A. piscivorus* comprised 77 (4.3%) of 1,782 snakes captured or observed, and 128 (12.6%) of 1,019 snakes at another site in the park (Dalrymple, Bernardino et al., 1991). Wharton (1969) reported that during the time water bird rookeries are active, cottonmouths may become very numerous under the nest trees.

Immature individuals made up 32.5% of those snakes examined by Burkett (1966) and about 45% of the Murphy's Pond population (Barbour, 1956a). Females comprised 53% of the adult specimens and in a group of 48 embryos (8 broods) examined by Burkett (1966).

Sistrurus catenatus (Rafinesque, 1818)
Massasauga
Plates 18–20

Recognition

The massasauga is a medium-sized (maximum length 100.3 cm) gray to light-brown rattlesnake with a row of 21–50 dark-brown to black dorsal blotches and three rows of small brown to black spots on each side of the body (occasional individuals are striped). Individuals from loamy plains are usually grayish brown, while those from Arizona appear more reddish. The black venter is mottled with yellow, cream, or white marks, or it may be nearly all black, and the tail is ringed with alternate dark and light bands. A dark, light-bordered stripe runs backward from the eye, and another dark middorsal, light-bordered stripe extends posteriorly on the back of the head. Midbody scale rows total 25(21–27). Ventral scales total 129–160, subcaudals 19–36, and the anal plate is complete. Dorsally on the head are nine enlarged plates: 2 internasals, 2 prefrontals, a large frontal, 2 supraoculars, and 2 parietals. The rostral scale is higher than wide. Laterally are 2 nasals, a loreal, 2 preoculars (the upper touches the postnasal), 3(2–4) postoculars, 1–2 suboculars, 11–12(9–14) supralabials, and 12–13(10–16) infralabials. The hemipenis is bifurcate with a divided sulcus spermaticus, about 33 recurved spines per lobe, and 23 fringes per lobe; some spines occur in the crotch. The hemipenis is illustrated in McCraine (1988). Only the fang occurs on the short maxilla; other teeth are 1–3 palatines, 5–7 pterygoidals, and 9–10 dentaries. Males have 129–155 ventrals, 24–36 subcaudals, and 5–11 dark tail bands; females have 132–160 ventrals, 19–29 subcaudals, and 3–8 dark tail bands. Tail length (exclusive of rattle) is 10–12.5% of total length in males, but only 7.5–9.0% in females (Wright, 1941; Minton, 1972).

Karyotype

The 36 chromosomes consist of 16 macrochromosomes (including 4 metacentric, 6 submetacentric, 4 subtelocentric), and 20 microchromosomes; sex determination is ZZ in males and ZW in females (Zimmerman and Kilpatrick, 1973).

Fossil record

Remains of *Sistrurus catenatus* have been found in deposits from the upper Pliocene (Blancan) of Kansas and Texas and from the Pleistocene (Blancan) of Kansas (Holman, 1979, 1981).

Distribution

The massasauga ranges from southern Ontario, central New York, and northwestern Pennsylvania, west to eastern Iowa, and southwest to western Texas, southern New Mexico, and southeastern Arizona. There are also Mexican populations in the Cuatro Ciénegas Basin, Coahuila, and near Aramberri, Nuevo León (McCoy and Minckley, 1969; Minckley and Rinne, 1972; Minton, 1983).

Geographic variation

Three subspecies are recognized (Minton, 1983). *Sistrurus catenatus catenatus* (Rafinesque, 1818), the eastern massasauga, is found from southern Ontario and central New York west to Iowa and eastern Missouri. It usually has 25 midbody scale rows, 129–157 ventrals, 24–33 subcaudals in males and 19–29 in females, 20–40 dorsal body blotches, and a venter that is predominately dark gray or black. Most adults are 55–80 cm long. *S. c. tergeminus* (Say, 1823),

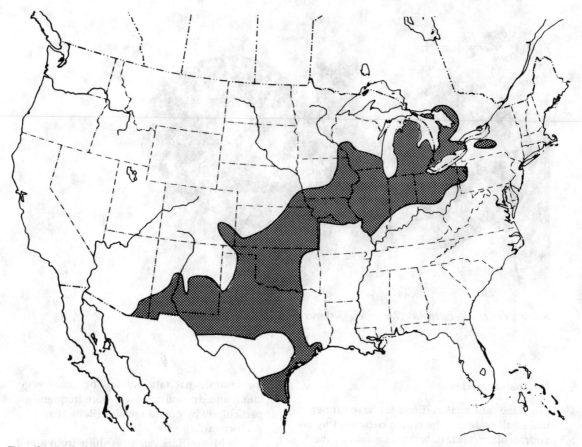

Distribution of *Sistrurus catenatus*.

the western massasauga, ranges from southwestern Iowa and northwestern Missouri southwest through extreme southeastern Nebraska, east and central Kansas and western Oklahoma to western Texas, and south from Oklahoma through east-central Texas to the Gulf Coast. This prairie race usually has 24 midbody scale rows, 138–160 ventrals, 27–34 subcaudals in males and 21–28 in females, 28–50 dorsal body blotches, and a whitish or cream-colored venter with dark lateral blotches. Most adults are 45–65 cm long. *S. c. edwardsi* (Baird and Girard, 1853), the desert massasauga, is a small race growing to about 55 cm that lives in southeastern Colorado, the Texas Panhandle, extreme southwestern Texas, and eastern and southern New Mexico in the

United States, and in the Cuatro Ciénegas Basin of Coahuila and at Aramberri, Nuevo León, in Mexico. This subspecies has an average of only 23 midbody scale rows, 137–152 ventrals, 28–36 subcaudals in males and 24–29 in females, 27–41 dorsal body blotches, and a whitish to cream-colored venter with only a few small dark spots.

A zone of intergradation between the subspecies *catenatus* and *tergeminus* occurs in south central Iowa and adjacent Missouri, and a second intergrading zone between *tergeminus* and *edwardsi* extends from southwestern Colorado through extreme western Oklahoma and the central Texas Panhandle (Minton, 1983). The total altitudinal range is from about sea level to 1,500 m.

Sistrurus catenatus catenatus (Roger W. Barbour).

Confusing species

The large rattlesnakes (*Crotalus*) have numerous small scales on the crown between the two supraocular scales. In *Sistrurus miliarius*, the prefrontal scales are in broad contact with the loreal scale. Hognose snakes (*Heterodon*) lack the loreal pit, rattle, and elliptical pupil.

Habitat

Over most of its range this pit viper is usually found in moist habitats such as swamps, marshes, bogs, wet meadows, or seasonally moist grasslands, but in the drier southwestern range in Arizona, New Mexico, and parts of western Texas it is restricted to habitats such as river bottoms, dry grasslands, and plains of mesquite, juniper, grass, yucca, creosote, and cacti. The massasauga's persistence in these arid regions may be correlated to its use of rodent burrows as retreats. Such burrows provide a humid microclimate that helps retard moisture loss from the snake's body.

In the spring and fall, massasaugas in western Pennsylvania inhabit low, poorly drained areas near the hibernacula, but during the summer, habitats with low or sparse vegetation and dry soil are used more frequently, particularly by gravid females (Reinert and Kodrich, 1982).

In Missouri, *S. catenatus* shifts from a prairie habitat in spring to upland old fields and deciduous woods in summer, and then returns to the prairie habitat in the fall (Seigel, 1986).

Behavior

In Ohio, Indiana, and Missouri this species is annually active from April to late October or early November (Conant, 1951; Minton, 1972; Seigel, 1986), farther south in Texas it may be active by mid-March (Greene and Oliver, 1965), and in the southwest and Mexico some are probably active in all months, but less so in the summer and winter. Most activity in Missouri takes place from April to mid-May, and again in October (Seigel, 1986).

Winter hibernation is spent in rock crevices, rodent and crayfish burrows, old stumps and rotten logs. Maple and Orr (1968) studied the overwintering behavior of *S. catenatus* in northeastern Ohio. By monitoring the temperature of various components of the hibernacula,

Sistrurus catenatus tergeminus (John H. Tashjian, Houston Zoo).

they concluded that the massasauga is capable of maintaining a cloacal temperature above the ambient temperature for 45 minutes, and that it can withstand a freezing body temperature for short periods without harm, but that it usually hibernates in wet crayfish holes at depths below the frost line.

During the spring and fall, massasaugas are more diurnal in their habits and often bask or actively forage. When the daytime temperatures become too warm in summer, they shift to a more crepuscular or nocturnal schedule. In northwestern Missouri, such a shift in the daily activity periods is seasonal (Seigel, 1986). In the spring most activity occurs from 1200 to 1600 (55%) with some activity from 1600 to 2000 (24%) and 0800 to 1200 (17%); little activity (4%) takes place from 2000 to 2400. In summer the snakes are most active from 1600 to 2000 (42%) and 1200 to 1600 (33%), but activity increases to 18% from 2000 to 2400. In the fall the most active period (70%) is 1200 to 1600. In Arizona most are crepuscular or nocturnal in the summer, to avoid the very warm daytime temperatures, and in Pennsylvania *S. catenatus* is active mainly from 0900 to 1500 (Reinert, 1978).

Twenty-five *S. catenatus* from western Pennsylvania were equipped with radio transmitters and their movements followed for up to 50 days (Reinert and Kodrich, 1982). The mean area and length of the home range were 9,794 m² and 89 m, respectively, and the mean distance moved per day was 9.1 m. No significant movement differences occurred between the sexes; however, gravid females had significantly shorter home ranges than did nongravid females.

Sisturus c. edwardsi is fairly adept at sidewinding on smooth or sandy surfaces (Lowe et al., 1986).

Massasaugas are accomplished swimmers and readily enter water.

Reproduction

Studies of the reproductive cycles of *Sisturus catenatus* from Illinois and Wisconsin by Wright (1941) and Keenlyne (1978) indicate that females became mature in their third year at a length of about three times that of birth. First-summer females from Wisconsin show no follicular growth, those in their second summer have follicles about 7 mm in diameter, and 50% of third-summer females are gravid and 25% are postpartum (a female collected on 31 August had yellow follicles about 25 cm long). Only about 3% of fourth-summer females are not gravid or have not already given birth. Only 7% third-summer or older females, and only 3% of fourth-summer or older females are nonreproductive. This strongly suggests an annual female reproductive cycle. However, Reinert (1981) and Seigel (1986) found evidence of a biennial reproductive cycle in females from western Pennsylvania and Missouri. Of 26 Pennsylvanian females examined in 1977 by Reinert (1981), 11 (42%) were nongravid adults and 3 (11.5%) were juveniles. The annual percentage of reproductive females in Pennsylvania was only 52%, and varied from 33 to 71% over three years in Missouri. The smallest gravid Pennsylvania female was 448 mm in total length. A significant size difference exists between females from Pennsylvania and those from Wisconsin, those from Wisconsin being larger, and perhaps this helped determine the breeding cycle. Seigel (1986) found a significant positive regression between snout-vent length and brood size in Missouri.

Published reports of matings seem to indicate the breeding period extends from March to November (Wright, 1941; Wright and Wright, 1957; Klauber, 1972; Chiszar, Scudder and Knight 1976; Reinert, 1981; Lowe et al., 1986), but Howard K. Reinert (pers. comm.) believes that this species has a late summer and fall mating season. Every literature record of an earlier mating that he checked was either based upon conjecture or on observation of mating in captivity.

Chiszar, Scudder and Knight (1976) described the courtship behavior of a captive pair. The entire sequence of behaviors took over six hours, and the courtship postures of the male and female followed the general pattern for *Crotalus* and other *Sistrurus* species. The front of the males's body was balanced on top of the female's back and he frequently rubbed her head and neck with longitudinal strokes of his chin (these chin rubs were accompanied by flexions of his entire body). Also, his tail was looped around her tail 5 cm anterior to the base of the rattle. He then massaged the last 5 cm of the female's tail with his tail loop by tightening the loop and stroking posteriorly until he touched her rattle. He then reversed the stroke until he arrived back at the site of his original grip. This process was re-

peated three times, although occasionally two or four stroking cycles occurred. Male courtship consisted of a series of chin-rubbing episodes, each of which was bounded at the beginning and end by three tail-stroking cycles during which chin rubbing was discontinued. Mean duration of 30 chin rubbing episodes was 56.3 seconds, the mean number of chin rubs per episode was 95.4, and the mean number of chin rubs per second was 1.68. The chin-rubbing rate was relatively constant and this behavior continued for the entire period of time between successive tail stroke cycles.

The ovoviviparous young are born during the period from late July to late September; the gestation period is about 100–115 days (Atkinson and Netting, 1927, thought it to be only about three months). If mating occurs in late summer or early autumn and the young are not born until the next summer, the sperm is probably stored in the female's oviducts and not used until the next spring. Embryos are not present in the fall or winter; Richard Seigel (pers. comm.) has found females in Missouri with developing follicles in October, but never embryos.

At birth, young *S. catenatus* are 140–252 mm in total length and weigh 8–10 g. As in other rattlesnakes, the newborn's rattle consists only of a scalelike button. Brood size is generally low, ranging from 2 (Stebbins, 1954) to 19 (Keenlyne, 1978) with broods of 8–10 young the norm. Some geographic variation may take place in brood size, with females from southern populations producing fewer young per broods (Fitch, 1985a).

In parturitions observed by Anderson (1965), from 1 to 10 minutes occurred between births, and the young ruptured the fetal membrane within the first few minutes. The first act of the neonates after emergence from the membrane was to stretch their jaws as if yawning. The little snakes would shake their tails and strike if annoyed. Births observed by Wright (1941) all occurred between 0630 and 1615. Young only a few days old have venom toxic enough to cause bitten mice to die within several hours (Conant, 1951).

A natural hybrid *Sistrurus catenatus* × *Crotalus horridus* has been reported by Bailey (1942); Howard Reinert examined this preserved specimen and found it truly intermediate in most characters.

Growth and longevity

Yearling massasaugas from Illinois are 39–43 cm in total length, an increase of about 65% from birth (Wright, 1941), and Missouri yearlings have snout-vent lengths of 30–40 cm (Seigel, 1986). Massasaugas with 50–54 cm snout-vent lengths are probably three to four years old (Wright, 1941; Keenlyne, 1978; Seigel, 1986).

The longevity record for this species is 14 years (Bowler, 1977).

Food and feeding

Most reports of the massasauga's food are based on single observations. The only detailed studies of prey were conducted on *Sistrurus c. catenatus* by Keenlyne and Beer (1973) in Wisconsin (59 specimens contained 91 food items), and by Seigel (1986) on *S. c. tergeminus* in Missouri (22 of 96 snakes contained food). In Wisconsin, no frogs were found, and nearly 95% of all prey were warm blooded; 85.7% of the entire diet consisted of the vole *Microtus pennsylvanicus*. Other prey and their percentages of occurrence were as follows: deer mice *Peromyscus leucopus*, 4.4; garter snakes *Thamnophis sirtalis*, 4.4; jumping mice *Zapus hudsonius*, 2.2; red-winged blackbird *Agelaius phoeniceus*, 1.1; masked shrew *Sorex cinereus*, 1.1; and an unidentified snake (probably *Thamnophis*), 1.1. Food items by sex and percentages of snakes containing prey were as follows: males, 83.6; nongravid females, 55.6; gravid females, 10.4. Stomach items in Missouri fell into two major classes: rodents (*Microtus, Peromyscus*) and snakes (*Storeria, Thamnophis*). Fowlie (1965) reported that, based on examination of stomachs, the desert massasauga, *S. c. edwardsi*, is primarily an amphibian predator; but Greene (1990), after examining museum specimens (mostly road-kills), concluded that they eat small lizards and pocket mice (*Perognathus*); and Lowe et al. (1986) reported that mice and lizards, especially whiptails (*Cnemidophorus*) and earless (*Holbrookia*), are taken.

Birds are also sometimes taken. Best (1978) thought the massasauga was a nest predator on field sparrows (*Spizella*), and Brush and Ferguson (1986) reported predation of lark sparrow (*Chondestes*) eggs by *S. catenatus*. Frogs and toads eaten include *Hyla, Rana* and *Bufo* (Klauber, 1972; Wright and Wright, 1957), and lizards (*Cnemidophorus, Gambelia*) have also been recorded (Klauber, 1972; Greene and Oliver, 1965). Other foods listed in the literature are insects, centipedes, crayfish, snakes, and fish (Wright and Wright 1957; Lardie, 1976; Klauber, 1972). Most captives will readily eat house mice (*Mus*).

Ruthven (in Klauber, 1972) thought that the other snakes devoured had probably been found dead. In support of this, Greene and Oliver (1965) found a massasauga in Texas attempting to swallow a recently road-killed hognose snake (*Heterodon*), and Schwammer (1983) observed one eating carrion in Colorado.

Endothermic prey is probably detected by the heat sensory facial pit, but sight and odors are also important feeding cues (Chiszar, Scudder and Knight, 1976; Chiszar, Scudder and Smith, 1979; Chiszar, Taylor et al., 1981). Movement is the primary cue in eliciting exploratory behavior (Scudder and Chiszar, 1977). Tail luring may also be important for young massasaugas to capture prey; Schuett et al. (1984) observed recently born young waving their tails back and forth over their heads to attract ranid frogs.

Most prey are usually struck and then eaten only after they die, but frogs may be swallowed alive.

Venom and bites

Massasauga fangs are short; Klauber (1939) reported lengths of 5.0–5.9 mm for *Sistrurus c. catenatus* and 4.5–5.5 mm for *S. c. tergeminus*.

The total dry venom yield for an adult massasauga is 25–35 mg (Minton and Minton, 1969); the average total dry venom yield is 31 mg per adult (Klauber, 1972); however, the yield is probably closer to 5–6 mg per bite. The minimum lethal dose for a 20-g mouse is 4.4 micrograms intraperitoneal and 105 micrograms subcutaneous (Minton and Minton, 1979); mice die in 3–23 minutes after being bitten (Swanson, 1930). The estimated human lethal dose is 30–40 mg (Minton and Minton, 1969).

The venom is largely hemolytic and causes much ecchymosis as capillary walls are destroyed, but neurotoxins may also be present. Other symptoms of bites include pain, discoloration, and swelling (although in one case numbness was reported) at the site of the bite, a cold sweat, faintness, nausea, tremors, and nervousness (Atkinson and Netting, 1927; Hutchison, 1929; LaPointe, 1933; Allen, 1956; Menne, 1959; Dodge and Folk, 1960; Klauber, 1972). Humans have died from massasauga bites (Lyon and Bishop, 1936; Stebbins 1954; Menne, 1959).

Predators and defense

Our knowledge of the natural predators of *Sistrurus catenatus* is scanty. Minton (1972) mentions a blue racer (*Coluber*) preserved in the act of swallowing an adult massasauga, and most likely other species of snakes take the young. Hawks (*Buteo*), large wading birds (*Ardea*, etc.), and carnivorous mammals (*Procyon, Mephitis, Taxidea, Canis, Felis, Lynx*) are also potential predators, and Chapman and Casto (1972) reported predation on *S. catenatus* by a loggerhead shrike (*Lanius*). However, habitat destruction and road kills by humans probably eliminate more of these snakes each year than all natural predators combined.

This snake is usually rather sluggish and mild mannered, only becoming aggressive after active disturbance. Richard Seigel (pers. comm.) has even stepped on one without it even rattling, and the only bite in over 10 years in the area about his study site in Missouri was of a drunk who handled one for 10 minutes. However, they are dangerous, and differences in disposition between individual massasaugas do occur. A large female that I found in Missouri was very alert and irritable, rattling her tail and striking whenever I came too near. The frequency of the sound produced by the rattle ranges from a mean low of 6.42 kHz to a mean high of 12.46 kHz; the mean sound band width is 6.05 kHz (Fenton and Licht, 1990).

Populations

In Ohio, 27 were once discovered and killed during the harvesting of about 15 hectares of wheat (Conant, 1951), and a single blast during the construction of the trans-Canada highway resulted in 117 dead massasaugas (Menne, 1959). In the past, fairly dense populations of these snakes probably occurred at suitable sites. However, habitat destruction has destroyed most colonies, and *Sistrurus c. catenatus* has been placed on threatened or endangered lists over most of its range (Breisch, 1984; Bushey, 1985; Seigel, 1986). This trend was noted as long ago as 1948 by Loomis, who stated that "cultivation seems to have greatly reduced its numbers." Suitable habitat must be preserved if this interesting species is to remain a part of our fauna.

Of 128 nonhatchling snakes examined in Missouri by Seigel (1986), 42 (33%) were adult females, 36 (28%) were adult males, and 50 (39%) were juveniles; a sex ratio of 0.85:1. The sex ratio of adults was 1:1.16, not significantly different from 1:1. In April–May most snakes found were adults, but juveniles dominated in August–October.

Remarks

Bonilla et al. (1971) found several differences in the chemical properties of the venom of *Sistrurus catenatus* and its congener *S. miliarius*. The venom of *S. miliarius* was yellow, that of *S. catenatus* colorless; their electrophoretic patterns were similar, but not identical; and the venom of *S. miliarius* contained a greater percentage of proteins.

Sistrurus miliarius (Linnaeus, 1766)
Pigmy rattlesnake
Plates 21–24

Recognition:

This is a small (maximum length, 80.3 cm) grayish rattlesnake with nine enlarged scales on the dorsal surface of the head (see *S. catenatus*), 1–3 rows of lateral dark spots along the body, a series of 22–45 dark-brown or black middorsal blotches, and a red to orange middorsal stripe. The ground color usually ranges from gray to tan. Individuals with reddish-orange to brick-red color patterns, however, occur along the northeastern edge of the range, chiefly in Beaufort and Hyde counties, North Carolina (Palmer, 1971). The venter is whitish to cream colored with a moderate to heavy pattern of black or dark-brown blotches. A dark-black or reddish-brown bar extends backward from the eye to beyond the corner of the mouth. Two dark longitudinal and often wavy stripes are present on the back of the head. Keeled scales occur in 19–25 rows at midbody and 15–19 rows near the vent. On the venter are 122–148 ventral scutes and 25–39 subcaudals with no sexual dimorphism; the anal plate is undivided. On the side of the head are 2 nasals, a loreal (lying between the postnasal and upper preocular), 2 preoculars, 3–6 postoculars, 4–5 temporals, 10(8–13) supralabials, and 11(9–14) infralabials. The hemipenis is similar to that described for *S. catenatus* and is illustrated in McCranie (1988). Only the fang occurs on the maxilla; other teeth present are 3 palatines, 7–9 pterygoidals, and 10–11 dentaries. Males have tails 10–15% of the total length with 7–14 dark bands; tails of females are 9–12% of the total length with 6–13 dark bands.

Karyotype

Each diploid cell has 36 chromosomes: 16 macrochromosomes (including 4 metacentric, 6 submetacentric, 4 subtelocentric), and 20 microchromosomes; sex determination is ZZ in males and ZW in females (Zimmerman and Kilpatrick, 1973).

Fossil record

Pleistocene fossils have been found at Irvingtonian (Meylan, 1982) and Rancholabrean (Holman, 1981) sites in Florida.

Distribution

The pigmy rattlesnake ranges from Hyde County, North Carolina, south to the Florida Keys, and west to eastern Oklahoma and central Texas (Palmer, 1978).

Geographic variation

Three subspecies have been described (Palmer, 1978). *Sistrurus miliarius miliarius* (Linneaus, 1766), the Carolina pigmy rattlesnake, is gray to reddish brown with one or two rows of lateral spots, usually 25 anterior and 23 midbody scale rows, and the ventral dark spots are at least two scutes wide. The head pattern is well marked. It ranges from Hyde County, North Carolina, southwestward to central Alabama. *Sistrurus m. barbouri* Gloyd, 1935, the dusky pigmy rattlesnake, is dark gray with three rows of lateral spots, usually 25 anterior and 23 midbody scale rows, and a heavily dark-spotted venter. The head pattern is obscured. This race ranges from extreme southwestern South Carolina south through peninsular Florida and west through southern Georgia, the Florida Panhandle, and southern Alabama to southeastern Mississippi. *Sistrurus m. streckeri* Gloyd, 1935, the western pigmy rattlesnake, is gray-brown to brown with one to two rows of

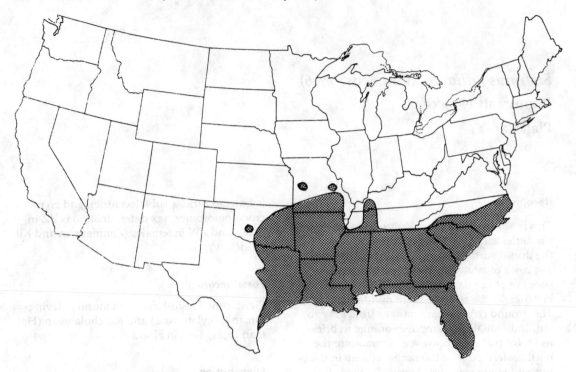

Distribution of *Sistrurus miliarius*.

Sistrurus miliarius barbouri (Roger W. Barbour, Collier County, Florida).

Sistrurus miliarius barbouri (Richard D. Bartlett, Florida).

lateral spots, usually 23 anterior and 21 midbody scale rows, and diffuse ventral blotches about one scale wide. The head has a well-marked pattern. It occurs from Land Between the Lakes in western Kentucky and Tennessee, southern Missouri, and eastern Oklahoma south to the Gulf Coast of Louisiana and central Texas.

Christman (1980) found that in Florida the number of ventral and subcaudal scales increases clinally to the south. Also, coastal populations have higher dorsal scale row and blotch counts, and larger, more rounded dorsal blotches than more inland Florida populations.

Confusing species

The larger rattlesnakes of the genus *Crotalus* have small scales between the supraoculars. *Sistrurus catenatus* is much larger and darker, and its upper preocular scales touch the postnasal scales. The hognose snakes, *Heterodon*, lack a rattle and facial pit, and have upturned rostral scales.

Habitat

The pigmy rattlesnake lives in a variety of habitats, but in the East none of these are very far from water. Mixed pine-hardwood forest, scrub pinewoods, sandhills, and wiregrass flatwoods are all occupied. In the Everglades, it is seldom encountered in pinewoods or other dry habitats, but flooding may force it to higher ground such as canal banks and roads (Duellman and Schwartz, 1958). Where the gopher tortoise (*Gopherus polyphemus*) is sympatric, this small snake may use its burrow as a retreat, and it is also known to reside in small mammal burrows (Lee, 1968). In Texas and Oklahoma, it is restricted to mesic grasslands.

Behavior

In Florida, southern Georgia, and South Carolina, the pigmy rattlesnake may be active in every month (Chamberlain, 1935; Hudnall, 1979; Dalrymple, Steiner et al., 1991), but farther north in North Carolina it is active from March to November (Palmer and Williamson, 1971). During the year, most northern *Sistrurus miliarius* are caught from June to September (Chamberlain, 1935; Palmer and Williamson, 1971), but in southern Florida, most are seen from July to November (Dalrymple, Steiner et al., 1991). Males, juveniles, and newborns are most active in October.

Sistrurus miliarius miliarius (Richard D. Bartlett, Moore County, North Carolina).

Little is known of their overwintering behavior. Neill (1948) found one in February emerging from a probable mammal burrow, and Palmer and Williamson (1971) reported that some had apparently hibernated in a small hole in a sawdust pile. Allen (in Klauber, 1972) found them in winter under logs in the drier areas of swamps.

During the warmer parts of the year in Florida this small rattlesnake is active from the late afternoon into the night (Hudnall, 1979). The mornings at this time are spent in basking. From November to February it is active primarily in the afternoon, but most activity is in the form of basking.

Apparently this species occupies a rather small home range. Hudnall (1979) monitored movements of marked *S. m. barbouri* in Palm Beach County, Florida. The maximum distance moved from the first point of capture ranged from 9 to 242 m. A male, recaptured five times, moved an average of 179.6 m between recaptures, while two males, recaptured three times, only averaged 81 and 89.5 m, respectively, between captures. Gravid females apparently move about relatively little; a gravid *S. m.*

streckeri in Texas was never found more than 2 m from its original capture point in seven observations between 1 July to 3 August, 1976 (Fleet and Kroll, 1978).

Although usually found on the ground, Klauber (1972) related an observation of one on a limb of a tree about 8 m high. Since it lives near water, *S. miliarius* is a good swimmer.

Male combat dances determine dominance in *S. miliarius* (Palmer and Williamson, 1971; Lindsey, 1979; pers. obs.) and are similar to those reported for other rattlesnakes. The males rise up, face each other, sometimes sway back and forth, and finally lunge toward each other and entwine the posterior 40–50% of their bodies (sometimes the entwining occurs before the snakes rise from the ground). Pushing with the raised anterior body and thrashing about may cause the combatants to fall to the ground, but they soon resume the raised, entwined position, and continue their fight. During the entire performance the posterior parts of their bodies are tightly entwined but there is no extrusion of the hemipenes. Their heads occasionally are facing or located one above the other, but most often are several cen-

Sistrurus miliarius streckeri (John H. Tashjian; Ray Folsom, Hermosa Beach, California).

timeters apart and situated at right angles to one another. Mouths are kept closed and no attempts are made to bite. Finally, after being pinned firmly to the ground, one snake untwists itself from the dominant male and crawls quickly away. Combat dances may last for only a few minutes to longer than two hours, and may be continuous or sporadic.

Reproduction

The only observation of copulation in nature was by Hamilton and Pollack (1955) who found a mated pair on 18 September. Montgomery and Schuett (1989) also reported September matings in captivity, but Verkerk (1987) observed that mating took place in January in captives that had just emerged from hibernation. As most young are born from July to September, viable sperm must be stored in the female's reproductive tract overwinter if spring matings do not occur (Montgomery and Schuett, 1989).

A brief note on male courtship behavior observed on 23 September 1988 has been published by Montgomery and Schuett (1989). When placed with the female, two males crawled about waving their tails. When she coiled both males moved over her with increased tongue flicking. She crawled away, followed by one male. The other male crawled to the first male and rubbed him with his snout. The first male unsuccessfully attempted intromission with the female, and then crawled from her toward the second male now coiled at the end of the cage. When the second male was reached, the first crawled over him as the second male first hid his head beneath his coils and then tried to crawl from beneath and away from the first male. This series of acts took 28 minutes (0955–1023). At 1200 hours all three snakes were coiled together. At 1630 the first male was observed copulating with the female with the right hemipenis inserted. Copulation continued past 2234. The female was isolated from the males after mating, and gave birth to 10 young on 14 July 1989, 294 days later.

A female *Sistrurus m. barbouri* collected on 7 January had enlarged ovarian follicles, another had partly developed embryos on 2 July, and a third had nearly full-term fetuses on 16 July; three other females gave birth on 15 July, 2 August, and 4 October, respectively (Iverson, 1978). Two Texas litters of *S. m. streckeri* were born on 2 and 12 August, respectively (Ford et al., 1990), but in southern Florida newborns are very common in July and August (Dalrymple, Steiner et al., 1991).

The ovoviparous young are born enclosed in a sheathlike membrane. Fleet and Kroll (1978) witnessed the birth of five young. The female was observed at 1804 hours with the first fetal sac partially protruding from her cloaca; it fully emerged one minute later. The second to fifth young were born 49, 30, 25, and 80 minutes apart, respectively. The female remained coiled during labor and each birth was characterized by a series of from 8 to 19 undulatory contractions. These peristaltic contractions passed tailward through the posterior two-thirds of the female's body, generally beginning two to five minutes apart and ending about 10 seconds apart. During contractions, the fetal bulge could be seen to move toward the vent. Near the end of the undulatory series, there was a series of up to five convulsive cloacal contractions and elevation of the tail. This was followed by emergence of the fetal sac which in three timed instances required three to four minutes. The neonates moved their heads vertically and uncoiled from within the membranes. Subsequently, two young moved their heads from side to side in scraping movements against the female's body, which lasted up to 25 seconds and continued at irregular intervals for 16 to 26 minutes. This behavior may remove mucus and debris from the mouth or loosen the natal skin. The fifth young had the shortest time between parturition and shedding and did not exhibit this behavior. Interspersed with the scraping behavior, the young gaped their jaws (1–4 times each) into a yawnlike position.

Litter size varies from three (Anderson, 1965) to 32 (Carpenter, 1960), but 7–10 young are most common. Fitch (1985a) analyzed available data on litter size and felt a north-south geographic variation was suggested, with northern *S. miliarius* producing fewer young. The young are 126–191 mm in total length at birth, weigh 1.8–5.0 g, and have cream, yellow or orange tails.

The preparturition weights of two female *S. m. streckeri* from Texas were 44.6 and 53.8 g, respectively, while their postparturition weights were 25.4 g (56.9%) and 22.0 g (40.9%), respectively. The first gave birth to six young which, along with their embryonic membrane and fluid, had a mass 50.2% that of the mother's prebirth weight. The other female had eight young which totaled 40.9% of her prebirth weight (Ford et al., 1990).

An apparent defense of young by a captive female was witnessed by Verkerk (1987). When disturbed, the young hid behind her back, while she rattled and tried to bite, but this may only have been normal fleeing behavior in the young and typical defense by the female.

Growth and longevity

No data on growth rates have been published on this species. A captive male *Sistrurus m. barbouri*, 235 mm in total length when collected, grew to 609 mm in two years in my laboratory.

A male *S. m. barbouri* at the Staten Island Zoo lived 15 years, 1 month, 28 days (Bowler, 1977).

Food and feeding

Sistrurus miliarius consumes a variety of small prey: insects, spiders, centipedes, frogs (*Acris, Rana*), toads (*Bufo*), lizards (*Anolis, Eumeces, Scincella*), snakes (*Diadophis, Nerodia, Thamnophis, Sistrurus miliarius*), nestling birds, and mice (*Peromyscus, Microtus*) (Clark, 1949; Hamilton and Pollack, 1955; Wright and Wright, 1957; Palmer and Williamson, 1971; Klauber, 1972; pers. obs.). Captives readily take house mice (*Mus*) and scincoid lizards. Chamberlain (1935) and Neill and Allen (1956) questioned insects and centipedes as food for this species, believing instead that these were secondarily ingested in the stomachs of vertebrate prey. However, the high frequency of reports of insects and centipedes in the stomachs of pigmy rattlesnakes indicates these are legitimate prey. Verkerk (1987) reported the young readily accept crickets.

Twelve of 16 *S. miliarius* from Georgia examined by Hamilton and Pollack (1955) contained prey remains. Reptiles (five lizards, one snake) were found in 50% of the stomachs, centipedes (*Scolopendra heros*, 50–85 mm) were found in 33%, and mammals (*Peromyscus, Microtus*) occurred in 17%.

Some prey may be ambushed, but most are probably actively sought. Many of these snakes that I have found seemed to be searching for prey, and I once observed a Florida pigmy rattler stalk, strike, and swallow an anole lizard. Neill (1960) felt that juvenile *S. miliarius* may use their yellowish tails as a lure for prey (see *S. catenatus*).

Venom and bites

The fangs of this small snake are correspondingly short, 5.2–6.3 mm (Klauber, 1939). Venom yields are also low. The usual total dry venom yield of an adult is 20–30 mg (Minton and Minton, 1969; Klauber, 1972). The largest *Sistrurus miliarius* yield no more than 35 mg of dried venom (Tennant, 1985). However, a typical bite probably injects only about 20 mg. The lethal dose for a 20-g mouse is 485 micrograms (Minton and Minton, 1969). The lethal dose for adult humans has not been calculated, but is probably more than the total capacity of the snake. Minton and Minton (1969) did not believe *S. miliarius* capable of inflicting a fatal bite to an adult human.

The venom seems rather virulent to mammals. Small mice bitten by an adult in my laboratory have died in 30–90 seconds. The venom apparently has little effect on rattlesnakes of the genus *Crotalus*. Munro (1947) reported that a smaller *Crotalus horridus* was bitten by a *S. m. streckeri,* but showed no symptoms of envenomation. However, a bitten *Micrurus fulvius* died in 24 hours (Wright and Wright, 1957).

Human envenomation by the pigmy rattlesnake is not uncommon, particulary in Florida; of 382 verified snakebite cases in Florida, 168 (44%) were inflicted by *S. miliarius* (Tu, 1977). The bite is more serious in children than adults, and Guidry (1953) reported that a small child required several weeks of hospitalization when bitten. Case histories of bites are given by Harris (1965) and Klauber (1972).

The venom contains no neurotoxic agents, but is strongly hemorrhagic. Bonilla et al. (1971) and Tu (1977) give detailed information on its chemistry.

The following symptoms have occurred in human bites: swelling, pain, weakness, giddiness, respiratory difficulty, hemorrhage, nausea, ecchymosis, and passage of bloody urine for two days (Hutchison, 1929).

Predators and defense

Sistrurus miliarius is frequently eaten by indigo (*Drymarchon*), king (*Lampropeltis*), and coral snakes (*Micrurus*), and adult pigmy rattlesnakes occasionally eat smaller members of their species (Allen and Neill, 1950b). Opossums (*Didelphis*), skunks (*Mephitis*), and domestic dogs (*Canis*) and cats (*Felis*) may also kill them, and large hawks (*Buteo*) probably prey on them. Most, though, are destroyed by habitat destruction and automobiles.

The pattern and grayish coloration of this species affords it some camouflage. This is especially true of the subspecies *S. m. barbouri* in areas where pine trees have been partially burned and ashes are abundant.

When accosted, this fiery tempered beast quickly coils, bobs its head, and strikes with little warning. Its small rattle can barely be heard.

Populations

In proper habitat, *Sistrurus miliarius* is often quite numerous. Bell (in Duellman and Schwartz, 1958) collected 27 individuals in eight days at one Florida site, and Viosca (in Dundee and Rossman, 1989) collected 103 on Delacrox Island, St. Bernard Parish, Louisiana, during a 12-day period when the snakes were forced onto levees by severe flooding. At one site in the Everglades National Park, 271 (15.2%) of the 1,782 snakes collected or observed were *S. miliarius,* and it was the most common of 21 species, but at another site in the park, only 25 (2.4%) of 1,019 snakes were this species (Dalrymple, Bernardino et al., 1991). The sex ratio in a brood of 32 young was 17 males and 15 females (Carpenter, 1960).

Remarks

Sistrurus is thought to be ancestral to *Crotalus* and to have originated in Mexico and later migrated north to the United States (Klauber, 1972). Data from a study of dorsal scale microdermoglyphics indicate *S. miliarius* is more closely related to *Crotalus pricei, C. cerastes* and *C. mitchelli* than to its congeners *S. catenatus* and the Mexican *S. ravus* (Stille, 1987).

Crotalus adamanteus Palisot de Beauvois, 1799
Eastern diamondback rattlesnake
Plate 25

Recognition

This rattlesnake achieves the largest size (maximum length, 244 cm; although most are 100–150 cm in total length) and bulk of any venomous snake in the United States. It is brown with a dorsal pattern of dark, yellow-bordered, diamondlike (rhomboid), broader than long blotches, several light stripes on its face, and a ringed brown and white tail. The venter is yellow to cream with some brownish mottling. A dark band, bordered front and back by a cream to yellow stripe, extends downward and backward from the eye to the supralabials. Several vertical light stripes are present on the rostrum. The pupil is elliptical, and there is a hole (pit) located between the nostril and eye. There are usually 27–29(25–31) midbody rows of keeled scales. On the underside of the body are 162–187 ventrals, 20–33 subcaudals, and an undivided anal plate. Most dorsal head scales are small, but the higher than wide rostral, 2 internasals, 4 canthals (between the internasals and supraoculars), and 2 supraoculars are enlarged. Prefrontals are absent, and usually 6–7 intersupraoculars are present. Laterally lie 2 nasal scales, several loreals, 2 preoculars, 2(3) postoculars, several suboculars, 14(12–17) supralabials, and 17(15–21) infralabials. There is usually prenasal-supralabial contact, but the postnasal is prevented from touching the upper preocular by the loreal scales. The short

Crotalus adamanteus (Roger W. Barbour, Collier County, Florida).

Crotalus adamanteus (John H. Tashjian, Gladys Porter Zoo).

hemipenis is bifurcate with a divided sulcus spermaticus and many recurved spines in the basal area. Spines also occur in the crotch between the lobes (see Klauber, 1972:696 for a photograph). As in all North American viperids, only the enlarged, hollow fang occurs on the shortened, rotational maxilla. Other teeth include 1–3 palatines, 7–11 pterygoidal and 9–10 dentaries.

Males have 165–176 ventrals, 26–33 subcaudals, and 5–10 dark rings on the tail; females have 162–187 ventrals, 20–27 subcaudals, and 3–6 dark rings on the tail.

Karyotype: Unknown.

Fossil record

Pleistocene remains of *Crotalus adamanteus* have been found at Irvingtonian and Rancholabrean sites in Florida (Holman, 1981; Meylan, 1982), and at a Rancholabrean site in Augusta County, Virginia, far north of its present range (Guilday, 1962). The fossil forms *C. a. pleistofloridensis* Brattstrom, 1953 and *C. giganteus* Brattstrom, 1953 have been synonymized with *C. adamanteus* (Auffenberg, 1963; Christman, 1975).

Distribution

The eastern diamondback rattlesnake ranges along the coastal plain from southeastern North Carolina to the Florida Keys and southeastern Mississippi and adjacent eastern Louisiana. McCranie (1980a) thought the Louisiana population possibly extinct, but Dundee and Rossman (1989) cited records from Washington and Tangipahoa parishes.

Geographic variation

Individuals from the Florida Keys have higher ventral counts (Christman, 1980), but no subspecies are recognized.

Confusing species

No other rattlesnake occurring east of the Mississippi River has the combination of a dorsal diamondlike pattern, light facial stripes, and a ringed tail.

Habitat

Crotalus adamanteus is best associated with dry, lowland palmetto or wiregrass flatwoods, pine, or pine-turkey oak woodlands. It usually

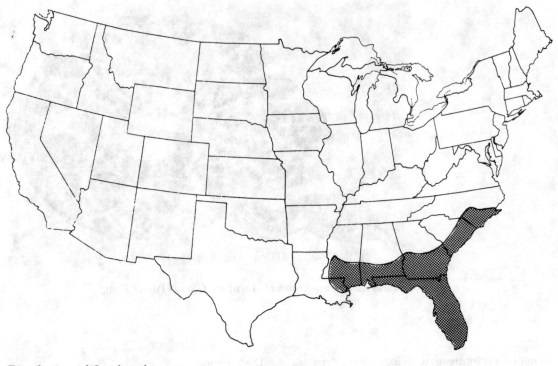

Distribution of *Crotalus adamanteus*.

avoids marshes and swamps, but occasionally lives along their borders, and it will swim across streams and narrow bodies of fresh water. It also occasionally swims or rafts across salt water to offshore islands.

Behavior

Where the winters become too cold, this large snake usually hibernates in mammal or gopher turtle (*Gopherus*) burrows, hollow logs or stumps, or among the roots of wind-felled trees. It does not live in uplands, and so does not usually have rock slides or crevices as winter retreats, as do *Crotalus horridus* and *Agkistrodon contortrix*. Probably, it also uses the same retreats to escape severe summer heat, although I have collected active Florida diamondbacks on very hot days. In Florida it may be active all year, but farther north it probably is restricted to an April–October annual cycle.

It is most commonly seen in the evening or early morning when rabbits, its favorite prey, are active; most of those I have caught were found in the morning. However, Ditmars (1936) has reported nocturnal activity, and many are run over by autos at night. Most daylight hours are spent in hiding or, if the air is cool, basking.

Activity is not restricted to ground level, and *C. adamanteus* occasionally climbs several meters off the ground into trees while pursuing prey (Klauber, 1972).

A male social dominance system is present in the form of a combat dance. When two males meet, they examine each other with much tongue flicking and staring, and then raise up the anterior 30–40% of their bodies and face each other. With heads bent at a sharp angle to the ground, the snakes entwine their anterior bodies and try to throw the other off balance and to the ground by pushing with the body and neck. This may last for only a few minutes to almost two hours (pers. obs.), and finally ends when one male (usually the smaller) is pinned to the ground, unwraps itself from the victor, and crawls away. Wagner (1962) observed a male become involved in a second combat almost immediately after winning the first bout. The incident took place at the Miami Serpentarium. Shortly, another rattlesnake approached the "victor" of the previous bout from the rear. The first snake started to crawl away, but the challenger caught up and crawled over him, their bodies being parallel. They proceeded to raise the anterior parts of their bodies off the ground, intertwining them while they were raised. When the two heads were about 30 cm from the ground they started leaning backward. Soon one or both lost balance and fell to the ground. The bout lasted for nearly an hour, but at the end the original male remained champion.

Reproduction

In South Carolina, *Crotalus adamanteus* has a spring mating season beginning in March (Kauffeld, 1969), but Meek (in Klauber, 1972) reported mating in mid-September, and Ashton and Ashton (1981) thought that copulation occurs in both the fall and spring.

The age and size at which sexual maturity is reached is not known, but Murphy and Shadduck (1978) reported that a male and female first attained at total body lengths of about 50 cm did not mate when the male was 135 cm and the female 110 cm, but that a successful copulation finally occurred two years later. Unfortunately their lengths at the time of copulation were not reported. The mating was observed on 30 January at 0800, and the snakes remained joined until 1700. Occasional pulsations near the cloaca of the male were seen, but the female remained passive. On 1 August, after a gestation period of 213 days, 10 young were born.

Mount (1975) reported that female of *C. adamanteus* from Alabama seek sheltered places and give birth in late summer or early fall. Judging from the duration of gestation above, this would make the mating period late winter or early spring in Alabama; however, a spring mating period is more likely. Known birth dates range from 16 July to 5 October (Klauber, 1972). Most young are born in retreats such as gopher tortoise burrows or hollow logs. It is not known whether female *C. adamanteus* reproduce annually or biennially.

The number of young per brood ranges from six (Curran, 1935) to 21 (Klauber, 1972); the mean for 28 litters reported in the literature was 12.8 young. At the time of emergence from the fetal membranes the neonates are usually 300–380 mm in total length and weigh 35–48.5 g (Murphy and Shadduck, 1978), but a brood from the Florida Panhandle averaged 388 mm (Dundee and Rossman, 1989).

A *C. adamanteus* has hybridized with *C. horridus atricaudatus* in Alabama (Klauber, 1972).

Growth and longevity

Christman (1975) reported measurements which showed that the length of a midbody vertebra was directly proportional to the overall length of several species of *Crotalus*. Prange and Christman (1976) studied this concept further in *C. adamanteus* and calculated correlation coefficients of regression of 0.99 for vertebral length plotted against body length ($y = 6.745x - 0.674$) and log body weight against log body length ($y = 3.108x + 2.766$).

Crotalus adamanteus has the potential of a rather long life. One kept by the late Louis Pistoia survived 22 years, 7 months and 3 days (Bowler, 1977).

Food and feeding

Young *Crotalus adamanteus* feed primarily on mice and rats, while adults seem to prefer rabbits (*Sylvilagus*) but will also take cotton rats (*Sigmodon*), white-footed mice (*Peromyscus*), squirrels (*Sciurus*), and birds—towhees (*Pipilo*), bobwhite quail (*Colinus*), turkey (*Meleagris*), pileated woodpeckers (*Dryocopus*), and king rail (*Rallus*) (Klauber, 1972; pers. obs.). The food need not be fresh; Funderberg (1968) reported that a large female contained a rabbit that was in such a state of decomposition and odor and with numerous broken bones that it had to have been a road kill from a nearby unpaved road.

Crotalus adamanteus may actively seek out prey by following their scent trails, but many animals are caught from ambush. These snakes often lie in wait for prey beside logs or among the roots of fallen trees. In fact, the first eastern diamondback I ever caught was lying between several cotton rat holes under a wind-felled tree. In addition to their odor, endothermic prey emit infrared heat waves which may be detected by the pit on the rattlesnake's face.

Usually only one envenomating bite is delivered. The wounded prey is released immediately after the strike and allowed to crawl off to die (Chiszar, Radcliffe, Byers et al., 1986). Biting of the prey sets off a chemosensory search image for the tasted animal. The snake will slowly pursue its victim until it finds it, and then examines the prey with its tongue to determine if the prey is dead. Swallowing then proceeds quickly, usually from the head end. Chemical searching may last for 2–62 minutes following a strike (Brock, 1981).

Venom and bites

The eastern diamondback has a well-developed solenoglyphous venom-delivery system (Klauber, 1972), with fangs to 27 mm in length (Telford, 1952). Its venom is strongly hemolytic and human deaths from severe untreated bites usually occur in 6–30 hours; the mortality rate may be 40% (Neill, 1957). Total dry venom yield from *Crotalus adamanteus* may be as high as 848 mg, but averages from about 492–666 mg per adult (Stadelman, 1929a; Klauber, 1972). The minimum lethal doses for a 350-g pigeon and a 20-g mouse are 0.2–0.3 mg and 0.04 mg, respectively (Githens and George, 1931; Githens and Wolff, 1939). The minimum lethal dose for a rabbit is 0.25 mg per kg of body weight (Boquet, 1948). The human lethal dose has been estimated as 100 mg (Dowling, 1975). Venom toxicity varies between individual snakes, even from the same litter (Mebs and Kornalik, 1984). The chemistry of venom from *C. adamanteus* is discussed in Minton (1974) and Tu (1977).

Symptoms of human envenomation by *C. adamanteus* include swelling, pain, weakness, giddiness, respiratory difficulty, hemorrhage, weak pulse or heart failure (or in some cases an increased pulse rate), enlarged glands, soreness, diarrhea (often bloody), collapse, shock, toxemia, and convulsion (Hutchison, 1929; Kitchens and Van Mierop, 1983). Necrosis around the bite occasionally involving much of the injured limb is fairly common, and sensory disturbances may occur, such as a sensation of

yellow vision (Minton, 1974). Bitten limbs often are at least partially crippled.

Wyeth Laboratories, Inc., Philadelphia, produces an antivenin to counteract the venom of *C. adamanteus,* and total doses of 130–170 ml may be needed in severe bites. Case histories of bites by *C. adamanteus* are related by Parrish and Thompson (1958) and Klauber (1972).

Predators and defense

Adults have little to fear except from humans. The young, however, have many enemies, including hogs (*Sus,* whose thick skin and subcutaneous fat make it difficult for the rattlesnake to effectively bite), carnivorous mammals (*Procyon, Ursus, Mephitis, Lutra, Canis, Felis, Lynx*), raptorial birds (*Bubo, Buteo, Caracara*), wood storks (*Mycteria*), other snakes (especially kingsnakes, *Lampropeltis,* black snakes, *Coluber,* indigo snakes, *Drymarchon,* and coral snakes, *Micrurus*), and river frog, *Rana heckscheri* (Neill, 1961). Humans are by far the worst enemy. Most will kill a rattlesnake on sight, and their automobiles and habitat alteration destroy many each year. Recently, commercial rattlesnake roundups have resulted in the death of many *Crotalus adamanteus* (Speake and Mount, 1973). The participants use gasoline to flush hibernating snakes from their retreats. Those snakes which emerge are either collected or killed, while many rattlesnakes and other innocuous animals probably die in the burrows as a result of inhaling the fumes. Rattlesnake meat is tasty, and occasionally they are killed for a gourmet meal.

It will lie quietly coiled when first discovered, displaying such a mild temperment for a rattlesnake that it has been referred to as "the gentleman of snakes" by Snellings (1986). However, if provoked severely, it coils, shakes its tail, and raises the head and neck into a striking position. If further annoyed it does not hesitate to strike, and if picked up will thrash about to deliver a bite. Overall, they are extremely dangerous and should be let alone.

The rattle is a highly diversionary defense mechanism, calling attention to the tail instead of the head. As far as humans are concerned, it is certainly an effective warning device. The mean sound frequency of rattling is 3.79 kHz, the mean highest sound frequency is 16.50

kHz, and the mean sound band width is 12.72 kHz (Fenton and Licht, 1990).

Populations

In suitable habitats, *Crotalus adamanteus* can be quite common. This is especially true in the vicinity of Corkscrew Swamp Sanctuary, Collier County, Florida (pers. obs.), and at Okeetee, South Carolina. A series of 280 snakes caught in 74 days of actual hunting over a period of six years at Okeetee included 60 (21%) *C. adamanteus* (Kauffeld, 1957). About 1,000 *C. adamanteus* were killed each year on seven hunting preserves in the Thomasville-Tallahassee area of Florida where a dollar bounty was being offered for each snake (Stoddard, 1942). The late Ross Allen reported that over a period of 28 years he had received 1,000–5,000 per year at his Florida snake exhibit with a grand total of 50,000 (Klauber, 1972)! At a site in the Everglades National Park, however, only 1.7% of the 1,782 snakes seen or collected over a three-year period (1984–1986) were this species, indicating a pos-

sible decline (Dalrymple, Bernardino et al., 1991).

Remarks

Gloyd (1940) thought that *Crotalus adamanteus* was a "climax form" derived from *C. atrox*, the western diamondback rattlesnake, which he believed closest to the ancestral type for the *atrox* group of rattlesnakes. Studies by Meylan (1982) support this theory, but Christman (1980) noted there is no reason to believe *C. adamanteus* arose on the Mexican Plateau and then migrated to Florida. Both Florida and the southwestern United States seem to be refuges where rattlesnakes, like *adamanteus* and *atrox*, evolved more slowly, retaining ancestral characters.

As can be seen from the above narrative, we are rather ignorant of many facets of the natural history of *C. adamanteus*. That so large and visible an animal is so poorly understood is remarkable. We must instigate ecological studies soon, as much of its habitat is disappearing, particularly in Florida, or it may be too late to assure the survivorship of this magnificent beast.

Crotalus atrox Baird and Girard, 1853
Western diamondback rattlesnake
Plate 26

Recognition

This is one of the largest venomous snakes (maximum length, 213 cm) in the United States, surpassed in length and bulk only by its eastern cousin *Crotalus adamanteus*, and individuals over 120 cm are not rare. Coupled with its great size is an aggressive nature, making *Crotalus atrox* a dangerous snake. Ground color is brown to gray, but some individuals are reddish, yellowish, or even melanistic (Pedro Almendariz lava field in Socorro and Sierra counties, New Mexico; Best and James, 1984). A series of 24–45 dark, light-bordered, diamond-shaped or hexagonal blotches occurs on the back, and some smaller dark blotches lie along the sides. The tail is distinctly pat-terned with alternating wide gray and black rings; usually 4–6(2–8) black rings are present. A pair of light stripes extends diagonally downward and backward from in front and in back of the eye to the supralabials. The posteriormost of these ends well in front of the corner of the mouth. Some dark pigment may be present on the top of the head. Ventrally the scutes are white, cream, or pink, with some fine dark mottling along the sides; the underside of the tail is often grayish. Midbody scale rows average 25(23–29). Beneath are 168–196 ventrals, 16–36 subcaudals, and an undivided anal plate. The rostral is higher than wide; it is followed by 2 small internasals, and behind these lie 11–32 additional scales. The 2 supraoculars are large; between them are 4–5(3–

Crotalus atrox (John H. Tashjian; University of Texas, El Paso).

8) intersupraocular scales. No prefrontals are present. Laterally, behind the 2 nasals are situated a loreal (occasionally the loreal is doubled), 2(3) preoculars, several suboculars, 3(2–4) postoculars, 15–16(12–18) supralabials, and 16–17(14–21) infralabials. The prenasal is normally in contact with the supralabials, and the postnasal usually touches the upper preocular. The bilobed hemipenis has a divided sulcus spermaticus, and on each lobe are a soft apical projection, about 64 spines, and 57 fringes (Klauber, 1972). Only a few spines lie in the crotch between the lobes. The maxilla bears only the fang, while the palatine, pterygoid, and dentary usually have 3, 8(7–9) and 9–10 teeth, respectively.

Males have 168–193 ventrals, 19–32 subcaudals, and 5–6(3–8) black tail rings; females have 173–196 ventrals, 16–36 subcaudals, and 4(2–6) black tail rings.

Karyotype

According to Baker et al. (1972), the karyotype of *Crotalus atrox* consists of 36 chromosomes: 16 macrochromosomes and 20 microchromosomes, with the sex determined as ZW (female) or ZZ (male), the Z chromosome submetacentric and the W chromosome subtelocentric. Zimmerman and Kilpatrick (1973) expanded and modified this description by reporting the presence of 4 metacentric, 6 submetacentric and 4 subtelocentric macrochromosomes, and that the Z is metacentric and the W submetacentric.

Fossil record

Upper Pliocene (Blancan?) remains have been found in Texas (Holman, 1979). Pleistocene records of fossil *Crotalus atrox* include: Irvingtonian, Texas; Rancholabrean, Arizona, California, Nevada, New Mexico, Texas, and Sonora, Mexico (Holman, 1981; Mead et al., 1984; Van Devender et al., 1985).

Distribution

This rattlesnake ranges from western Arkansas westward through eastern and south-central Oklahoma, Texas (except the extreme east and the northern panhandle), central and southern New Mexico and Arizona, possibly extreme southern Nevada (Clark County; see Emmer-

son, 1982), southeastern California (Riverside and Imperial counties), and southward in Mexico to extreme northeastern Baja California and northern Sinaloa in the west, and northern Veracruz, Hidalgo and Querétaro in the east. Most of the range lies below 1,500 m.

Crotalus atrox was once accidentally introduced into Wisconsin, where a small population persisted for several years (Gloyd, 1940).

Geographic variation

Although some color variation occurs between populations, no subspecies are recognized.

Confusing species

Crotalus scutulatus has very narrow dark tail rings and two to three large intersupraoculars, and the most posterior, diagonal white stripe on the side of the face ends in front of and above the corner of the mouth. *C. ruber* is more reddish, lacks dark spots in the body blotches, has 29 midbody scale rows, and the first pair of infralabials is usually transversely subdivided. In *C. viridis*, the tail rings are less pronounced and the light diagonal stripe behind the eye extends above the corner of the mouth. *C. mitchellii* is usually speckled with black, has faded tail rings, and a series of small scales separating the rostral from the prenasals. *C. molossus* has a unicolored, black tail.

Habitat

This is a snake of varied arid or seasonally dry habitats such as deserts, grasslands, shrublands, scrub woods, or open coniferous forests, with typical plants ranging from cacti and thornbushes to mesquite, paloverde, grasses, and even riparian oak groves. It is often found among scattered rocks and boulders, or rock outcrops with crevices.

Behavior

The earliest record for an active *Crotalus atrox* is 9 March (Minton, 1958). However, over most of its range, it can be found above ground from April to late September or early October, although Klauber (1972) reported one active as late as 18 December, and Wright and Wright (1957) listed other late autumn rec-

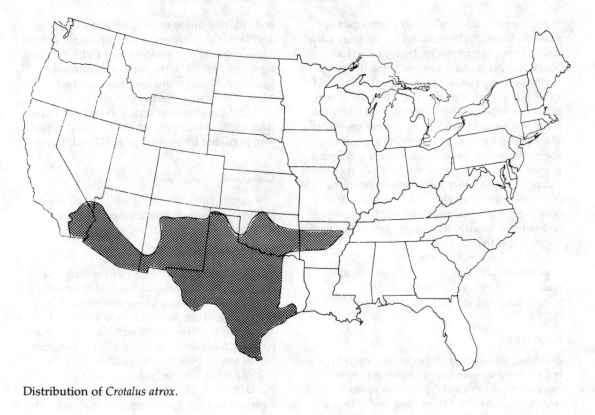

Distribution of *Crotalus atrox*.

ords. In Chihuahua, Mexico, Reynolds (1982) found this species to be active mainly from July to September, with a decided peak in August. Precipitation there is highest in July, and *C. atrox* seems to reach its greatest activity about a month later, possibly owing to increases in rodent populations.

Winter is spent in hibernation, most often down a rock crevice or in a cave located on a south-facing slope, but also sometimes in animal burrows. At one Texas site, hibernating snakes had apparently crawled into the cracks formed in the heavy black soil during drying in the previous hot summer (Klauber, 1972), and in Arizona some have been found hibernating under wood piles (Gates, 1957). Although some weight is lost (most likely due to dehydration) during the fasting period of hibernation, little change occurs in most blood constituents. The level of cholesterol and NaCl drop while other values remain essentially the same (Martin and Bagby, 1973). Hibernation is

adaptive, and saves the snake from having to cope with both falling air temperatures and, in some cases, decreases in prey numbers. If the weather improves, some *C. atrox* will bask at the mouth of their retreats around noon on warm winter days.

Crotalus atrox is one of our rattlesnake species that regularly congregates at specific dens for overwintering. Such dens may serve over 100 rattlesnakes, and other species of snakes and mammals may occasionally join them. Klauber (1972) told of several winter dens that usually had as many as 200 *C. atrox* present, although recently the numbers of rattlesnakes at such dens have decreased. A period of winter cooling or inactivity is needed for successful breeding in some species of snakes, and the same may be true of *C. atrox* (Tryon, 1985).

Western diamondbacks are more diurnally active during the spring and fall; then they forage most often in the morning or early evening, with some basking in the morning. In the summer they are chiefly crepuscular or nocturnal, resting under some sheltering rock or shading vegetation during the day.

Daily activity is determined by air temperature (Landreth, 1973). Brattstrom (1965) recorded body temperatures of 21–34°C (mean 27.4) from active *C. atrox*, and Cowles and Bogert (1944) found an active snake with a body temprature of 18°C, but they thought that body temperatures of 27–30°C brought about persistent activity (fresh captured snakes with body temperatures of 14°C were thought to be active only because of abnormal disturbance). The critical thermal maximum of one *C. atrox* was 39°C (Cowles and Bogert, 1944), but this is lower than expected, as Mosauer and Lazier (1933) reported that a *C. atrox* finally died at a body temperature of 46.5°C after 10.5 minutes of full exposure to the sun. The snake had been severely affected in the last minutes before death; grasping and convulsively wriggling. All body processes, including the frequency of rattling, increase with increasing body temperature (Martin and Bagby, 1972).

Most movement is by the typical serpent method of laterally undulating the body, but on loose sand a caterpillar-like crawl may be employed. *Crotalus atrox* may make annual migrations of from 0.7 to 3.5 km to and from winter dens (Landreth, 1973). During these migrations the snakes do not wander randomly, but instead follow directed courses. After directional trials in an arena, Landreth (1973) concluded that *C. atrox* used solar cues (sun compass?) to determine long-distance directional goals, but thought that olfaction probably played some role when near the den. In the spring, males moved an average of 102.4 m per day, but only averaged 61.2 m and 54.3 m in the summer and fall. Females made average daily movements of 82.4 m in the spring, 46.1 m the summer, and 46.3 m in the fall. However, three different females made unexplained midsummer, overnight trips of 72.4–105.6 m. Average winter movements were of little distance from the den (males, 3.5 m; females, 2.7 m).

Like most other rattlesnakes, *C. atrox* can climb, and has been seen on branches as high as 1.2 m above the ground (Klauber, 1972). One I kept in the laboratory continually climbed onto the shield of a heating lamp sus-pended from the ceiling of its cage. This snake is also an accomplished swimmer, and Klauber (1972) related several instances of its aquatic activities. It has even been seen swimming at sea about 32 km from land (Shaw and Campbell, 1974).

Male western diamondback rattlesnakes participate in combat bouts in both the wild (Lowe, 1942, 1948a) and in captivity (Armstrong and Murphy, 1979). These bouts or "dances" are of similar pattern in all large species of *Crotalus*, and the reader is referred to the description presented for that by *C. ruber*.

Reproduction

The age and size at which *Crotalus atrox* reaches sexual maturity are speculative. During a study of the female reproductive cycle of *C. atrox* from northwestern Texas, Tinkle (1962) found the shortest mature female to be 800 mm in snout-vent length. The largest definitely immature female was 936 mm, with the exception of a 1005-mm female which appeared immature. Most females longer than 900 mm in snout-vent length were mature. Klauber (1972) reported a gravid female of 742 mm in snout-vent length, Minton (1958) found one that was 889 mm long, and Simons (1986) reported another gravid female of 800 mm in total length. The heaviest immature female weighed by Tinkle (1962) was 450 g, while the lightest mature female weighed 320 g; most adults weighed more than 500 g. Females from Texas in their third hibernation were mature and would have reproduced the following fall at an age of about three years (Tinkle, 1962). Maturity seems to take place in 30–36 months.

Females reproduce biennially (Tinkle, 1962). Those which are reproductive store more fat than females in their nonreproductive year. Follicle enlargement begins after emergence from hibernation in the previous nonreproductive year. Follicles reach a maximum length of about 30 mm with considerable yolk deposition prior to the next hibernating period, but little follicular enlargement occurs over the winter. In the spring of the second (the reproductive) year of the cycle, the follicles further enlarge and the ova are ruptured and fertilized early in the summer. The developing young are carried through the summer, and are usually born in August or September.

No follicles enlarge when the female is gravid, and postpartum females have follicles no larger than 10 mm. Fat volume slowly increases after parturition.

In Chihuahua, Mexico, male *C. atrox* experience early spermiogenesis in July and full spermiogenesis during August and September, and are probably reproductive during the entire summer (Jacob et al., 1987). Testes mass does not vary significantly during these months, but seminiferous tubule diameter is significantly greater in August. Snout-vent lengths of mature males examined by Jacob et al. (1987) were 491–1,110 mm.

In captivity, courtship and mating activity have been observed in almost all months. Wild copulating *C. atrox* have been found in spring as early as 25 March (Bogert, 1942) and as late as 14 May (Wright and Wright, 1957). A second late summer or early fall breeding period may also take place; Taylor (1935) reported a 29 August mating in Arizona. Olfaction may play an important role in mate selection at the den. A typical mating sequence is as follows. A male senses the presence of a female through olfaction and begins rapid tongue flicking (1/ sec). He then raises the fore portion of his body about 50–60 cm and begins to jerk spasmodically (increasing to 1/sec). The male crawls to the female and closely examines her body with his tongue, particularly her back and sides, and increases the rate of spasmotic body jerking to about 2/sec. He touches his chin to the female's back and begins to rub it sideways at irregular intervals, and brings his entire body in contact with hers. The female usually raises her tail and opens her cloacal vent at this stage, and the male positions his vents next to her open vent; this may take considerably manipulation on his part. When the vents are finally aligned one hemipenis is inserted. The female may begin rhythmic body spasms at this time, but usually lies still once insertion has taken place. The male may pulsate his anterior tail region during the copulation. The actual mating act may take as little time as 15 minutes (possibly due to my disturbing the pair) to as long as four to five hours (Armstrong and Murphy, 1979, reported one mating lasted eight hours).

Birth dates in captivity have ranged from June to October (Armstrong and Murphy, 1979), but it is more likely that parturition in the wild takes place in August or September (Price, 1988). Wiley (1929) described the birth process during captive parturition.

The average number of young in 61 litters reported in the literature is 10.6(4–25). Litter size is positively correlated with female body length (Tinkle, 1962). Newborn young average 278 mm (214–339) in total length, and are more boldly marked than adults.

Price (1988) related several instances in which an adult female was found near newborn young caught in a pitfall trap, and thought this possibly an example of maternal behavior.

Crotalus atrox may mate with other species of rattlesnakes when in captivity (*C. molossus, C. adamanteus, C. scutulatus;* Klauber, 1978; Aird et al., 1989), but a study by Jacob (1977) to determine if *C. atrox* and *C. scutulatus* naturally interbred provided no positive evidence of hybridization.

Growth and longevity

A 503-mm (62-g), wild female *C. atrox* grew 38.1 mm (17 g) in 43 days between captures, and another wild 774-mm (258-g) female grew 50.8 mm (8 g) in 59 days (Laughlin and Wilks, 1962).

A captive *C. atrox*, wild caught as a juvenile, lived 25 years, 10 months, and 27 days (Bowler, 1977), and William W. Lamar (pers. comm.) has called my attention to an individual collected as a 90-cm adult at the Okeene Rattlesnake Roundup in 1964 that was still alive in January 1991.

Food and feeding

Crotalus atrox has been the subject of many reports and two extensive studies of its feeding habits. As a result, we have a better understanding of its prey than for most other rattlesnakes. Warm-blooded animals are preferred, but reptiles, insects, and possibly even amphibians are occasionally taken. Known wild prey includes: mammals—shrews (*Cryptotis*), cricetid mice and rats (*Baiomys, Microtus, Neotoma, Peromyscus, Onychomys, Reithrodontomys, Sigmodon*), pocket mice (*Perognathus*), kangaroo rats (*Dipodomys*), murid rats (*Rattus*), pocket gophers (*Geomys*), ground squirrels (*Spermophilus*), fox squirrels (*Sciurus*), cottontails (*Sylvilagus*), and jackrabbits (*Lepus*); birds (including eggs and nestlings)—quail (*Lophortyx*), gulls (*Larus*), terns (*Sterna*), black skimmers (*Rynchops*), owls (*Athene*), mockingbirds (*Mimus*), doves (*Columbigallina*), towhees (*Pipilo*), horned larks (*Eremophila*), and sparrows (*Amphispiza, Melospiza*); snakes (*Crotalus atrox*); lizards (*Cnemidophorus, Coleonyx, Crotaphytus, Eumeces, Phrynosoma, Sceloporus*); insects—lubber grasshopper (*Brachystola*), beetles and ants; and possibly a toad (*Bufo*) (Hermann, 1950; Woodin, 1953; Stebbins, 1954, 1985; Fouquette and Lindsay, 1955; Gates, 1957; Wright and Wright, 1957; Minton, 1958; Cottam et al., 1959; Smith and Hensley, 1958; Klauber, 1972; Conant, 1975; King, 1975; Beavers, 1976; Reynolds and Scott, 1982; Best and James 1984; Quinn, 1985; Tennant, 1985; Vermersch and Kuntz, 1986). The beetles and ants were found in a specimen that also contained mammal hairs and the remains of an iguanid lizard (Klauber, 1972) and so may have been secondarily ingested with other prey. The toad specimen was reported by Centerwall to Klauber (1972), but it may have been the remains of a lizard. In captivity, this snake does well on house mice (*Mus*), and Klauber (1972) reported a captive *C. atrox* ate a *C. molossus* that may have died before it was eaten. Wiley (1929) fed neonates small bits of beef and liver until they were large enough to take prey.

Beavers (1976) examined the digestive tracts of 205 Texas *C. atrox*; 78 (38.1%) contained food. Small mammals made up the largest diet by weight (94.8%) and frequency of occurrence (86.7%) (birds, 7.6%, 13.3%; lizards, 2.9%, 11.1%). Of the mammals, pocket mice (*Perognathus*) had been eaten by 39% of the snakes, harvest mice (*Reithrodontomys*) by 9%, and white-footed mice (*Peromyscus*) by almost 7%. In a similar study of 43 western diamondbacks from Chihuahua, Mexico, Reynolds and Scott (1982) also found that mammals were the predominant foods (85.7% occurrence), with pocket mice having been eaten by 26.5%, white-footed mice by 16.3%, and ground squirrels (*Spermophilus*) by 10.2%. Reptiles were found in 12.2% of the digestive tracts, and birds in only 2.0%. Prey was selected on the basis of size; animals either too

large or too small were not eaten, nor were those that could seriously harm the snake.

Carrion is probably taken when available. Captives usually are willing to consume prekilled mice. In laboratory tests using dead or decaying rodents conducted by Gillingham and Baker (1981), *C. atrox* readily accepted the carrion; the snakes apparently used olfaction to find putrescent mice, even those buried in gravel, but failed to find fresh-killed mice. Gillingham and Baker (1981) thought this strong evidence that this species possesses a scavenging feeding strategy. Being opportunistic, *C. atrox* probably eats what it can find, dead or alive, providing it is of the proper species and body size.

Crotalus atrox are nocturnal hunters for most of the year, but forage more often during the day in the spring and fall. Prey are detected either by infrared reception or by olfaction, and two hunting strategies are used: active foraging and ambushing. When a prey odor trail is discovered by a hunting snake, the rate of tongue flicking increases as it gains information about the freshness of the trail (Gillingham and Clark, 1981). If reasonably fresh, the odor trail is followed, the snake frequently tasting the air with its tongue, and such chemosensory searching may continue for as long as 2.5 hours (Chiszar, O'Connell et al., 1985). When the warm-blooded prey is finally either detected visually or by infrared reception, the rate of tongue flicking slows.

When in ambush, *C. atrox* usually lies quietly beside a frequently used rodent trail. At such times it may lay a coil over its rattle, perhaps to muffle it (Thayer, 1988).

Once the prey is within striking range, it is quickly bitten, and then left to wander away to die. At its leisure, the rattlesnake follows the odor trail of the dying animal, and once having discovered its body, cautiously examines it with its tongue to make sure the prey is dead. When satisfied the animal is no longer a threat, swallowing rapidly proceeds; usually the prey is taken down head-first.

Venom and bites

While the venom of *Crotalus atrox* is certainly potent enough to rapidly kill its prey, many small animals probably die almost instantly from being struck by such long fangs as possessed by this snake. From 51 *C. atrox* with body lengths of 100–140 cm, Klauber (1939) recorded fang lengths of 9.6–12.9 mm, second only to *C. adamanteus* which had fangs to 15.8 mm.

The venom of *C. atrox* is highly hemorrhagic. Almost 53% of its enzymes are concerned with lytic breakdown of the circulatory system, but another 17% are neurotoxic in action, and 30% are digestive proteases (Tennant, 1985). A neurotoxin analagous to the Mojave toxin of *C. scutulatus* is present in the venom of some *C. atrox* (2 of 12 samples examined by Weinstein et al., 1985), making these individuals more dangerous than most. Hemorrhaging from the breakdown of vascular tissues occurs rapidly, being evident as soon as six minutes after introduction of the venom (Soto et al., 1989). In addition, the venom contains strong hemaglutinizing enzymes (which cause clumping of the red blood cells; Minton, 1956).

Gregory-Dwyer et al. (1986) found no seasonal variations in the isoelectric properties of the proteins in venom from individual *C. atrox* kept under controlled conditions and tested monthly for 20 months; however, intraspecific differences in concentrations of certain proteins were evident. The venom is quite stable if stored in closed containers in the dark; Russell et al. (1960) reported that venom thus kept lost relatively little of its potency after 16–17 years.

Newborn western diamondbacks, like other crotalids, possess fangs and active venom glands that can deliver a serious bite. However, the venom yield per bite is small. The yields of venom naturally increase with the size of the snake. Klauber (1972) has estimated that a 60-cm *C. atrox* could yield about 0.10 ml of venom per bite (0.19 ml high yield), while one 150 cm long would inject 1.27 ml per bite (1.88 high yield), and presented a table showing the yields for snakes of lengths between these two extremes. A typical adult *C. atrox* may contain 200–300 mg of dried venom, and the lethal dose for a human is estimated to be about 100 mg (Minton and Minton, 1969). One 165.5-cm *C. atrox* yielded a total of 1,145 mg of dried venom (3.9 ml liquid) (Klauber, 1972). So, venom yields for this species are often very high. Lethal doses for some other animals are: pigeons, 0.09–0.35 mg (Githens and

George, 1931); mice, 0.12 mg (Githens and Wolff, 1939); and rats, 0.025 mg (per g rat body weight; Billing, 1930).

Bites of humans by *C. atrox* in the Southwest are not uncommon and often are serious (Hutchison, 1929). Without antivenin treatment, death may occur, and this species has probably been responsible for more human deaths than any other snake in the United States. Symptoms reported resulting from bites by *C. atrox* include: intense burning pain, swelling, discoloration of tissues, edema, ecchymosis, hemorrhage, necrosis, hematemesis, hemolytic anemia, lowered blood pressure, lowered heart rate, increased heart rate, fever, sweating, weakness, giddiness, nausea, vomiting, breathing difficulties, and secondary gangrene infection (Hutchison, 1929; Russell, 1960). Obviously this snake is not one to be messed with! It also lives near sites of human activity and bites are a common result. Case histories of human envenomation are presented in Ehrlich (1928), Crimmins (1927), Hutchison (1929, 1930), Fidler et al. (1938), Young (1940), Klauber (1972), and La Rivers (1976).

Predators and defense

Adult western diamondback rattlesnakes have few, if any, natural predators; however, newborn and juvenile *Crotalus atrox* probably are eaten by a number of other carnivorous animals: hawks (*Buteo*), owls (*Bubo*), roadrunners (*Geococcyx*), coyotes (*Canis*), badgers (*Taxidea*), skunks (*Mephitis*), bobcats (*Lynx*), peccaries (*Tayassu*), and ophiophagous snakes: kingsnakes (*Lampropeltis*), whipsnakes (*Masticophis*), and other rattlesnakes (*Crotalus*). Despite this long list of potential predators, relatively few actual reports of predation on *C. atrox* are available. Blair (1954) observed what he thought to be an attempted attack by a bobcat (*Lynx rufus*) in Texas. A small *C. atrox* has been taken from the stomach of a bullfrog (*Rana catesbeiana*) by Clarkson and de Vos (1986). Wilson (1954) published photographs of an attack on a *C. atrox* by an indigo snake (*Drymarchon corais*), and Shaw and Campbell (1974) related several instances of predation by western whipsnakes (*Masticophis*). Klauber (1972) reported a case of captive cannibalism of a juvenile by another juvenile (*C. atrox* are not immune to the venom of their own species).

Indirect evidence of kingsnake (*Lampropeltis getulus*) predation has been presented by Cowles and Phelan (1958), who recorded a fear reaction increase in heart rate from 40 to 60 beats/minute when *C. atrox* was subjected to the kingsnake's odor.

The western diamondback sometimes enters chicken coops to feed on the chicks (or possibly eggs), but this can be risky for the snake. Crimmins (1931) observed chickens carrying young *C. atrox* in their bills, and after he killed one snake and chopped it up, the chickens and muscovy ducks in the pen fought over the pieces. If a *C. atrox* enters a pen through a chicken-wire fence and then feeds, it may become hopelessly stuck in the fence when it tries to leave (Campbell, 1950).

Humans are still the major predator of *C. atrox*. The automobile, habitat destruction, and wanton killing (especially at dens) has severely decimated populations in some areas, and possibly one of the worst destructive actions by humans is in the form of the popular "rattlesnake roundups" in Oklahoma, Texas, and other southwestern states, where *C. atrox* is usually by far the most common snake captured (Campbell, Formanowicz et al., 1989).

A large, irate western diamondback is an awesome adversary. It will coil, raise its neck and head as high as 50 cm above its coils, continually rattle its tail, strike, and sometimes even advance toward you, presumably to get within better striking range. Even the newborn young instinctively coil and strike. Once disturbed, some *C. atrox* may not soon calm down; Martin and Bagby (1972) reported that one rattled continuously for three hours. The mean lowest and highest frequencies of rattling are 2.36 and 15.31 kHz, respectively; the mean sound bandwidth is 12.95 kHz (Fenton and Licht, 1990). The worst case scenario has been presented above, and it must be remembered that individual variation in temperment occurs; some *C. atrox* may remain remarkably placid when disturbed. However, this animal must be considered extremely dangerous.

Some *C. atrox* spray musk from anal scent glands when disturbed (particularly if handled), and this secretion has been considered a possible deterrent to predators. Studies by Weldon and Fagre (1989) on the response of dogs (*Canis familiaris*) and coyotes (*Canis latrans*) to the scent gland secretions of *C. atrox* have indicated this may not be true. The canids were not repulsed.

Populations

Crotalus atrox is one of the most common snakes in the southwestern United States, and is surely the most numerous venomous snake over much of the area. In western Texas it may be found in dense populations (although not as dense as in the past). About 3,500 individuals were collected at Floresville, Wilson County, Texas, between 1 June and 1 September 1926 (Crimmins, 1927), and a rancher killed 1,200 while clearing about 4,049 ha of cactus and brush in Shackelford County, Texas, a density of 0.3 *C. atrox* per hectare (Wood, in Klauber, 1972). It is extremely abundant in Arizona: of 425 rattlesnakes collected in central Arizona by Klauber (1972) 218 (51.3%) were *C. atrox*. However, Vitt and Ohmart (1978) noted that due to its high trophic position, it existed at relatively low densities, at least, in the lower Colorado River area.

Boyer (1957) recorded a sex ratio of 117 females to 97 males (1.00:1.21) in 214 *C. atrox* from the Wichita Mountain Wildlife Refuge in southwestern Oklahoma, but this is not significantly different from 1:1.

Remarks

In his 1940 monograph on the rattlesnakes, Gloyd proposed that *Crotalus atrox* formed a complex based on morphology and scalation with the closely related species *C. adamanteus*, *C. ruber*, *C. exsul*, and *C. tortugensis*. This arrangement was followed, at least for those species occuring in the United States, by Brattstrom (1964) and Klauber (1972). Data from electrophoretic studies of venom proteins by Foote and MacMahon (1977) have strengthened this arrangement, but have also showed a closer linkage to *C. viridis* than had been recognized.

Although it is a dangerous animal, *C. atrox* has several economically important features. It is an excellent "mouser," eating many rodents each year that might otherwise destroy or damage crops or stored grains, and potentially spread human diseases (such a bubonic plague, tuleremia, and lyme disease). The meat of this snake is also tasty and has been favorably compared to chicken (although it is more stringy) and there is a thriving export industry to the Orient, but, fortunately, there is currently not a great demand for the meat in North America. Finally, this snake is one of the favorite exhibitions in captivity, and probably all major zoos in North America and Europe display them. Would it not be a shame if *C. atrox* was to slowly disappear from the wild so that only these captive, relatively short-lived (25 years) specimens were all that remains of North America's King of the Rattlesnakes?

Crotalus cerastes Hallowell, 1854
Sidewinder
Plates 27–29

Recognition:

This small (maximum length, 82.4 cm), pale rattlesnake, generally of deserts, has a pointed, hornlike projection over each eye. The dorsum is pinkish, cream colored, tan, or gray with a series of 28–47 darker tan, yellowish-brown, orangish, or gray blotches. Three longitudinal rows of dark spots (often faded) extend down each side, and a light-bordered, dark stripe runs diagonally backward from the orbit to the corner of the mouth. Some dark spots or streaks may occur on the dorsal surface of the head behind the eyes. The tail may bear two to six rings, the last of which are usually the darkest. The venter is white to cream, and may bear dark pigment. The keeled body scales occur in 21–23(19–25) rows at midbody, and a strongly keeled, tuberculate spinal ridge is present. The ventral scutes occur in a series of 132–154, and the subcaudals in 14–26; the anal plate is not subdivided. The rostral scale is wider than high, and the 2 internasal scales which follow are moderate in size. Behind these lie 12–34 scales before the 4–6 intersupraoculars. No prefrontals are present, and the supraoculars are elevated into pointed, hornlike structures over the eyes. Laterally are 2 nasals (the prenasal touches the supralabials), 1(2) loreals (the loreal extends to the orbit splitting the proculars), 2 preoculars, 2–3 suboculars, 2 postoculars, 12–13(10–15) supralabials, and 12–13(10–17) infralabials. The hemipenis is composed of two short, thick lobes each with approximately 54 spines and 21 fringes, and numerous spines in the area between the lobes (Klauber, 1972). The dental formula is 1 maxillary (the fang), 3(2–4) palatine; 8(7–9) pterygoidal, and 9–10(8–11) dentary teeth.

This is generally thought to be the only species of North American *Crotalus* with females larger than males (which seldom are longer than 60 cm), but this may not be true. Stephen M. Secor, who is studying this snake in the eastern Mojave Desert, has never noticed a distinct size dimorphism; the largest sidewinder he collected was a 66.7-cm male, and of the 11 snakes greater than 55 cm that he captured, 5 were males and 6 were females. Males have 132–151 ventrals and 18–26 subcaudals; females have 136–154 ventrals and 14–21 subcaudals.

Karyotype

Zimmerman and Kilpatrick (1973) have described the karyotype as 2n = 36: 16 macrochromosomes (4 metacentric, 6 submetacentric, 4 subtelocentric, the Z sexual chromosome metacentric, and the W sexual chromosome subtelocentric), and 20 microchromosomes.

Fossil record

Pleistocene sidewinder remains have been found in Rancholabrean deposits in Arizona (Van Devender and Mead, 1978; Van Devender et al., 1991).

Distribution

Crotalus cerastes ranges from extreme southwestern Utah, southern Nevada, and southeastern California southward through southeastern Arizona to northwestern Sonora, eastern Baja California del Norte, and Isla Tiburón in Mexico.

Geographic variation

The species is composed of three poorly defined subspecies. *Crotalus cerastes cerastes* Hal-

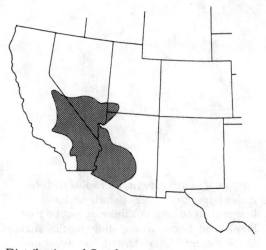

Distribution of *Crotalus cerastes*.

lowell, 1854, the Mojave Desert sidewinder, has the proximal rattle-matrix lobe brown, usually 141 or fewer ventrals in males and 144 or fewer in females, 22 subcaudals in males, 17 or more in females, and 21 scale rows at midbody. It is restricted to the Mojave Desert of California (Mono, Inyo, Kern, Los Angeles, and San Bernardino counties) and adjacent Mohave County, Arizona, southern Nevada (Esmeralde, Ivye, Lincoln and Clark counties), and southwestern Utah (Washington County). *C. c laterorepens* Klauber 1944, the Colorado Desert sidewinder, has the proximal rattle-matrix lobe black, usually 142 or more ventrals in males and 146 or more in females, 21–22 subcaudals in males and 17 in females, and 23 scale rows at midbody. This race ranges in the Colorado Desert from southeastern California (Riverside, San Diego, and Imperial counties) and southwestern Yuma County, Arizona, south in Mexico to Baja California and the panhandle of Sonora. *C. c. cercobombus* Savage and Cliff, 1953, the Sonoran sidewinder, has a black proximal rattle-matrix lobe, usually 141 or fewer ventrals in males, 145 or fewer in females, 20 subcaudals in males, 16 in females, and 21 midbody scale rows. It is found from southcentral Arizona (eastern Yuma, Maricopa, Pinal, and Pima counties) southward to western Sonora and Isla Tiburón, Mexico.

According to the table describing scalation in Klauber (1972: facing p. 124), the above scale counts are not adequate characters for differentiating the races, and in some individuals the tail color may also be a questionable trait. A thorough review is needed of the geographic variation in this snake.

Confusing species

No other snake in the southwestern United States has elevated hornlike projections over its eyes.

Habitat

As indicated by the common names of the subspecies, this is a desert dweller. It seems most common where there is loose sand in low-lying areas, but sometimes inhabits rocky, or gravelly sites within deserts. What little vegetation there is within its habitat usually consists of creosote bush, mesquite, paloverde, and various cacti. Its habitats extend in altitude from below sea level to elevations of about 1,800 m.

Behavior

In San Diego County, California, the sidewinder's annual period of activity may be prolonged; Klauber (1972) collected or observed it from February into November. More southern populations in Mexico may be active all year. Farther inland in the United States, *Crotalus cerastes* usually first appears from mid-March to the early days of April, depending on suitable air temperatures (daytime surface temperatures of 32–35°C, Stephen M. Secor, pers. comm.). Most spring surface activity takes place in May and June, and another activity peak occurs in late September or October. Winter dormancy begins in October and all are underground by December (Secor, pers. comm.). In the spring when surface temperatures are no higher than 35°C, *C. cerastes* will stay on the surface all day before seeking shelter at sunset. It avoids the summer daytime heat by entering animal burrows or by burying beneath the sand so that at most only the outer coil is exposed. It enters a tight coil and then edges or nudges the sand outward from beneath the body or uses the head and neck to pull sand over its coils to form a saucer-shaped craterlike depression in which it lies with its back flush with the surrounding surface (Cowles and Bogert, 1944; Cowles, 1945; Secor, pers. comm.). Any appreciable drift of sand tends to submerge the snake by either surface movement of rolling grains or by the deposition of wind-blown material, and wind often blows away tracks leading to the spot. These roosts are usually in the shade of some bush.

Winter is usually spent alone within a rodent burrow. Of 15 snakes monitored by Stephen Secor, 14 hibernated in burrows of the kangaroo rat (*Dipodomys merriami*) and the other in a desert tortoise (*Gopherus agassizi*) burrow. Cowles (1941) speculated that *C. cerastes* might hibernate at depths below 30 cm, and Secor recorded hibernation depths of 20–70 cm..

Crotalus cerastes is usually considered a nocturnal snake; almost all summer activity takes place at night, but some diurnal movements occur in early spring and in the fall, particularly during the morning or late in the afternoon. Nocturnal activity takes place from 2000 to 0500 hours, with the peak period occurring between 1000 and 0100 (Moore, 1978).

Sidewinders are nocturnal despite widely varying ambient temperature conditions (Moore, 1978). The normal activity range in body temperatures is 6.3–40.8°C (Moore, 1978; Secor, pers. comm.). The mean body temperature varies from month to month (November 18.4°C, August 31.4°), but averages 25.8°C. Active sidewinders studied by Cunningham (1966) had a mean body temperature of 25.5°C (14.8–37.0) and temperatures of those found coiled in the morning after sunrise were 27.8–37°C. Brattstrom (1965) reported a mean voluntary temperature of 26.2°C (20.6–33.5), and Cowles and Bogert (1944) reported a voluntary minimum temperature of 17.5°C, optimum of 31.5°C, and critical maximum of 41.6°C. Brattstrom thought the critical minimum to be −2°C. Basking sometimes occurs in the morning or late afternoon, probably following a meal (the region of the body containing the food bolus is usually exposed directly to the sun), and, at night, *C. cerastes* near asphalt roads will crawl onto the highway to warm themselves by conduction of the residual heat remaining there (this leads to high mortality).

A sidewinder can often be found coiled near the mouth of a rodent burrow. It usually maintains this position until its body temperature starts to rise in midmorning, and then either retreats into the burrow or crawls to a nearby shaded spot. Due to circulatory adjustment, tightly coiled, inactive sidewinders may be able to conserve heat more effectively than when uncoiled (Moore, 1978).

With increasing temperature, the oxygen capacity, red cell counts, amount of hemoglobin present, and hematocrit are lowered, while the blood pH rises. The hemoglobin system appears well adapted to the thermal variability of the sidewinder's environment, helping to maintain lower oxygen levels, resulting in lower metabolism and internal heat production, at higher ambient temperatures. Photoperiod does not seem to play a major role in blood adaptations (MacMahon and Hammer, 1975a,b).

Sidewinders may maintain rather large home ranges, probably owing to the energy needs caused by widespread desert prey. In the eastern Mojave Desert, male *C. cerastes* have home ranges of 8.3–51.2 ha, and females maintain home ranges of 3.7–21.3 ha (Secor, 1991). Secor (1991) found that in the spring juveniles move an average of 100.2 m between captures, subadults 172.6 m, adult females 76.3 m, and adult males 122.5 m. Movements for these same groups in the summer and fall are 163.7, 202.3, 73.0 and 99.1 m, and 140.8, 188.0, 78.5 and 237.5 m, respectively. One of Secor's subadults moved 963 m between captures, and Brown (1970) reported they can travel over 1 km in a single evening, but such long-distance movements probably are not frequent.

Although *C. cerastes* is capable of performing all four typical methods of crawling in snakes, its sandy habitat often dictates that it resort to the specialized method from which it derives its name. Sidwinding is essentially a series of lateral looping movments in which only vertical force is applied and usually no more than two parts of the body touch the sand at any one time (Klauber, 1972; Jayne, 1986). However, when the neck first makes contact, three points (neck, body, and tail) are in contact with the sand (Secor, pers. comm.). The sidewinding snake moves diagonally forward rather quickly. Separate J-shaped tracks, each paralleling the other, and angling in the direction of movement, result as the snake literally skips across the loose sand. The weight seems to be shifted from head to tail as it moves. Jayne (1986) observed that sidewinding is sometimes alternated with typical lateral undulatory crawling, even on loose sand, and reported that maximum mean forward velocities of 0.75–1.7 total body lengths per second could be achieved. Mosauer (1935) reported that *C. cerastes* could voluntary sidewind to speeds of 0.14 m/sec, and that if harrassed could reach a speed of 0.91 m/sec . Typically nightime sidewinding is slow, 0.04–0.07 m/sec (Secor, pers. comm.). According to Stephen Secor, lateral undulation is used when moving through brush or plant debris, while rectilinear crawling is typically used during courtship and the first 1–3 m of nighttime movement.

Sidewinding seems ideally suited for movement in sandy deserts, but also allows some thermal regulation, as only two points of the body are in contact with the hot (or cold at night) sand.

Sidewinders occasionally climb above ground into desert vegetation. Baldwin (in Klauber, 1972) wrote that a disturbed *C. cerastes* crawled into a small bush about 30 cm high and later transferred to an even higher one. Armstrong and Murphy (1979) found a Mexican sidewinder 30 cm above ground in a creosote bush, and Cunningham (1955) reported that captives coiled 60–90 cm above the floor on the branches of a bush placed in their cage. Several times recently released sidewinders climbed into a bush, and Stephen Secor thought this a response to the disturbance of being handled. The only undisturbed sidewinder he found off the ground was a subadult coiled 10 cm high in the base of a burrweed.

When placed in water *C. cerastes* swims readily with a lateral undulatory stroke, head and neck held well out of the water (Klauber, 1972).

Male combat behavior has been recorded only once in *Crotalus cerastes* (Lowe and Norris, 1950), and that episode included a sequence of biting not known in other North American crotalids. The snakes raised the anterior 33% of their bodies suddenly to force or throw one another to the substrate, disengaged, and repeated the same stereotyped aggressive behavior over and over again uninterrupted for approximately 90 minutes. In addition to muscular action in the anterior portions of the bodies,

where the major contacts were made, entwining movements occasionally occurred at the posterior parts of the bodies and at the tails. The posterior parts became involved in the same general twisting and entwining activity as the rest of the body and were under similar muscular tension. The posterior movements appeared to be involved in gaining leverage. The snakes then suddenly changed tactics while parrying in the upright position and vigorously bit each other in the neck, approximately 50 mm posterior to the head. They then separated, waited motionlessly for approximately two minutes, and again advanced toward each other raising their necks and the anterior portions of their bodies with heads held as before at approximately 45° to the horizontal. Again, instead of the usual entwining, they suddenly bit into each others' necks. One fang of one combatant did not penetrate during the second biting and no venom was seen extruded from it, and no venom was seen on either snake; it appears doubtful that either male ejected venom during the biting. This bout took place in April, and both males were in breeding condition.

Reproduction

The size and age of maturity have not been determined for either sex. The smallest gravid females of each subspecies examined by Klauber (1972) were: *Crotalus cerastes cerastes*, 485 mm; *C. c. cercobombus*, 440 mm; and *C. c. laterorepens*, 478 mm. Wright and Wright (1957) reported a gravid *C. c. laterorepens* was 434 mm (attributed to Klauber). The shortest male found in copulation by Stephen Secor was 49.5 mm in total length.

Observed copulations by wild *C. cerastes* have mostly occurred in the spring, 8 April–4 June (Stebbins, 1954; Wright and Wright, 1957; Klauber, 1972; Secor, pers. comm.), but Lowe (1942) reported mating activity on 20 September and 18 October, and Secor observed mating episodes on 7 and 22 October. All episodes occurred in the morning, and Klauber (1972) thought that coitus had acutally begun the night before, as even after discovery, some mated pairs remained joined for over seven hours.

Perkins (in Klauber, 1972) has given the best description of the courtship and mating behavior of *C. cerastes*. He collected a pair of *C.*

c. laterorepens on 9 May. The next day the male courted the female with jerky head movements along her body. On 11 May they were found to be in copulation. The male nudged the female at any part of her body that his head touched, and often held his tail straight up, waving it slowly in a circle, but no pulsating or pumping motion occurred. The male dragged the female halfway across the cage. There was considerable motion in the entire body of the male, moving around and over the female, while still nudging with his head, but there was no tail motion. There was then little movement for about 15 minutes. Some tail waving and nudging followed, but not necessarily at the same time. Later, both snakes crawled slowly about dragging the other. A pronounced pulsating or pumping motion at a rate of 1/sec then occurred in the swollen cloacal region of the female. It appeared that the pulsating motion was being imparted by the female. Pulsations continued at this rate for about 10 seconds, then a little faster or slower until they stopped entirely. Afterward the female dragged the male almost continuously (he seemed to sidewind backward). The female seemed to be trying to separate. On the following morning they were separated. The observed duration of the coitus was almost eight hours. As a result of this mating eight young were born on the night of October 13–14, 155 days later. During copulation the male's tail may be wrapped around that of the female (Lowe, 1942).

Known birth dates range from 17 August to 28 November, but most occur in October (Wright and Wright, 1957; Klauber, 1972; Secor, pers. comm.); 5–18 young (typically 7–12) are born in each litter. Mean total body length for 22 newborn young (measured, or from the literature) was 180.3 mm (161–203).

Four young (one nonviable) were produced from a captive mating of a male *C. c. laterorepens* and a female *C. s. scutulatus* (Powell et al., 1990). The living neonates were patterned more like *C. scutulatus* but had head scalation typical of *C. cerastes* and moved by sidewinding.

Growth and longevity

Growth data are sparse. In his studies, Stephen Secor has found a growth rate of about 0.53 mm/day, which apparently slows in larger individuals. Klauber (1972) commented that the Mojave Desert sidewinder (*C. c. cerastes*) is a shorter animal than the Colorado Desert sidewinder (*C. c. laterorepens*), and that this may be correlated with a shorter growing season.

The sidewinder seems not to have as long a life span as some other species of *Crotalus*; Bowler (1977) listed the following age records for captives: *C. c. cerastes*, 8 years, 1 month, 11 days; *C. c. cercobombus*, 10 years, 8 months, 21 days; and *C. c. laterorepens*, 13 years, 9 months; and Secor wrote that he thought the normal life span of wild sidewinders is between 5 and 7 years.

Food and feeding

Adult sidewinders feed mostly on a variety of rodents and lizards; neonates and small juveniles rely almost exclusively on lizards (*Cnemidophorus, Uta,* or small *Uma*) or pocket mice (*Perognathus;* Secor, pers. comm.), although possibly small snakes may also be eaten. As the snakes grow they gradually take larger lizards and rodents.

The most extensive study of prey taken by *Crotalus cerastes* was that of Funk (1965) who examined 226 snakes. Mammals, birds, lizards, and small snakes were present in 171 (76%) digestive tracts; 88 (51.5%) contained mammals, 73 (42.7%) contained lizards, 5 (2.9%) had eaten birds, and 5 (2.9%) contained snakes. Mammals eaten included ground squirrels (*Ammospermophilus, Spermophilus*), kangaroo rats (*Dipodomys*), pocket mice (*Perognathus*), wood rats (*Neotoma*), house mice (*Mus*), harvest mice (*Reithrodontomys*), white-footed mice (*Peromyscus*) pocket gophers, (*Thomomys*), and shrews (*Notiosorex*). The most common mammals of the 91 eaten were 31 *Perognathus* (34.1%), 24 *Dipodomys* (26.4%), and 10 *Peromyscus* (11%). Reptiles taken were of the lizard genera *Callisaurus, Cnemidophorus, Coleonyx, Dipsosaurus, Gambelia, Phrynosoma, Uma, Urosaurus,* and *Uta,* and the snakes *Arizona, Chionactis, Crotalus* (*cerastes*) and *Sonora. Cnemidophorus tigris* (27 individuals, 35.1%) and *Uma notata* (14, 18.2%) were the most common prey. Only four birds were among the food: the black-throated sparrow (*Amphispiza*), cactus wren (*Campylorhynchus*), lark sparrow (*Chondestes*), and house sparrow (*Passer*). Also listed as prey in the literature are the yellow warbler (*Dendroica*) and a "caterpillar" (Klauber, 1972). The caterpillar was found in a juvenile snake that also contained a lizard (*Uma*), and Klauber (1972) thought that the lizard had either first eaten the caterpillar, or had been captured while eating it.

Carrion is sometimes consumed. Cunningham (1959) reported that a captive ate a mouse dead for at least two days, and Klauber (1972) found a sidewinder eating a kangaroo rat (*Dipodomys*) that obviously had been a traffic victim.

Although some *C. cerastes* may occasionally forage, most prey is struck from ambush. This is particularly true of small rodents, and especially quick lizards. One of the most common behavioral traits of the sidewinder is its habit of waiting patiently, partially concealed by sand, outside the entrances of rodent and lizard burrows. Evidence that foraging may be of less importance than ambushing, and also that olfactory prey detection may possibly play a more minor role than visual or infrared prey detection, has been provided by Chiszar and Radcliffe (1977), who recorded no significantly different response from naive neonate sidewinders to cotton swabs soaked either in water or in extracts of both prey and non-prey surface substances, or to skin patches from prey and non-prey animals. In contrast, Stephen Secor has observed that olfaction seems more important in prey detection than vision or heat detection, as sidewinders usually tongue flick before striking and then use olfaction to trail wounded prey. Lizards are usually retained in the mouth when struck, but rodents are released after the strike.

Prey species evolve with their predators, and some develop important avoidance behavior to specific predators. Such is the case with some kangaroo rats (*Dipodomys*) and *C. cerastes*. The rats kick loose sand with their hind feet at suspect sand mounds which may conceal the snake, particularly near their burrows. While the truth of this has been questioned, it apparently does occur, as witnessed by Fowlie (1965). A kangaroo rat aroused by his field party made off for a nearby clump of creosote bush, and while giving chase, a member of the party was nearly bitten by a partially buried sidewinder resting under the creosote bush at the mouth of a rodent burrow. It required no time at all to find that kangaroo rats and sidewinders were throughout the area. When aroused, a rat would dash to a nearby creosote bush and stand perfectly motionless. On finding itself near a buried sidewinder, the rat would first tremble and shake and then jump up and down in a very peculiar mannner, hopping here and there in front of the snake. The rat made little attempt to flee and next began a peculiar sand kicking maneuver that finally caused the sidewinder to take refuge in the brush at the base of the clump.

Venom and bites

Although individuals of *Crotalus cerastes* are generally small, the fangs of this species are relatively long; 518–767-mm adults had fangs 5.0–8.1 mm long (Klauber, 1939).

The venom is of moderate toxicity and *C. cerastes* must be considered dangerous; however the amount of venom injected per bite seems low. Amaral (1928) reported an average venom yield per bite of 0.06 ml (0.018 g dried) for adult *C. cerastes*; Klauber (1972) a total yield of 33 mg of dried venom per adult, with a maximum yield of 63 mg; and Dowling (1975) yields of 20–35 mg. The minimum lethal venom dose for 350-g pigeons is 0.9–0.15 mg (Githens and George, 1931); for 22-g mice, 0.06 mg (Macht, 1937), and for 20-g mice, 0.17 mg (Githens and Wolff, 1939). The lethal dose for mice injected subcutaneously is 5.5 mg/kg (Minton, 1956). The human lethal dose is 40 mg (Dowling, 1975). The venom does not lose its lethal potency for mice and cats after 26–27 years of storage, although it does take longer to produce complete neuromuscular blockage than does fresh venom (Russell et al., 1960). Sidewinder venom shows much protease activity (MacKessy, 1988) which aids in predigestion of prey.

Russell (1960) reported that in his experience with treating snake bites, those of the sidewinder have tended to be relatively mild, but that this may be due in part to the snake's smaller size resulting in the injection of a smaller volume of venom. He thought the venom to be very toxic, as humans have died from it. Hutchison (1929) listed the following symptoms as typical of a sidewinder bite: swelling, pain, weakness, giddiness, paralysis, and an increase in body temperature. To this may be added necrosis of the tissue at the bite site, which may be deep (Minton, 1956). Lowell

(1957) experienced pain, swelling, an itching sensation, dizziness and slight nausea (after administration of antivenin), discoloration of the bitten finger, and tenderness at the site of the bite for several days after being bitten. When bitten, Stephen Secor had his fingers, hand and arm swell to almost twice normal size, but discoloration was restricted to the area about the bite, and pain in the hand and arm occurred only on the day of the accident. The swelling subsided in two to four days.

Predators and defense

Natural predators (particularly of the young) include bobcats (*Lynx*), badgers (*Taxidea*), skunks (*Mephitis*), raccoons (*Procyon*), coatis (*Nasua*), coyotes (*Canis*), foxes (*Vulpes*), peccaries (*Tayassu*), roadrunners (*Geococcyx*), hawks (*Buteo*), kestrels (*Falco*), owls (*Bubo, Athene*), ravens (*Corvus*), shrikes (*Lanius*), other snake-eating serpents (*Lampropeltis, Masticophis*), and leopard lizards (*Gambelia*) (Secor, pers. comm.). *Crotalus cerastes* may even prey on smaller individuals of its own species (Funk, 1965). During Secor's studies in the eastern Mojave Desert, several radio-equipped snakes were dug out of their winter retreats and preyed on by coyotes, and this canid may be a leading cause of winter mortality. Humans often kill it on sight, or go out of their way to run over sidewinders crossing roads.

The sidewinder is a nervous snake that will often remain quietly coiled when first discovered. If prodded it will rattle and usually try to crawl away, but, if prevented from doing so, will then put up a spirited fight, coiling, rattling, jerking the head, alternately laterally compressing and inflating the body, and striking viciously. The frequency of rattling may range from a mean low of 5.69–5.81 kHz to a mean high of 13.9–15.1 kHz; the mean band width is from 8.24 to 9.32 kHz (Fenton and Licht, 1990).

If confronted by a kingsnake, *Crotalus cerastes* will form a broad loop or bend in its body which it then lifts off the ground. Approximately 25–33% of its body is elevated from this position, and the loop is then used to strike a blow at the kingsnake much as a human would gouge with the elbow. As it approaches, the sidewinder strikes downward with such force as to render the blow a very effective defense mechanism (Cowles, 1938).

The sidewinder is capable of changing its body melanism to somewhat match that of the substratum (Klauber, 1931; Neill, 1951). While Neill (1951) thought this a thermal adjustment, it also may provide some protection from overhead avian predators.

Populations

Brown (1970) marked 72 adults, but recaptured only eight during an ecological study in the Mojave Desert, and estimated the density to be less than 1/ha, but Secor captured 177 *C. cerastes* and recaptured 69 at this same site and thought the density closer to 1/ha. However, the species may be locally common; Fowlie (1965) found 47 on one moonlit night on a stretch of road between Yuma and Gila Bend, Arizona, and Armstrong and Murphy (1979) have seen as many as 30 per night while driving roads.

Of 116 sidewinders captured by Secor (pers. comm.), 59 were females and 57 males, a 1:1 sex ratio. The ratio of mature to immature snakes was 0.33:1.

Remarks

Because of its specialized adaptations for desert life, *C. cerastes* is generally thought to be a relatively, newly derived species. However, its affinities are somewhat clouded. Gloyd (1940) could not determine its relationships, while Brattstrom (1964) thought it closest to *C. mitchelli, C. tigris,* and *C. viridis,* and Klauber (1972) grouped it with *C. durissus, C. molossus,* and *C. horridus.* Recent studies on venom proteins by Foote and MacMahon (1977) show it a near relative of *C. pricei, C. willardi, C. lepidus, C. triseriatus,* and *Sistrurus ravus,* but those on dorsal scale microdermatoglyphics by Stille (1987) place it with *C. mitchelli* and *C. pricei.*

Cohen and Myres (1970) suggested that the supraocular horns function as eyelids that protect the eyes of *C. cerastes* while the snake crawls through burrows entangled with such obstructions as creosote roots, rocks, and gravel which could abrade the eye.

Crotalus horridus Linnaeus, 1758
Timber rattlesnake
Plates 30–32

Recognition

This large rattlesnake grows to 189.2 cm, has 15–34 transverse chevronlike or V-shaped bands on the body, a gray, dark-brown or black unpatterned tail, and numerous small scales between the supraocular scales on top of the head. The ground color varies from gray to yellow or dark brown, and some individuals (especially in the Northeast) may be totally black. Martin (1988) reported that light-phase rattlesnakes make up 58% of the population in the Shenandoah National Park of Virginia. Elsewhere, at other sites studied by Martin, the proportion of light-phase snakes varies from 40 to 92.5%. The dark phase tends to predominate at higher, cooler, wetter, more densely forested sites in the Shenandoah National Park and in other areas of the Appalachian Mountains, particularly in the Northeast. In the South, timber rattlesnakes have a red to reddish-orange middorsal stripe. A dark stripe extending from the eye backward to beyond the corner of the mouth may be present on the side of the head, and some individuals may have round occipital spots. The venter is pink, white, cream, or yellow with small, dark, stipplelike marks; on it are 158–183 ventrals, 13–30 subcaudals, and an undivided anal plate. At midbody there are 23–24(21–26) rows of keeled scales. Dorsal head scales include a higher-than-wide rostral, 2 internasals, 4 canthals (between the internasals and supraoculars), and 2 supraoculars. Lateral head scalation consists of a prenasal, a postnasal (separated from the preoculars by loreals), 1–3 loreals, 2 preoculars, 2–6 postoculars, several suboculars, 13–14(10–17) supralabials, and 14–15(11–19) infralabials. The bifurcated hemipenis has a divided sulcus spermaticus, more than 70 long, thin, recurved spines, and

30 or more fringes per lobe, but no spines in the crotch between the lobes. The dental formula is maxillary 1 (the fang), palatine 2–3, pterygoid 8–11, and dentary 10–13.

Males grow generally larger than females (Brown, 1990). Males have 158–177 ventrals, 20–30 subcaudals, and tail lengths 6–10% of the total length. Females have 163–183 ventrals, 13–26 subcaudals, and tails 4–8% as long as the total length.

No significant differences in the frequency of occurrence of dark or light color phases has been found between the sexes of *C. h. horridus* (Schaeffer, 1969). A differences in head pattern, however, exists in this subspecies (Storment, 1990). Females retain the dark postorbital stripe throughout life (even though the intensity may decrease), but this stripe becomes infused with white and yellow spots and is obscured in mature males.

Karyotype

The karyotype is 2n = 36: 16 macrochromosomes (including 4 metacentric, 6 submetacentric, 4 subtelocentric) and 20 microchromosomes. Chromosomal sex determination is ZZ in males and ZW in females (Zimmerman and Kilpatrick, 1973).

Fossil record

Pleistocene fossils of *C. horridus* are from the Irvingtonian of Maryland and West Virginia and the Rancholabrean of Alabama, Arkansas, Georgia, Indiana, Missouri, Pennsylvania, Tennessee, and Virginia (Holman, 1981, 1982; Holman and Grady, 1989; Holman et al., 1990; Richards, 1990).

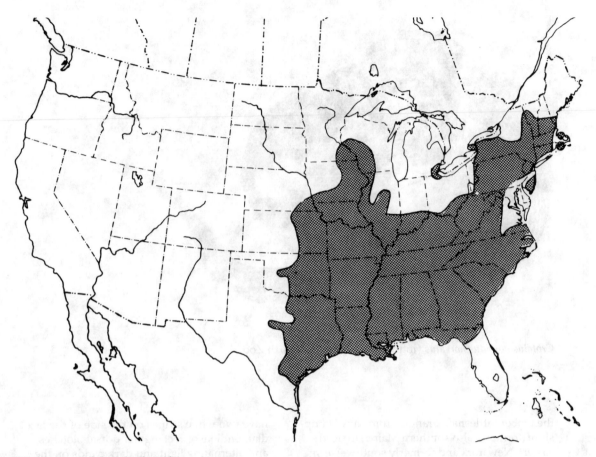

Distribution of *Crotalus horridus*.

Distribution

Crotalus horridus formerly occurred in southern Maine and now ranges from eastern New Hampshire west through northeastern and southern New York to extreme southwestern Wisconsin and adjacent southeastern Minnesota, and south to northern Florida and eastern Texas. In Canada, it once lived in southwestern Ontario, but the last specimen taken in Ontario was from Niagara Glen in 1941, so this species has apparently been extirpated in that country (Cook, 1984).

Geographic variation

Pisani et al. (1973), after studying scutellation and character variation in *Crotalus horridus*, con-

cluded that no valid subspecies occur, but many of their specimens were from western areas where intergradation between the two described subspecies, *C. h. horridus* and *C. h. atricaudatus*, is known to occur. Brown and Ernst (1986) repeated Pisani et al.'s analyses of morphological characters and added several others relating to adult size and pattern in a study of eastern *C. horridus* where little intergradation is known. They demonstrated a separation between the populations east of the Appalachians into the two subspecies using adult size and pattern differences in conjunction with the number of dorsal scale rows and ventral scales. In view of Brown and Ernst's results, I recognize as valid the two described subspecies of *C. horridus*.

Crotalus horridus horridus Linnaeus, 1758,

Crotalus horridus horridus (John H. Tashjian, Ft. Worth Zoo).

the timber rattlesnake, ranges from New Hampshire (formerly also southern Maine) to northeastern New York and formerly southwestern Ontario, west to Illinois and southwestern Wisconsin and southeastern Minnesota, and south to northern Georgia, northwestern Arkansas and northeastern Texas. It is yellow to gray or black in ground color, lacks a distinct middorsal stripe, and has 23(21–26) midbody scale rows and 15–34 dorsal body bands. *Crotalus h. atricaudatus* Latreille, in Sonnini and Latreille, 1801, the canebrake rattlesnake, ranges from southeastern Virginia along the Atlantic Coastal Plain to northern Florida, westward to central Texas, and northward in the Mississippi Valley to southern Illinois. It is pinkish brown or gray in ground color, has a distinct reddish-orange middorsal stripe, 25(21–25) midbody scale rows, and 21–29 dorsal body bands.

Confusing species

The eastern diamondback rattlesnake, *Crotalus adamanteus*, western diamondback rattlesnake, *C. atrox*, and western rattlesnake, *C. viridis*, all have two white stripes on the side of the face, diamondlike or rectangular dorsal blotches, and alternating light and dark bands on the tail. The smaller pigmy rattlesnakes and massasaugas of the genus *Sistrurus* have nine enlarged platelike scales on top of the head.

Habitat

Crotalus h. horridus inhabits upland wooded areas, usually with nearby rocky ledges or rock slides. *C. h. atricaudatus* is more at home in lowland thickets, pinewoods, and cane thickets, or along swamp borders. In the Pine Barrens of southern New Jersey, male and nongravid female *C. h. horridus* utilize forested habitats with greater than 50% canopy closure, thick surface vegetation (approximately 75%), and few fallen logs, while gravid females use less densely forested sites with approximately 25% canopy closure, an equal mixture of vegetation and leaf litter covering the ground, frequent fallen logs, and a generally warmer condition (Reinert and Zappalorti, 1988a). The habitat preferred by gravid females is of lesser

Crotalus horridus atricaudatus (Roger W. Barbour, Aransas County, Texas).

extent, being mostly restricted to road borders in southern New Jersey. Reinert (1984a,b) found a similar use of more open, generally warmer microhabitats by gravid females in Pennsylvania. *C. h. horridus,* although primarily a forest dweller, apparently requires open sites for successful breeding.

According to Reinert (1984b), *C. h. horridus* uses a gradient of habitats in Pennsylvania from mature forest sites with numerous fallen logs to young forest sites with predominant leaf litter cover. Dark-phase individuals prefer the former while yellow-phase rattlesnakes are more frequently found in the latter. This apparent habitat separation of the two hue morphs probably has to do primarily with background color matching for camouflage from both would-be prey and predators.

Behavior

Annual activity usually occurs in the period from mid-April to late September, but some individuals, especially those in the more southern populations, may emerge from hibernation in March or retreat into hibernacula in Octo-

ber. The duration of dormancy usually depends on how warm are the early spring or fall days. Early and late records for this species are 5 March (Wright and Wright, 1957) and 5 November (Martin 1988). In the spring and fall *Crotalus horridus* is primarily active during the daylight hours, but when the days become hot in summer it shifts to a more crepuscular or nocturnal activity cycle.

Formerly, *C. h. horridus* congregated in large groups of as many as 50–100 snakes at suitable hibernacula, usually located in rock crevices or talus slopes on south-facing slopes (Wright and Wright, 1957). Timber rattlesnakes from many kilometers around would crawl over set pathways to reach these sites (Neill, 1948). Heavily populated dens still exist in remote or protected areas, but many have been reduced or extirpated. *C. h. atricaudatus* hibernates individually or in much smaller groups in mammal burrows, old stumps, or shallow rock crevices. In fact, this subspecies may never have hibernated communally or in extensive numbers at any particular site. Neonates and adults may use the same overwintering site. Apparently the young snakes initially find the dens by following adult odor trails to the site (Brown and MacLean, 1983; Reinert and Zappalorti, 1988b).

In Virginia, communal dens of *C. h. horridus* may contain as many as 200 snakes (Brown, 1990). The mean date of entrance into hibernacula is 12 October (1–18 October), but stragglers continue to arrive until 5 November after other *C. horridus* have gone underground (Martin, 1988). Some temporary and sporadic emergence usually occurs in the spring before general emergence. Commonly the first rattlesnake is seen about 14 April, but the annual range of first emergence is from 8 March to 2 May depending on the warmth of the early spring. The most common date of general emergence is 1 May (18 April–12 May). Oldfield and Keyler (1989), in a limited study in Wisconsin, recorded the first timber rattlesnakes on 1 May and the last on 11 September.

The minimum and maximum voluntary temperatures of five *C. horridus* recorded by Brattstrom (1965) were 21.2 and 31.7°C, respectively, but Brown et al. (1982) found active New York snakes at 12.5 and 33.3°C. Body temperatures of hibernating *C. horridus* in New York averaged 10.5°C (4.3–15.7) from September through May (Brown, 1982). The mean rate of body-temperature decline of these snakes was 0.5°C per week through February, in March the temperature stabilized at 4.3°C, and then rose by 0.6°C per week in April and May. *C. horridus* from New York averaged body temperatures of 30.1° on the surface on clear days but only 27.8°C when on the surface on cloudy days or underground on clear days (Brown et al, 1982). In Wisconsin, the mean body temperature of nine active snakes was 29.6°C; the lowest body temperature was 21.5°C for one coiled beneath a rock (Oldfield and Keyler, 1989).

The annual climatic variables for active *C. horridus* in New Jersey were recorded by Reinert and Zappalorti (1988a). Mean data for 119 males, 50 nongravid females, and 88 gravid females, respectively, are as follows: surface temperature 25.6, 25.9, 29.1°C; air temperature 24.8, 25.8, 28.3°C; soil temperature 18.2, 18.1, 21.0°C; soil moisture 35, 35, 19%; surface relative humidity 67.3, 65.4, 65.5%; air relative humidity 59.9, 59.4, 59.0%; illumination at snake 5,382, 4,198, 9,041 lux. Obviously gravid females seek warmer, drier, more illuminated (open) microhabitats.

Movements from the hibernation den to the summer feeding range of radio-equipped New York adults were recorded by Brown et al. (1982) and Brown (1987). Average dispersal was 504 m (females 280 m, males 1,400 m), and the snakes gained a mean 102 m in elevation. The mean maximum migrations (using the five longest records for each sex) were 4.07 km by males and 2.05 km by females (Brown, in Macartney et al., 1988). The maximum single migratory movement from the den by a male was 7.2 km and 3.7 km by a nongravid female (W. S. Brown, pers. comm.). Not all females migrate long distances; a gravid female moved only 39 m between 20 June and 5 August (Brown et al., 1982). Based on recaptures, average migration distances in Virginia are estimated at 2.45 km for adult males, 2.16 km for nongravid females, 0.5 km for gravid females, and 1.73 km for juveniles (Martin, 1990). Little movement by gravid females has also been reported by Fitch (1958), Galligan and Dunson (1979), Martin (1988), and Reinert and Zappalorti (1988a). Two of the females tracked by Brown et al. (1982) appeared to use the same migratory routes returning to the den in autumn as they had in leaving it in the spring. Timber rattlesnakes may remain at one spot for a period of time, then move a considerable distance, only to once again settle down for some time (Galligan and Dunson, 1982). Perhaps such erratic movements are stimulated by prey availability.

In New Jersey, radiotelemetry studies showed that males have the largest activity ranges (mean length 1,463 m), but the sizes of their ranges positively correlate with the number of days they were monitored (Reinert and Zappalorti, 1988a). This was not true of either nongravid females (mean length, 995 m) or gravid females (mean length, 665 m). Time series analyses indicated that movement patterns of males and nongravid females consisted of constantly shifting, nonoverlapping home ranges. In most cases, these snakes moved during the active season in a looping pattern that returned them to the same hibernaculum from which they had departed in the spring. Reinert and Zappalorti (1988a) showed that gravid females have more static overlapping activity ranges and shorter dispersal distances from the hibernaculum.

The average home range radii by sex for

Virginia *C. horridus* based on recaptures were: adult males, 2.47 km; nongravid adult females, 2.18 km; gravid females, 0.48 km; and juveniles less than three years old, 1.74 km (Martin, 1988). The greatest distance traveled from the hibernaculum was about 6 km.

Adult male *C. horridus* are known to occasionally engage in combat dances, and these usually occur in April or May (Sutherland, 1958; Anderson, 1965; Collins, 1974; Klauber, 1972; pers. obs.), but McIlhenny (in Klauber, 1972) reported that they always occur in the fall in Louisiana. These bouts may last from about 30 minutes to more than two hours. The males face each other, put the sides of their heads together, gradually raise the anterior 40–50% of their bodies, and entwine their necks. When raised as high as possible without losing balance, a shoving match begins in which each snake tries to bring its entire weight down across the neck of the opponent, forcing it to the ground. If the first pushing attempt fails, this stereotyped behavior is continued until one snake's head and neck are pinned (sometimes with considerable force) and held on the ground until the defeated snake breaks contact and crawls away.

Reproduction

Female *Crotalus horridus atricaudatus* from South Carolina bear their initial litter in their sixth year at a snout-vent length of more than 100 cm and a weight of more than 700 g; South Carolina males are mature at snout-vent lengths of 90–100 cm and probably reach this state in their fourth year (Gibbons, 1972). Four years is also probably the age of maturity in Kansas (Fitch, 1985b). In Wisconsin, a 67.7-cm (snout to base of rattle) female *C. h. horridus* contained follicles, and those longer than 89.4 cm produced eggs or contained embryos (Keenlyne, 1978). Female *C. h. horridus* in Pennsylvania are mature before they reach a snout-vent length of 77 cm and weigh 430 g (Galligan and Dunson, 1979). Data gathered by Brown (1991) on females from New York suggest an estimated snout-vent length of 84 cm and an estimated mass of 460 g at maturity, and a mean snout-vent length of 92.7 cm for all reproductive-sized females; vitellogenic females weighed 561–1,141 g, gravid females 514–1,414 g, and postpartum females 329–837

g. Brown's overall sample indicated a female range at first reproduction of 7–11 years (mean, 9.3); 61% of his females reproduced for the first time at 9 or 10 years of age. In Virginia, females mature at 74 cm or longer in four to five years (Martin, 1988). *C. h. horridus* is the shorter subspecies, and perhaps tdhis explains its maturation at a shorter length.

Biennial, triennial, and quadrennial female reproductive cycles have been reported. Keenlyne (1978) and Galligan and Dunson (1979) thought female reproduction is probably biennial; but Gibbons (1972) thought some from South Carolina were possibly on a triennial cycle, and Fitch (1985b) and Brown (1987) reported triennial female cycles from Kansas and New York. Computer programs written by Brown (1991) to calculate reproductive cycle lengths using reproductive states of mature New York females recaptured in years subsequent to their original marking indicate several supported observations: (1) no direct evidence for a biennial cycle was obtained by captures of females in alternate years; most females (57%) reproduced triennially and many (27%) quadrennially; (2) gravid females with their sedentary selection of thermally optimal gestation sites showed a strong tendency to remain in the open outcrop knolls in the vicinity of the den; and (3) not capturing a female in a certain year suggests that she did not occupy a typical gestation site but made longer migratory movements typical of nonreproductive females.

Over an eight-year period, 27% of all New York females and 51% of the mature females sampled by Brown (1991) were gravid. Proportions of gravid females varied significantly from year to year, and years of highest reproduction (1986, 1988) were preceded by years of lower reproduction (1985, 1987). Similarly, the low year of 1981 was followed by rising rates of female reproduction in 1982–1984. The annual proportion of gravid females varied from 25% in 1985 to 77% in 1988. Brown's data suggest considerable fluctuations in yearly reproductive output. Of 43 mature females collected during organized rattlesnake hunts in Pennsylvania between 1985 and 1987, 36 (83.7%) were gravid (Reinert, 1991).

Female reproduction in *C. h. horridus* from deciduous forests in New York and Virginia was compared by Brown and Martin (1990). Proportions of mature females gravid each year ranged from 27 to 75% in New York and from 18 to 51% in Virginia. There were wide annual fluctuations in this parameter which were not synchronous between the two states. Most females in both populations reproduced triennially, 57% in New York and 58% in Virginia, and a substantial number of females reproduced quadrennially, 27% in New York and 28% in Virginia. A few Virginia females even showed a five-year cycle. Biennial cycles were not observed in New York, but 8% of the females had such a cycle in Virginia. Age of first breeding was 5–10 years in Virginia, with most females (47%) first reproducing at 8–9 years; in New York, females first bred at 7–11 years, with most (62%) at 9–10 years of age. Mean age at first litter production was 8.1 years in Virginia and 9.3 years in New York.

In Wisconsin, both follicular development and yolking begin in late July (Keenlyne, 1978). In South Carolina, follicles are unyolked in the spring (30 April–6 June), and females contain 4.24% fat, but in the fall (2–9 October) the follicles are yolked and the female fat content is 5.82% of body weight (Gibbons, 1972). South Carolina females may contain embryos in the spring (27 April–6 June) and a fat content of 3.17%, but after parturition in late summer (24 August–8 September) the fat content is only 0.84%. If fat reserves are low, female reproduction may be delayed one or more years, and some females may bear their first young as late as six to nine years of age (Martin, 1988).

It has previously been thought that *C. horridus* breeds in the spring before dispersing from the hibernaculum, but this is in error. Most mating occurs from July to mid-September with the sperm being stored until June of the next year (Brown, 1987; 1991; Martin 1988, 1990). Keenlyne (1978), however, thought the mating period in Wisconsin to include late summer, fall, and the next spring, and Anderson (1965) found a copulating pair on 27 April in Missouri. During courtship the male actively pursues the female and, on reaching her, repeatedly strokes her neck with his chin. The male positions himself alongside the female and stimulates her with quick, rapid jerks of his head and body, and then curls his tail beneath hers until the vents touch and inserts his hemipenis. Some pumping motions of the male's tail near the vent may occur during the copulation, and a mating may last for several hours (captive, pers. obs.).

Newborn young can follow the scent trail of conspecific young, adult males, and females (Brown and MacLean, 1983). It is strongly possible that odor plays an important role in finding and determining the opposite sex, and as a guidance mechanism for newborn young to find a communal hibernaculum (Brown and MacLean, 1983; Reinert and Zappalorti, 1988b).

Crotalus horridus is ovoviviparous. The young are usually born between late August and early October. In the southern part of the range it is possible that female *C. h. atricaudatus* may give birth earlier; Kauffeld (in Klauber, 1972) reported that a captive female canebrake rattlesnake produced a litter on 20 July. Birthing rookeries are sometimes located at the hibernaculum, but in Virginia they average 0.04 km away, and parturition has occurred as far away from the overwintering den as 0.5 km (Martin, 1988). In New Jersey, parturition has taken place as far as 1 km away from the hibernacula (Reinert and Zappalorti, 1988b).

The birthing process, as observed by Trapido (1939), takes from 5 to 25 minutes, and the interval between extrusion of the young may be 11 minutes to just less than an hour. The young are born about 20 minutes apart, and may remain in their fetal membranes for 10–60 minutes (Minton, 1972). Neonates are patterned like the adults, but gray in hue. They are 271–350 mm in total length, weigh about 22–25(16–32) g, and have a buttonlike scale at the end of the tail which is exposed after the first shedding of the skin, which usually take place in 7–10 days. They are dangerous even at this small size, having fangs 2.6–3.8 mm long (Stewart et al., 1960) and a ready supply of venom.

Most broods consist of 6–10 young, but litters of 3 (Wright and Wright, 1957) to 19 (Martof et al., 1980) are known. In Kansas compliments of 5–14 have been recorded, but averaged 8.5 young, of which 6.8 were born alive (Fitch, 1985b). In Virginia, broods average 8(4–16) young (Martin, 1990).

The female and her young remain together for 7–10 days, then all disperse (Martin, 1988, 1990). Neonates are believed to trail adults away from the birthing area and, later in the fall, to the hibernaculum (Reinert and Zappalorti, 1988b); some have even been found traveling with adults during these movements (Reinert and Zappalorti, 1988b). In Virginia, only about 30% of the young of the year overwinter at the communal hibernaculum; an unknown percentage use alternate ancestral hibernacula that are limited (probably by the narrowness of the crevices) to use by smaller snakes (Martin, 1988).

Growth and longevity

Minton (1972) found a 369-mm juvenile on 16 July in Indiana that must have been from a very late brood the previous fall. He estimated that a snake 81 cm in length was almost two years old, and four others were 79 and 95 cm in their second year and 100 and 110 cm in their third year (W. S. Brown, pers. comm., believes these ages are probably vastly underestimated). In Virginia, juveniles averaged 432 mm at one year, 584 mm at two years, and 706 mm at three years (Martin, 1988). Fitch (1985b) reported the following ranges in snout-vent length for tentative age cohorts of *C. horridus* from Kansas: first year (button), 298–413 mm; first year (2 rattle segments), 495 mm; second year (3 segments), 548–665 mm; second year (4 segments), 577–670 mm, third year (5 segments), 504–802 mm; third year (6 segments), 814–855 mm; fourth year (7 segments), 800–870 mm; fourth year (8 segments), 812–995 mm; fifth year (9–10 segments), 598–1,031 mm; sixth year (11–12 segments), 922–1,082 mm; seventh year, 1,132–1,230 mm; eighth year, 1,020–1,175 mm, and ninth year, 1,038–1,196 mm. South Carolina *C. horridus* measured by Gibbons (1972) were 35–43 cm in snout-vent length shortly after birth, and by the next June had reached 50–60 cm; two-year-olds were 65–75 cm in snout-vent length and three-year-olds 80–90 cm. Growth can be accelerated with increased feeding; a captive grew from 343 mm total length to 756 mm in only eight months (Schwab, 1988).

The longevity record for this species is 30 years, 2 months and 1 day (Bowler, 1977).

Food and feeding

Warm-blooded prey is preferred: mammals—bats (species not given), shrews (*Blarina, Cryptotis, Sorex*), moles (*Scalopus*), mice (*Clethrionomys, Microtus, Mus, Napaeozapus, Peromyscus, Synaptomys, Zapus*), rats (*Neotoma, Rattus, Sigmodon*), pocket gophers (*Geomys*), chipmunks (*Tamias*), young woodchucks (*Marmota*), squirrels (*Sciurus, Tamiasciurus*), rabbits (*Sylvilagus*), weasels (*Mustela*), and skunks (*Mephitis*)—and birds (eggs, young and adults)—domestic chickens (*Gallus*), bobwhite quail (*Colinus*), turkey (*Meleagris*), ruffed grouse (*Bonasa*), yellow-billed cuckoo (*Coccyzus*), towhee (*Pipilo*), brown thrasher (*Toxostoma*), wood thrush (*Hylocichla*), cedar waxwing (*Bombycilla*), ovenbird (*Seiurus*), black-throated blue warbler (*Dendroica*), field sparrow (*Spizella*), grasshopper sparrow (*Ammodramus*), and white-throated sparrow (*Zonotrichia*) (Uhler et al., 1939; Barbour, 1950; Wright and Wright, 1957; Minton, 1972; Klauber, 1972; Fitch, 1982; pers. obs.). However, other prey are also taken: lizards (*Eumeces*), snakes (*Coluber, Thamnophis*), frogs (*Rana?*), toads (*Bufo*), and insects (Surface, 1906; Hamilton and Pollack, 1955; Myers, 1956; Wright and Wright, 1957; Klauber, 1972; Fitch, 1982). Barbour (1950) took a snail shell from the stomach of a *C. horridus*, but thought it probably had been secondarily consumed in a chipmunk pouch.

A typical eastern timber rattlesnake may eat only 6 to 20 meals a year (Brown, 1987), and Fitch (1982) estimated that each timber rattlesnake from Kansas consumes 2.5 times its body weight in prey each year.

Not all prey is taken alive. Swanson (1952) reported a timber rattlesnake regurgitated a brown rat (*Rattus norvegicus?*) which, from the smell, had obviously been eaten as carrion. He also noted that a large *C. horridus* once disgorged a half-grown rabbit (*Sylvilagus?*) which contained many maggots. Nicoletto (1985) reported the apparent taking of a dead *Peromyscus leucopus* by a timber rattlesnake.

The stomach contents of 141 Virginia timber rattlesnakes were analyzed by Uhler et al. (1939). Mice composed 38% of the diet, squirrels and chipmunks 25%, rabbits 18%, shrews 5%, and birds (mostly songbirds) 13%. One specimen contained a bat. In Pennsylvania, the food is composed of 94% mammals (Surface, 1906).

The feeding behavior of female *C. horridus* is strongly related to their reproductive condition (Keenlyne, 1972; Reinert et al., 1984). Gravid females feed very little, if at all, while those with maturing follicles eat more often.

Crotalus horridus is a "sit and wait" predator, taking up an ambushing position at a site where small mammal prey probably will pass (Reinert et al., 1984). A favored ambush position consists of coiling adjacent to a fallen log with the head positioned perpendicular to the log's long axis. Often the chin is rested on the side of the log. Some prey may be actively sought, however, as evidenced by the abundant records of this snake climbing into trees, presumably seeking birds or squirrels. Striking of prey starts a chemical search with much tongue flicking by the snake (Chiszar, O'Connell et al., 1985). Since white-footed mice (*Peromyscus*) and chipmunks (*Tamias*) are the preferred foods in the East (Brown, 1987; Martin, 1988), feeding must occur both day and night. Neill (1960) has suggested that the young use their tails as a lure for small prey, but this has not been ascertained.

Venom and bites

Adult *C. horridus* have fangs 8.7–10.4 mm long (Klauber, 1939), and replacement fangs for those lost or broken are already present in the newborn (Barton, 1950).

The venom is strongly hemolytic, and human fatalities have resulted from bites (Hutchison, 1929; Barbour, 1950; Guidry, 1953). The activity of some venom enzymes tends to increase with the age and size of the snake (Bonilla et al., 1973). Venom chemistry is discussed by Johnson, Hoppe et al., 1968; Bonilla et al., 1973; and Tu, 1977).

Bites of humans by *C. horridus* have resulted in the following symptoms: swelling, pain, weakness, giddiness, breathing difficulty, myonecrosis, blood coagulation, hemorrhage, weak pulse and lowered blood pressure, increased heart rate, heart failure, nausea and vomiting, ecchymosis, paralysis, unconsciousness or stupor, shock, gastric disturbance, diarrhea, and heart pain (Hutchison, 1929; Kitchens et al., 1987). A case history of a fatal bite delivered by the decapitated head of a *C. h. atricaudatus* is presented by Kitchens et al. (1987); other case histories of bites are given by Parrish and Thompson (1958), Klauber (1972) and Brown (1987).

A typical *C. horridus* contains 100–200 mg of dried venom (Minton and Minton, 1969; Klauber, 1972). The maximum known dry yield of venom is 229 mg (Klauber, 1972). Wet venom volumes may range from 0.23 to 0.71 ml per snake (Minton, 1953). The LD_{50} for a 20-g mouse is 52–62 micrograms intravenous, 14–145 intraperitoneal, and 161–183 subcutaneous (Minton and Minton, 1969). The estimated lethal dose for a human adult is believed to be 75–100 mg (Minton and Minton, 1969).

Predators and defense

The young have many predators: hawks (*Buteo*), bobcats (*Lynx*), coyotes and dogs (*Canis*), foxes (*Vulpes*), skunks (*Mephitis*), cottonmouths (*Agkistrodon*), racers (*Coluber*), kingsnakes (*Lampropeltis*), and indigo snakes (*Drymarchon*). Adults have few enemies except humans. A few adults may be trampled by deer (Minton, 1972), but far more die on our highways, are blown apart by shotguns, or perish from destruction of their habitat (Tyning, 1987). Collecting for the live animal trade, bounty hunting (largely in the past), and, in Pennsylvania, licensed snake hunting—all at levels that are nonsustainable—are major threats to the timber rattlesnake throughout its range (W. S. Brown, pers. comm.).

Reinert (1991) surveyed 13 organized rattlesnake hunts in Pennsylvania between 1985 and 1987, and collected data on 139 *C. h. horridus*. The sample included 90 males and 49 females, 36 (83.7%) of the 43 mature females were gravid. Gravid females outnumbered nonreproductive females by nearly 6:1. The most plausible explanation of this is that gravid individuals have a greater susceptibility to being captured by hunters than their nongravid counterparts. This is undoubtedly a result of the preference of gravid females for more open, rocky habitats than nonreproductive females (see above), and hunters that Reinert interviewed indicated that this type of habitat was most frequently searched for rattlesnakes. As can be seen, organized hunts may seriously damage the reproductive capacity of a population by removing a large proportion of the gravid females in any biennial or triennial cycle. Reinert also reported that visible injuries, ranging from minor skin lesions to death, were observed in 40 (28.8%) of the snakes he examined. Injury rate was associated with the method of capture. Capture by nooses and hooked sticks resulted in injury rates of 83.3% and 33.8%, respectively, while tongs resulted in only a 10.2% injury rate. When the snakes were released after the events, they were not always returned to the site or even area of capture, and, since *C. h. horridus* displays den site fidelity, this probably resulted in further death of individuals. Reinert determined that organized snake hunts are of limited value as a resource management tool, and that they have a negative impact upon individual rattlesnakes and rattlesnake populations. He recommends prohibition or strict regulation of the events, and I concur.

If given a chance, a disturbed *C. horridus* will retreat, but, if prevented from escape, it will form a loose coil with head raised (as opposed to that of *C. atrox* or *C. viridis*) and strike (sometimes with mouth closed) when an intruder gets too close, rattling all the time (mean sound frequency of rattling: low 2.91–9.08 kHz, high 7.10–17.92 kHz, sound band with 4.16–11.62 kHz; Fenton and Licht, 1990). Personally, I have found them to be rather mild tempered for rattlesnakes, especially when compared to some of the western species of *Crotalus* or to *Sistrurus miliarius*, but they are still extremely dangerous.

Populations

Crotalus horridus may occur in large numbers in suitable habitats. Harwig (1966) has seen up to 17 at one time at a 6-m rock in Pennsylvania, and, in the 10-year period 1956–1965, recorded 1,628 individuals (however, this equates to only 163 snakes/year, and the number of field days/year was not given). He also has observed as many as 40–80 newborn on one September day.

Virginia dens have populations of 10–205 snakes, and the population in that state increases by an average of 63% with the birth of the young in the fall (Martin, 1988). From 1973 to 1988, Martin (1988, 1990) made observations at 509 Virginia sites and collected 5,195 *C. h. horridus*, including 1,271 neonates.

The density at Fitch's (1982) study site in Kansas was estimated to be 0.3 snakes/ha, a biomass of 0.16 kg/ha.

Sex ratios of *C. horridus* from Pennsylvania snake hunts were 1.79:1 (111 males to 62 fe-

males; Galligan and Dunson, 1979) and 1.84:1 (90 males to 49 females; Reinert, 1991). The sex ratio for 45 timber rattlesnakes from Kansas was 1.69:1 (Fitch, 1982). The sex ratio at birth for New Jersey litters was 1:1 (Odum, 1979), and Oldfield and Keyler (1989) noted a 1:1 adult ratio in Wisconsin, but this is in contrast to that reported by other investigators. Edgren (1948) reported a 2:5 male to female ratio in a Minnesota litter, and two Pennsylvania litters examined by Galligan and Dunson (1979) contained nine males and five females.

Fitch (1982) recorded a 1:1.5 juvenile to adult ratio in his Kansas population.

Survivorship of young-age classes may be low. Martin (1988, 1990) estimated that overwinter mortality was 50% for all age classes, and 61% for young of the year. The death rate for the second year was estimated at 40%, for the third year 25%, for the fourth year 17.5%, and 10% from the fifth to 14th year for females and fifth to 17th year for males. Only about 17% of the total population was composed of snakes 15 years of age or older. Fitch (1985b) estimated the percentage survival through the fifth year in Kansas was 17.3%.

Remarks

Crotalus horridus is a member of the *durissus* subgroup of *Crotalus*, and in North America is closest morphologically to *C. molossus* (Gloyd, 1940). Recent electrophoretic studies on venom by Foote and MacMahon (1977) and on scale microdermoglyphics by Stille (1987) have confirmed this relationship.

According to William S. Brown (pers. comm.) when proper mark-recapture estimates are combined with its low reproductive replacement rate and long generation time, *C. horridus* will come to be regarded as a classical K-selected species. It likely has low adult mortality, with a relatively low turnover making it very susceptible to small amounts of exploitation. Given its demographic characteristics, there is a good case to be made for giving the timber rattlesnake total protection throughout its range. Some states have recognized its plight by listing it as threatened or endangered (e.g., New York, Vermont, New Jersey), because of its fragmented distribution and small numbers of often isolated denning colonies.

Crotalus lepidus (Kennicott, 1861)
Rock rattlesnake
Plates 33, 34

Recognition

This gray, greenish, pink, or tan rattlesnake (maximum length, 83 cm) has a pattern of 13–24 narrow, dark-brown or black, irregularly spaced bands across its back and 1–6 similar bands on the tail. These bands are often bordered by paler pigments (particularly those anterior), and small spots or irregular gray mottling may occur between the bands. A pinkish-brown stripe runs backward from under the eye to the corner of the mouth. The venter is pink to light tan anteriorly, but more gray toward the tail. There are 21–26 (usually 23) rows of keeled scales at midbody. The venter has 147–172 ventral scutes, 16–33 subcaudals, and an undivided anal plate. Dorsal head scalation consists of a rostral (usually wider than long), 2 internasals, 5–15 scales in the internasal-prefrontal area (the internasals touch medially and a canthal is usually present on each side, but prefrontals are absent), 2 supraoculars, and 2(1–4) intersupraoculars. On the sides of the face are 2 nasals (the prenasal touches a supralabial, but the postnasal does not contact the upper preocular), 1(1–2) loreal, 2–5 prefoveals (a series of small scales anterior to the preoculars), 4(3–4) preoculars, 3–5 postoculars, several suboculars, 12–13(14–15) supralabials, and 11–12(9–13) infralabials. The shortened, bifurcate hemipenis has about 33 spines and 19 fringes per lobe; the sulcus spermaticus is divided. No

Crotalus lepidus lepidus (Richard D. Bartlett).

spines are present in the crotch between the lobes. In addition to the maxillary fang, the following teeth are present: 2–3 palatines, 7–10 pterygoids, and 7–10 dentaries.

Males have 147–172 ventrals and 20–33 subcaudals; females have 149–170 ventrals and 16–24 subcaudals. The male tail is 7.0–10.1% of the total body length; that of the female, 6.0–8.5%. The subspecies *C. l. klauberi* is sexually dichromatic: males are green and females gray (Jacob and Altenbach, 1977). This difference may have evolved for background color-matching through natural selection influenced by predation pressure, and is a result of a sex-linked gene that either balances the percentage of color morphs in the population or functions in sex recognition.

Karyotype

Crotalus lepidus has 36 diploid chromosomes: 16 macrochromosomes and 20 microchro-

mosomes. Sex determination is ZW, with the heteromorphism occurring in the fourth pair of macrochromosomes in the female (see Baker et al., 1972, for an illustration of the female karyotype).

Fossil record

A Pleistocene (Blancan) vertebra from Cochise County, Arizona, has tentatively been assigned to this species (Brattstrom, 1954a; Holman, 1981), and Brattstrom (1954a) reported that another vertebra from the Pliocene (again probably Blancan) of Arizona is possibly from *C. lepidus*. It agrees with those from the rock rattlesnake in all characters, but is extremely fragmented.

Distribution

The rock rattlesnake ranges from southwestern Texas, southern New Mexico and southeastern

Distribution of *Crotalus lepidus*.

Crotalus lepidus lepidus (John H. Tashjian, San Diego Zoo).

Arizona, southward in Mexico through Coahuila and western Nuevo Léon, Chihuahua, and northeastern Sonora to southwestern Tamaulipas, western San Luis Potosi, Aguascalientes, northern Jalisco, Zacatecas, and southeastern Sinaloa. It probably also occurs in eastern Nayarit (Campbell and Lamar, 1989).

Geographic variation

Four subspecies have been described, but only two live in the United States. *Crotalus lepidus lepidus* (Kennicott, 1861), the mottled rock rattlesnake, has a dark-gray or brown stripe extending from the orbit to the corner of the mouth, an absence of dark blotches on the dorsal neck, a mottled dorsal pattern with the crossbands faded and only slightly distinct from the pale ground color, and a relatively dark venter. It is found from southwestern Texas (except the El Paso area) and southeastern New Mexico southward on the Mexican Plateau to San Luis Potosi. *C. l. klauberi* Gloyd, 1936, the banded rock rattlesnake, often lacks a dark stripe from orbit to corner of mouth, has two dark blotches on the nape of the neck, a darker ground color crossed by well-defined

dark bands with little mottling between the bands, and a generally light-colored venter. A mint-green and black male *C. l. klauberi* is one of our more beautiful rattlesnakes. This race occurs from extreme western Texas, southwestern New Mexico and southeastern Arizona, southwestward across the Mexican Plateau to northern Jalisco.

There is much variability in coloration and pattern in *C. lepidus*. That related to sexual dicromatism is mentioned above, but interpopulational differences may be great. Ground color variation in *C. l. lepidus* seems to be the result of adaptation to the dominant substrate colors of the microhabitat. Vincent (1982a,b) compared the Texas populations of mottled rock rattlesnakes from the Davis Mountains and Big Bend region with those from the Stockton-Edwards Plateau. Uniformly dark bands on a pink or buff ground color characterize the former population which lives among darkly colored volcanic rocks, while snakes from the latter population live on more lightly colored limestones and are various shades of gray combined with a pronounced anterior fading of the bands. Campbell and Lamar (1989) discussed variation further, including that in the Mexican races.

Crotalus lepidus klauberi (John H. Tashjian, Dallas Zoo).

Confusing species

In the United States *C. lepidus* can be distinguished from all other sympatric rattlesnakes by its vertically divided upper preocular scale.

Habitat

The rock rattlesnake inhabits arid to semiarid areas at medium elevations (600–2,285 m) in the United States. It is most often associated with rocky areas, particularly talus slopes, rock-strewn hillsides, outcroppings with crevices, and dry arroyos in brushlands to pine forests, but has also been found in mesquite-grasslands and deserts (only *C. l. lepidus*). In such areas this snake has particular retreats (usually under rocks or in animal burrows) to which it quickly crawls when disturbed.

Behavior

Although the rock rattlesnake is common in some parts of its range, very little has been published concerning the natural history of the species.

Wright and Wright (1957) gave the annual activity period of *C. l. lepidus* as 1 May to 11 November. However, Armstrong and Murphy (1979) found both subspecies active through-out the year in Mexico if the temperature was warm, but reported that few were seen before or after the rainy season. Seventeen were found on one day after the first major rainfall.

Most activity is diurnal, but some forage on warm nights. *C. l. lepidus* is usually found in the open during the cool early morning hours, and in the shade until 0930 or 1000 hours (Telotte, in Conant, 1955). The hottest hours of the day are spent undercover, with surface activity resuming in the evening. Probably *C. lepidus* is active more daylight hours in the spring and fall. Hibernation takes place beneath the frost line under rocks, in crevices, or within old stumps.

Temperature data are scanty, but, living at relatively high altitudes, it probably is adapted to cooler temperatures than lowland rattlesnakes. Basking or otherwise active *C. lepidus* have been found at air temperatures of 24°C to about 35°C (Klauber, 1972). Five rock rattlesnakes acclimated to 24°C had heart and breathing rates averaging 23.4 (5.7–50.0) beats/minute and 1.76 (0.5–8.6) ventilations/minute (Jacob, 1980).

The rock rattlesnake does not seem to be much of a climber, but, like other snakes, is a natural swimmer (Klauber, 1972).

Male *C. lepidus* may participate in ritualized combat. Carpenter et al. (1976) filmed such a bout, and described it in great detail. The sequence of these actions and postures is a form of aggression for social communication. The typical actions, postures, and features of the sequence may occur simultaneously and involve investigations, ascent and descent of the anterior body and head, vertical display with swaying of the anterior trunk, bending over backward and the head bent sharply at the neck. Topping occurs as one male bends his trunk over that of the other male and pushes, causing the two snakes to fall to the ground. This is usually followed by quick recovery and return to displaying. The primary aggressor exhibits higher and more vertical posturing and more exaggerated neck flexures while his posterior trunk region is usually laid over that of the less dominant male. Topping occurs when the less aggressive male also assumes the exaggerated postures of the more dominant male. Eventually one male will establish dominance, and the defeated male will break contact and retreat.

Reproduction

The smallest pregnant females measured by Klauber (1972) were a 39.0-cm *C. l. klauberi* and a 44.1-cm *C. l. lepidus*, but Armstong and Murphy (1979) reported a 43.5-cm probable female *lepidus* gave birth to six young, and Kauffeld (1943a) related that a 42.9-cm *klauberi* had four young. An intergrade 41.2-cm female *C. l. maculosus* × *l. klauberi* is also known to have given birth (Harris and Simmons, 1972). So, sexual maturity in females is probably attained at a total length of about 40 cm. Comparable maturity data are lacking for males, as are also data on both sexual cycles.

Courtship and mating activities have been observed in captive *C. l. klauberi* on 21 February, 14–28 September, and 11 October (Armstrong and Murphy, 1979). The male directed rapid head bobs (3–5/5 sec) on the female's back, and tongue flicking occurred at the same rate.

A female *C. l. lepidus* obtained by Stebbins (1954) on 1 July contained five large ova that measured approximately 34 × 17 mm.

The young are normally born during late July and August, but Harris and Simmons (1972) reported a birth on 14 April, and Arm-

strong and Murphy (1979) another on 5 June. These early births were by Mexican females.

Sixteen litters in the literature averaged 3.56 young, and ranged from one (Harris and Simmons, 1972) to eight young (Stebbins, 1954). Newborn young average 187 mm (160–229) in total length and weigh 5–6 g. They have yellowish tails.

Growth and longevity

Data on the growth rate of wild individuals have not been published, but some indication of potential growth is available from two sets of captives. Kauffeld (1943b) reported that a male *C. l. klauberi* 200 mm long at birth grew to 312 mm in 118 days. The male fed readily, as opposed to a female that had to be force fed and grew only 32 mm. Woodin (1953) reported that young *klauberi* 205–210 mm when collected grew to 214 and 218 mm in 23 days.

Captive longevity records for this species include 23 years, 3 months and 24 days for *C. l. klauberi*, and 17 years, 11 months and 10 days for *C. l. lepidus* (Bowler, 1977).

Food and feeding

Crotalus lepidus seems to prefer reptilian food over mammals or amphibians. Rock rattlesnakes taken in the wild are known to have eaten lizards of the genera *Sceloporus*, *Urosaurus*, *Phrynosoma*, *Holbrookia*, and *Cnemidophorus* and the snake *Gyalopion canum*, and captives have eaten skinks (*Eumeces*) and the snake *Virginia striatula* (Campbell, 1934; Kauffeld, 1943a; Marr, 1944; Milstead et al., 1950; Woodin, 1953; Stebbins, 1954; Conant, 1955; Axtell, 1959; Klauber, 1972). Mice (Woodin, 1953), the frog *Syrrhophis marnocki* (Milstead et al., 1950), and a large grasshopper (Conant, 1955) have also been found in wild *C. lepidus*, and Falck (1940) reported the consumption of mice of various kinds, several frogs, and three tiger salamanders (*Ambystoma tigrinum*) by captives. Although lizards are favored, captives will also readily take small house mice (*Mus*). Williamson (1971) and Harris and Simmons (1977) related instances of cannibalism in which both adults and young consumed cagemates.

Juvenile *C. lepidus* have yellow tails, which

may be waved as a lure to attract lizards (Kauffeld, 1943b; Neill, 1960). Lizards are struck on the body, and usually retained until comatose. Swallowing quickly follows. Mice struck are usually released and later trailed. However, captives fed exclusively on mice and then used in studies of their trailing behavior by Chiszar, Radcliffe, Feiler et al. (1986) scored significantly below the prairie rattlesnake, *C. v. viridis*, which in nature depends more heavily on rodent prey. In trailing, *C. lepidus* used chemoreception (definitely the vomeronasal system, and possibly the nasal system) to find the rodent carcasses.

Few species of mice of suitable size live at the altitudes inhabited by *C. lepidus*, while lizards of various types are usually plentiful, and this may have resulted in natural selection favoring lizard predation.

Venom and bites

Five *C. l. klauberi* 520–595 mm in total length had fangs 3.2–3.6 mm long (Klauber, 1939).

A *C. lepidus* secreted 0.1 ml (0.03 g) of venom (Amaral, 1928), and fresh adults may yield as much as 129 mg of dried venom (Klauber, 1972). The minimum lethal dose of venom for a 20-g mouse is 0.02 mg (Githens and Wolff, 1939), and for a typical pigeon, 0.01 mg (Githens and George, 1931). The venom is extremely potent in its hemorrhagic activity (Soto et al., 1989), and this snake should be treated with much caution. It is not one to be carelessly handled, as learned by the late W. W. Wright who was bitten by a *C. l. lepidus* he was holding when distracted. Both fangs pierced his thumb. His arm swelled considerably and his lymph glands were affected, but he fully recovered (Wright and Wright, 1957). Klauber (1972) told of another human envenomation by *C. l. klauberi*. One fang entered the middle finger. By the next day the swelling had reached the forearm, and on the second day, the shoulder. An intense and continual burning pain developed at the site of the puncture; it began on the day following the bite and became virtually unbearable on the second day, but then lessened. Within a day or so after the bite, a blood blister about 12.5 mm × 19.0 mm formed at the site of the bite. The swelling continued for five weeks, with numbness and tingling sensations.

Predators and defense

No records of predation have been published except that by Tennant (1985) that a *C. lepidus* was eaten by a copperhead *Agkistrodon contortrix*, and those of cannibalism in captivity mentioned above (Williamson, 1971; Harris and Simmons, 1977). However, ophiophagous snakes, birds of prey, and carnivorous mammals must eat at least the juveniles. Humans seem to cause the most destruction of *C. lepidus*. Johnson and Mills (1982) listed mining, grazing, road-building, scientific and pet collecting, and recreational or urban development as threats (although minor) in Arizona. Most of these involve habitat destruction or alteration, but with the exception of grazing and road construction, are of local impact. Mining may actually benefit the snake in some cases by creating talus slopes with favorable vegetation, exposure, and lizard populations.

The rock rattlesnake is rather calm in disposition. When an enemy is detected, it often lies quietly in place, where its coloration may blend into the background so well as to conceal it. If further disturbed, it usually tries to crawl beneath some shelter, rattling all the while, but if trapped will coil and strike, and will attempt to bite if handled (see above).

Populations

Crotalus lepidus may be locally common, with large populations occurring on some talus slopes. It is known from more than 40 localities in southeastern Arizona (Johnson and Mills, 1982), and although no quantitative population estimates are available it appears to be relatively common, well protected on federal lands, and largely unthreatened in that state.

Remarks

Electrophoretic studies of venom proteins by Foote and MacMahon (1977) indicate that *C. lepidus* is most closely related to *C. triseriatus*. This same relationship was previously concluded on the basis of morphological similarities by Gloyd (1940).

Crotalus mitchelli (Cope, 1861)
Speckled rattlesnake
Plates 35, 36

Recognition:

In this rattlesnake (maximum length, 13.7 cm), the rostral scale is separated from the prenasal by at least one but usually several scales, and the preocular scale is subdivided. The ground color is variable, ranging from gray, cream, yellowish pink, tan, or brown to black (Pisgah lava flow, San Bernadino County, California). White and black specks adorn each keeled body scale. Usually some indication of dark bands or diamonds is present on the back, and the tail bears three to nine dark rings. The venter is white, pink, cream or yellowish brown with some grayish-black blotches at the sides or borders of the ventrals. The head is marked dorsally with dark flecks or spots. Midbody scales usually occur in 25(21–27) rows, ventrals total 166–187, and there are 16–28 subcaudals; the anal plate is not divided. Dorsally, the head scalation includes a rostal (which is wider than high), as many as 50 scales in the internasal and prefrontal area (no prefrontal scales are present), 2 supraoculars and 1–8 intersupraoculars; laterally are a prenasal (which may contact the first supralabial), a postnasal, (which does not touch the upper preocular), 2(0–5) loreals, 2–3 preoculars, 2–3 postoculars, several suboculars, 12–19 supralabials, and 12–19 infralabials. The bifurcate hemipenis has a divided sulcus spermaticus and 1–3 spines in the crotch; each lobe has 40–60 spines and 31 fringes (see photograph in Klauber, 1972). Dentition other than the maxillary fang includes 2–3 palatine teeth, 7–10 pterygoid teeth, and 7–10 teeth on the dentary.

Males have 166–185 ventrals, 20–28 subcaudals and 4–9 dark tail rings; females have 168–187 ventrals, 16–24 subcaudals, and 3–6 dark tail rings.

Karyotype

The karyotype is 2n = 36: 16 macrochromsomes and 20 microchromosomes (Zimmerman and Kilpatrick, 1973). Sex determination is ZW for females and ZZ for males.

Fossil record

Fossil vertebrae dating from the Pleistocene Rancholabrean have been found in Gypsum Cave, Nevada (Brattstrom, 1954a,b; Holman, 1981), and another Rancholabrean series of fragmented vertebrae from the Lower Grand Canyon, Arizona, may be either from *C. mitchelli* or *C. viridis* (Van Devender et al., 1977).

Distribution

The speckled rattlesnake ranges from southern Nevada and extreme southwestern Utah south-

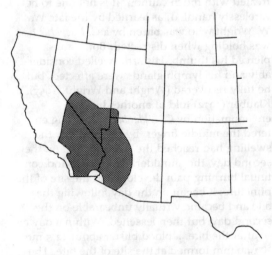

Distribution of *Crotalus mitchelli*.

Crotalus mitchelli pyrrhus (John H. Tashjian, San Diego Zoo).

ward through western Arizona and southern California, barely reaching northwestern Sonora, Mexico, but occurring throughout most of Baja California.

Geographic variation

Five subspecies are recognized, but only two live in the United States. *Crotalus mitchelli pyrrhus* (Cope, 1867), the southwestern speckled rattlesnake, occurs in southwestern Nevada and adjacent eastern California. It is white, gray, pink, or tan to orange-red with dark crossbands that are often divided by light pigment, has small scales separating the rostral and prenasal scales, and no pitting or furrowing on its smooth-edged supraocular scales. *C. m. stephensi* Klauber, 1930, the Panamint rattlesnake, ranges from southwestern Utah through western Arizona to northwestern Sonora, Mexico, and through southern California and the northern half of Baja California. It is tan or gray with light-brown blotches or crossbands that are bordered by light pigment, lacks small scales between the rostral and prenasals, and its supraoculars have pitted or furrowed outer edges.

Confusing species

Crotalus cerastes has a prominent horn above each eye, while *C. atrox* and *C. scutulatus* have more intense dorsal and tail markings.

Habitat

The speckled rattlesnake is primarily a desert dweller, where it most often occupies the hottest, driest, rocky areas such as canyons, foothills, buttes, and erosion gullies. It has also been found in thickets of chaparral, creasote bush, sagebrush, thornscrub, and pinyon-juniper woodlands, and sometimes lives in areas with loose sand. Its range spans elevations of a few hundred to at least 2,100 m.

Behavior

Active *Crotalus mitchelli* have been found as early as 28 February and as late as December (Klauber, in Wright and Wright, 1957), but April into early October seems to be the most active period. In the southwestern part of the range, activity peaks during the late summer rainy season (Armstrong and Murphy, 1979). Although it may be diurnally active in the

Crotalus mitchelli stephensi (Richard D. Bartlett).

spring or fall, foraging during most of its annual period (June to September) is nocturnal,
predominantly between dusk and shortly after
midnight (Klauber, 1972; Moore, 1978). *C.
mitchelli* has a unimodal activity pattern. It initially becomes active on the surface in April,
and the number of hours of activity increases
in each successive month through September
(Moore, 1978). Activity is then reduced and finally ceases between December and March.
Since surface temperatures sometimes fall rapidly after sunset in its desert habitat, *C. mitchelli* may be limited to only a few suitable hours
of activity each night.

Recorded body temperatures of active *C.
mitchelli* have ranged from 18.8 to 39.3°C; the
overall preferred body temperature (April to
December) seems to be about 31°C (Moore,
1978). The critical minimum temperature of
this snake is −2°C (Brattstrom, 1965). Tightly
coiled, inactive *C. mitchelli* may be able to conserve heat more effectively through circulatory
adjustments than uncoiled snakes (Moore,
1978).

Speckled rattlesnakes may bask in the
morning, particularly after feeding, but the hottest period of the summer day is spent undercover beneath rocks or bushes, or in rock crevices, caves, or animal burrows. In the winter

most northern populations hibernate, usually
below the frost line in crevices or animal burrows, but caves or abandoned mines may also
be used. While usually solitary in the summer,
these snakes often congregate at suitable hibernation dens. A hibernaculum may contain as
many as 20 to 180 snakes (Klauber, 1972).

Crotalus mitchelli may occasionally climb as
high as 90 cm into vegetation (Klauber, 1972).
It is also a natural swimmer (Klauber, 1972).

Reproduction:

The shortest gravid female *C. m. pyrrhus* and
C. m. stephensi known were 57.3 and 67.4 cm,
respectively (Klauber, 1972). The male and female reproductive cycles of this species have
not been studied, but Cunningham (1959) reported that a female *C. m. pyrrhus* collected 10
June in San Bernardino County, California, contained small eggs.

A pair of *C. m. pyrrhus* was found mating
in the afternoon of 18 April by J. W. Warren (in
Brattstrom, 1965). Both had body temperatures
of 31.8°C (air, 31.8; soil, 31.0). Armstrong and
Murphy (1979) observed a captive pair of the
Mexican subspecies *C. m. mitchelli* copulating
from 0800 to 1500 hours on 13 October. The
right hemipenis was inserted, and a prominent

bulge was evident which extended 20 scale rows anterior of the female's vent. The female gave birth to one young and an infertile egg mass on 29 June.

At the Houston Zoological Gardens, a pair of *C. m. mitchelli* has copulated on 8 January, 3 March, 25 March, and 13 May over a six-year period (Peterson, 1983). The 3 March 1981 copulation led to the birth of seven young 142 days later on 28 July.

Parturition in nature usually occurs from July through September. Broods may contain 2–11 young, (Stebbins, 1985), but 4–8 young are more normal (Klauber, 1972). Newborn young range from 203 to 304 mm in length.

Growth and longevity

Nothing has been published on the growth rate of this species. Bowler (1977) reported captive life spans of 15 years, 6 months, 1 day for *C. m. pyrrhus* and 12 years, 5 months, 23 days for *C. m. stephensi*.

Food and feeding

Adult speckled rattlesnakes seem to prefer warm-blooded prey. Klauber (1972) reported that of specimens containing food he examined, 18 had eaten mammals, nine had eaten lizards, one a bird, and another both a mouse and a lizard. The mammals most often consumed are ground squirrels (*Ammospermophilus, Spermophilus*), wood rats (*Neotoma*), kangaroo rats (*Dipodomys*), pocket mice (*Perognathus*), and white-footed mice (*Peromyscus*), but Shaw (in Klauber, 1972) found a large *C. mitchelli* that had taken a nearly grown cottontail rabbit (*Sylvilagus auduboni*). Mammal prey may be eaten fresh or as carrion. Although birds are often listed as prey, the only one yet identified has been the goldfinch (*Carduelis;* Batchelder, in Klauber, 1972). Some adults will also prey on lizards, and these reptiles are apparently the primary foods of juveniles. Lizards known to be eaten include side-blotched (*Uta*), fence (*Sceloporus*), chuckwalla (*Sauromalus*), zebra-tailed (*Callisaurus*), skink (*Eumeces*), and whip-tailed (*Cnemidophorus*).

Porter (1983) has related three instances of cannibalism by captive juvenile *C. m. stephensi*.

The speckled rattlesnake may capture its food by either actively foraging or lying in ambush.

Venom and bites

Five *C. m. stephensi*, 74.6–88.5 cm in length, examined by Klauber (1939) had fangs 5.6–7.1 mm long, while five *C. m. pyrrhus*, 82.6–111.4 cm long, had fangs 7.8–10.8 mm.

The venom is very potent, particularly for birds; the minimum lethal dose for a 350-g pigeon is only 0.002–0.04 mg dried venom, and for mice 0.05–0.12 mg (Klauber, 1972). The lethality of the dried venom is not even diminished with time; Russell et al. (1960) found it had not weakened after 26–27 years storage. A typical adult *C. mitchelli* may contain as much as 227 mg of dried venom (Klauber, 1972); however, on the average, young of *C. mitchelli* yield only 0.06 mg (0.18 ml) of venom per bite, while adults may inject 0.1 mg (0.3 ml), and old snakes 0.16 mg (0.48 ml) (Amaral, 1928).

Bites to humans may result in considerable swelling (edema), discoloration, and pain about the area bitten (Klauber, 1972). When treated with antivenin, recovery is usually complete and uneventful.

Predators and defense

Cunningham (1959) reported that notes accompanying a *C. m. pyrrhus* in the UCLA collection describe a bobcat (*Lynx rufus*) attack on the specimen, and Zweifel (in Stebbins, 1954) reported a gray fox (*Urocyon cinereoargenteus*) ate a speckled rattlesnake. Probably other mammals, such as skunks (*Mephitis*) and coyotes (*Canis*), birds of prey (*Buteo* and *Bubo*), and ophiophagous snakes (*Lampropeltis, Pituophos, Crotalus*) also prey on *C. mitchelli*. Shaw and Campbell (1974) noted that crows (*Corvus*) eat rattlesnake carrion, and so may also attack live ones.

Temperament varies among individual *C. mitchelli;* some may lie quietly with only an occasional rattle when first discovered, but most seem very nervous, and, if they cannot escape, will coil, raise a loop of the body from the ground, inflate the trunk, and strike, while rattling continuously. The subspecies *stephensi* seems to be less excitable than *pyrrhus*.

Remarks

Gloyd (1940), Brattstrom (1964), and Klauber (1972) thought *C. mitchelli* to be most closely related to *C. tigris* and *C. viridis*. However, electrophoretic studies of rattlesnake venoms by Foote and MacMahon (1977) indicate *C. ti-gris* has a closer relationship with the tropical *C. durissus*, and that *C. mitchelli* is closest to *C. pricei* and *C. cerastes*.

Competition or some other factor of habitat exclusion may occur between *C. mitchelli* and *C. atrox* and *C. scutulatus*, since the other two rattlesnakes will occupy its microhabitat when it is absent.

Crotalus molossus (Baird and Girard, 1853)
Blacktail rattlesnake
Plates 37, 38

Recognition

This is one of our most beautiful rattlesnakes. Its body (maximum length, 126 cm) is olive gray, greenish yellow, yellow, or even reddish brown with 20–41 dark-brown or black blotches (sometimes diamond shaped) or irregular crossbands, a black tail, and often a black snout. The darkest individuals are usually from areas with dark substrates. A dark band lies between the eyes, and a light-bordered, dark stripe extends diagonally downward from the eye to the corner of the mouth. The venter is white to cream anteriorly, but more greenish toward the tail. The keeled body scales occur in 27(23–31) rows at midbody. Ventral scutes total 166–201, subcaudals 16–30; the anal plate is entire. Behind the slightly higher than wide rostral scute lie 2 large internasals and 2 pairs of canthals. The internasals touch the rostral, the prenasal rarely touches the supralabials, the postnasal rarely touches the upper preocular, and the second pair of canthals is situated between the large supraoculars. Prefrontals are usually present, and 2–5 intersupraoculars lie between them. On the side of the face, the 2 nasal scales are followed by 2–4(1–9) loreals, 2(3) preoculars, several suboculars, 5(3–7) postoculars, 17–18(13–20) supralabials, and

Crotalus molossus molossus (Roger W. Barbour, Arizona).

Crotalus molossus molossus (John H. Tashjian, Gladys Porter Zoo).

17–18(14–21) infralabials. The short hemipenis is illustrated in Klauber (1972). It is bilobed with a divided sulcus spermaticus, about 68 spines and 21 fringes per lobe, and no spines in the crotch between the lobes. The pterygoid bears 8–9(6–10) teeth, the palatine 2(1–3), and the dentary 10(9–11). The maxilla has only the movable fang.

Males have 166–199 ventrals, 22–30 subcaudals, and tails comprising 5.8–8.6% of the total body length; females have 177–201 ventrals, 16–25 subcaudals, and tails that are 4.6–6.7% of the total body length.

Karyotype

Like other species of North American *Crotalus*, *C. molossus* has a diploid chromosome total of 36: 16 macrochromsomes (including 4 metacentric, 6 submetacentric, and 4 subtelocentric) and 20 microchromosomes, and sex determination by either ZW or ZZ (Baker et al., 1971, 1972; Zimmerman and Kilpatrick, 1973). Baker et al. (1971, 1972) reported that the Z macrochromosome is submetacentric and the W macrochromosome is subtelocentric; however, Zimmerman and Kilpatrick (1973) thought these two chromosomes to be metacentric and submetacentric, respectively.

Fossil record: None.

Distribution

Crotalus molossus is found from central and west-central Texas northwest through the southern half of New Mexico to northern and extreme western Arizona, southward to the southern edge of the Mexican Plateau and Mesa del Sur, Oaxaca. It also occurs on the Tiburón and San Esteban islands in the Gulf of California.

Geographic variation

Three suspecies are currently recognized, but only the nominate race, *Crotalus molossus molossus* Baird and Girard, 1853, the northern blacktail rattlesnake, inhabits the United States. It is described above.

Campbell and Lamar (1989) reported clinal variation in hemipenial morphology. Blacktail rattlesnakes from the United States and adjacent Mexico have straight, thick, stubby lobes with more than 50 small spines at the base of each lobe. Those from farther south have slender lobes and fewer spines.

Crotalus molossus from the uplands of western Texas are generally darker than those from

Crotalus molossus molossus (Richard D. Bartlett).

Distribution of *Crotalus molossus*.

New Mexico and Arizona which tend to be more yellowish in color.

Confusing species

No other adult rattlesnake within its range has an entirely black tail, six or fewer large scales in the internasal-prefrontal area, and lighter scales within the dark body blotches.

Habitat

Crotalus molossus is most common in upland pine-oak or boreal forests, but has been taken at altitudes of 300–3,750 m. It seems to prefer rocky sites such as talus slopes, the sides of canyons, or crevices in outcrops and caves. At lower elevations it may be found in mesquite-grassland, chaparral, or even desert situations. Those living on dark lava flows may be melanistic, apparently adapted to matching the substrate color (Lewis, 1949; Prieto and Jacobson, 1968).

Behavior

The winters become quite cold at the higher elevations occupied by this snake, restricting activity at that time of the year to only the warmest days. Blacktail rattlesnakes are more typically seen in the late summer or fall in Arizona (Greene, 1990), but some are surface active in the spring (Steve W. Gotte, pers. comm.). The earliest recorded dates of appearance are 4 April (Wright and Wright, 1957) and 21 April (Gates, 1957); Lowe et al. (1986) reported surface activity in February and March, but gave no dates. The latest observations recorded are in October (Wright and Wright, 1957; Greene, 1990), late November when there was a thin covering of snow on the ground (Kauffeld, 1943a), and December (Lowe et al., 1986). In Chihuahua, they are active from July through September (Reynolds, 1982). Hibernation takes place below the frost line in animal burrows, rock crevices, caves, and possibly abandoned mines.

Spring and autumn activity is mostly during the day, but *Crotalus molossus* may shift to a more nocturnal pattern during the hotter summer months. It usually remains concealed during cold or overcast days, but blacktails may not remain constantly torpid during the win-

ter: instead some may bask on warm days in January and February (Greene, 1990). *C. molossus* is often surface active after summer rains. Perhaps they drink water from puddles at such times. Greene (1990) observed one drinking water from the film seeping over a rock face on an October afternoon.

The thermal ecology of *C. molossus* has not been adequately studied. Bogert (in Klauber, 1972) maintained *C. molossus* at 4°C for about 10 days and the snake survived, so its critical thermal minimum is below this temperature. On the other hand, Wright and Wright (1957) reported that a specimen exposed for 10 minutes to bright summer sunlight died. Blacktail rattlesnakes detected in the morning are usually basking, apparently trying to warm after the chilling effects of night air temperatures. At 24°C, *C. molossus* had apneic and ventilatory heart rates of 11.26(5.8–23.1) and 13.87(8.6–31.6) beats per minute, respectively, and a breathing rate of 1.36(0.01–8.8) breaths per minute (Jacob, 1980).

Not all activity is restricted to the ground, as *C. molossus* is an accomplished climber. Allen (1933) found them in bushes about 2 m above ground, Klauber (1972) related several observations of *C. molossus* climbing in trees to heights of 2.5–2.7 m, Campbell and Lamar (1989) also encountered these snakes several meters high in trees, and Hastings (in Klauber, 1972) saw one climb a stone wall. Lowe et al. (1986) reported that *C. molossus* may sleep above ground in shrubs and trees.

This snake swims readily when placed in water (Klauber, 1972).

Greene (1990) observed males in late summer dominance combat, sometimes with a female nearby. As in other *Crotalus*, the combat is a stereotyped wrestling match rather than a violent fight. Each male attempts to pin his opponent's head to the ground, and the loser crawls away undamaged.

Reproduction

Surprisingly little is known of the reproductive biology of this fairly common rattlesnake. The smallest mature female *C. molossus molossus* known had a total length of 703 mm (Klauber, 1972). A female collected on 5 July had well-developed embryos (Gates, 1957), so the young of *C. m. molossus* are probably born in

August. Five young of the Mexican subspecies *C. m. nigrescens* were born in captivity on 9 June (Armstrong and Murphy, 1979).

A possible mating attempt by a male *C. m. molossus* on the dead body of a female occurred on 7 August (Wright and Wright, 1957), and a captive mating between a male *C. molossus* and a female *C. atrox* was reported by Davis (1936).

Observed matings of *C. m. nigrescens* have taken place in the wild on 1 February, and in captivity on 2 March and 28 May (Armstrong and Murphy, 1979). The 1 February copulation took place at 0920, during which the diameter of the female's cloaca was distended 15 mm owing to the male's hemipenis. A noticeable bulge in the female's body extended 35 mm anterior to her vent. The shoulder spines of the hemipenis were visible and the organ was colored dark purple. The copulation lasted 105 minutes, during which both partners engaged in intermittent head bobbing and tongue flicking.

A heterosexual pair of *C. molossus* was radiotagged by Greene (1990) after they were found in a wood rat nest. The pair traveled and basked together for several weeks, but by early fall they had moved apart and entered separate hibernacula in the same rocky bluff. Evidently, the male remains with the female after mating, perhaps guarding her from mating with another male (Greene, 1990).

Seventeen litters of *C. m. molossus* reported in the literature averaged 6.7(3–16) young (Kauffeld, 1943a; Wright and Wright, 1957; Klauber, 1972; Lowe et al., 1986); the subspecies *C. m. nigrescens* may also produce as many as 16 young (Dunkle and Smith, 1937). Neonate *C. m. molossus* average 272 mm (229–315) in total body length, weigh 11–28 g, have discernable dark tail bands, and are born in July or August.

Growth and longevity

Neither wild nor captive growth rates have been reported. An adult male lived 15 years, 6 months and 14 days (Bowler, 1977).

Food and feeding

Like other rattlesnakes, *Crotalus molossus* probably captures its prey by both ambush and active hunting. In the spring and fall most foraging is done in the morning or late afternoon, but nocturnal hunting is the mode in the summer. Prey is struck and then later trailed, found, and swallowed after it has died (pers. obs.).

Small mammals are the favorite foods, but birds and lizards may occasionally be taken, and suitable carrion is probably also consumed.

Reynolds and Scott (1982) studied the food habits of 12 *C. molossus* in Chihuahua, Mexico. Eighty-three percent of the prey taken was mammalian: pocket mice (*Perognathus*), 33.3%; kangaroo rats (*Dipodomys*), 16.7%; white-footed mice (*Peromyscus*), 25.0%; and wood rats (*Neotoma*), 8.3%. They also found bird remains (16.7%) in *C. molossus*. *Dipodomys, Peromyscus, Neotoma, Eutamias, Sciurus* and *Sylvilagus* have also been found in *C. molossus* from the United States (Milstead et al., 1950; Woodin, 1953; Gehlbach, 1956; Minton, 1958; Klauber, 1972; Lowe et al., 1986; Greene, 1990). *Neotoma* and *Sylvilagus* are the typical prey in the Arizona population studied by Greene (1990). A 101-cm *C. molossus* examined by Funk (1964a) contained a 40-cm Gila monster, *Heloderma suspectum,* and Klauber (1972) found lizard scales in the digestive tract of a Mexican specimen. The diet may also include snakes and frogs, and young snakes may feed on insects (Vermersch and Kuntz, 1986).

Venom and bites

Adults have fangs 9.6–13.5 mm long (Klauber, 1939), and, although the venom is relatively mild, the bites can be serious.

The venom has strong hemorrhagic properties. Minton (1958) reported the venom of moderate toxicity, the mouse LD_{50} being 17.4 mg/kg. The minimum lethal dose is 0.06 mg of dried venom for a 22-g mouse (Macht, 1937), and 0.14 mg/kg for a 20-g mouse (Githens and Wolff, 1939). The minimum lethal dose for a 350-g pigeon is 0.40 mg of dried venom (Githens and George, 1931). *Crotalus molossus* yields an average of 0.60 ml (0.18 g dried) venom when milked (Amaral, 1928), and the average total yield in dried venom per fresh adult is about 286 mg (Klauber, 1972); the maximum known yield is 540 mg. After 32 days, the liquid venom yield is only 47% of the first milking (Klauber, 1972). Gregory-Dwyer et al.

(1986) found no variation in isoelectric patterns of the venom from individual *C. molossus* over a 20-month period.

In two cases of human envenomation by *C. m. molossus*, there was marked swelling and ecchymosis of the bitten extremity and thrombocytopenia and, in one case, hypofibrinogenemia (Hardy et al., 1982). Both victims recovered after treatment with antivenin, crystalloid solutions, fresh frozen plasma, and cryoprecipitates. Studies show the venom has fibrinolytic and platelet-aggregating properties; a coagulent effect although present is much less marked.

Predators and defense

The only record of predation on *C. molossus* is that of one being eaten (possibly as carrion) in captivity by *C. atrox* (Klauber, 1972). Humans are the worst natural enemy, often killing this snake on sight, but also destroying many on the roads or through habitat destruction. Natural enemies, particulary of the young, probably include bullsnakes (*Pituophis*), kingsnakes (*Lampropeltis*), hawks (*Buteo*), owls (*Bubo, Asio*), skunks (*Mephitis*), coyotes (*Canis*), bobcats (*Lynx*), and peccaries (*Tayassu*).

This is usually a mild-mannered snake that will often remain calmly coiled either quietly or with occasional rattling, or will try to crawl away into some shelter. However, some will begin rattling when a human is still some distance from them, and a few will actually aggressively strike when first disturbed. Armstrong and Murphy (1979) observed an open mouth defensive posture in *C. m. nigrescens* during which the mouths were held open for more than five minutes while the snakes were provoked.

Populations

While this snake may be common at some localities, no study of its population structure or dynamics has been published.

Of 425 rattlesnakes collected in central Arizona, only 27 (6.4%) were *C. molossus* (Klauber, 1972).

Remarks

Gloyd (1940) hypothesized that *C. molossus* is part of a complex that includes *C. horridus* and *C. durissus*, but that it is most closely related to the Mexican species *C. basiliscus*. Electrophoretic studies of rattlesnake venom by Foote and MacMahon (1977) show *molossus* venom closest to *C. scutulatus, C. tigris*, and *C. horridus* (*C. basiliscus* was not tested).

Crotalus pricei (Van Denburgh, 1895)
Twin-spotted rattlesnake
Plate 39

Recognition

This small (maximum length, 66 cm), slender, gray to pale-brown, occasionally reddish rattlesnake has a dorsal pattern consisting of two longitudinal rows of 39–64 (usually about 50) dark spots. These spots may be either brown or black, and may be united across the back or alternate, with those nearest the tail usually forming crossbands. Smaller dark spots may occur along the sides, and normally a dark stripe extends from below the eye backward along the cheek. The basal segment of the rattle is orange to red. The throat and venter are gray or light brown; the posteriormost ventrals contain much black mottling. Body scales are keeled and usually lie in 21–23 rows at midbody and about 17 rows just before the tail. Ventrals total 135–171, and subcaudals 18–33; the anal plate is undivided. *Pricei* is unique among *Crotalus* species in having the first supralabial curving dorsally behind the postnasal scale to contact a small scale (a prefoveal) lying between the postnasal and loreal scales. Supralabials total 8–10, infralabials 8–12. Other head scalation includes a rostral (usually wider than long), 2 internasals, 1(2) canthal, 2 supraoculars, 2–3 intersupraoculars, 4–11 scales in the internasal-prefrontal area (prefrontals are usually absent), 2 nasals (the prenasal touches the supralabials, but the postnasal does not touch the upper preocular), 1–2 loreals, 2–3 preoculars, 3–4 postoculars, and several suboculars. The loreal does not touch any supralabial. The bifurcate hemipenis is short and has a divided sulcus spermaticus, about 52 spines and 22 fringes per lobe, and many spines in its crotch (Klauber, 1972). The dental formula consists of a maxillary fang, 3 palatine teeth, 6–7 pterygoid teeth, and 9–10 dentary teeth.

Males have 135–162 ventrals and 21–33 subcaudals, females 143–171 ventrals and 18–27 subcaudals.

Karyotype and fossil record: Unknown.

Distribution

In the United States, *Crotalus pricei* occurs in only five isolated populations on mountain sides (Santa Rita, Huachuca, Pinaleno [Graham Peak], Dos Cabezas, and Chiricahua) in southeastern Arizona. Its main range lies in Mexico from northeastern Sonora and western Chihuahua southward through western Durango to northern Nayarit. A second Mexican center for distribution, representing the range of the eastern subspecies, *C. p. miquihuanas*,

Distribution of *Crotalus pricei*.

135

occurs from southeastern Coahuila and south-western Nuevo Léon to southern Tamaulipas. Probably the range of *C. pricei* was continuous during the Pleistocene (Rancholabrean) glaciation, but later drying of the intervening lowlands restricted the scrub pine-oak wood-lands to high altitudes, isolating the Arizona populations from each other and those from Mexico.

Geographic variation

Two subspecies are recognized, but only the nominate race, *Crotalus pricei pricei* Van Denburgh, 1895, the twin-spotted rattlesnake, occurs in the United States.

Confusing species

Crotalus willardi has a sharp, slightly upturned snout, white bars across its back, and 25 or more midbody scale rows. *C. lepidus* has widely spaced black transverse bars or blotches on its back, no dark cheek stripe, and usually more than 10 supralabials.

Habitat

In Arizona, this snake lives only at relatively high altitudes, 1,900–3,200 m. There it occu-pies the rocky areas (particulary talus slopes) of canyons and ridges vegetated with scrub-brush, pine-oak, or coniferous woodlands. Wa-ter is probably obtained by drinking that trick-ling over rocks (Kauffeld, 1943b).

Behavior

Nothing is known of the annual cycle of this species, but surely it must spend the colder months hibernating. Armstrong and Murphy (1979) have collected *C. p. pricei* in Arizona on 17 March and in Mexico from May through September, sometimes under adverse weather conditions.

In captivity, *C. pricei* is most active in the late afternoon and at night (Kauffeld, 1943b). In nature, the nights at the higher elevations occupied by this snake are usually cold, and, although it may be nocturnally active on the warmest nights, most activity seems to occur during the day, particularly after 1100. This is especially true immediately following precipita-

tion, when it is commonly found basking, sometimes in pairs. Humid days also bring this snake out from its retreats under rocks and rock crevices.

Active *C. pricei* have been found at air tem-peratures ranging from 11°C (Armstrong and Murphy, 1979) to 27°C (Wright and Wright, 1957), and body temperatures of five *C. pricei* sent to Brattstrom (1965) by Frederick Gehlbach averaged 21.1°C (18.0–23.8). From these few data, it can be tentatively surmised that this small rattlesnake is adapted to rather low ambient temperatures.

Crotalus pricei swims with the same typical lateral undulations used in crawling, and it does not elevate its rattle above the water (Klauber, 1972).

Reproduction

The smallest gravid female *Crotalu pricei* mea-sured by Klauber (1972) was 301 mm; nothing is known of the size needed for males to attain maturity. Armstrong and Murphy (1979) re-ported that a captive mating observed by Jona-than A. Campbell took place on 9 July; how-ever, no specific description of the courtship or mating behavior has been published.

Like other *Crotalus*, the twin-spotted rattle-snake is ovoviviparous and bears live young in July or August. Litters range from three to nine (Armstrong and Murphy, 1979), although five to six is more common. Short females probably produce the smallest litters. Neonates are 127–203 mm long (Lowe et al., 1986), and weigh between 2.4 and 5.6 g.

Growth and longevity

Five captive *Crotalus pricei* grew an average of 98 mm in their first 125 days (Kauffeld, 1943b). The greatest increase was in the largest at birth, from 170 to 280 mm. Their weights increased an average of 11.5 g during this period.

A wild-caught, 76-cm adult lived more than 10 years in captivity (Kauffeld, 1969).

Food and feeding

Sceloporine lizards seem the primary prey of *Crotalus pricei.* Wild-caught adults have been found to have consumed both *Sceloporus jarrovi* (Woodin, 1953) and *S. poinsetti* (Armstrong and Murphy, 1979), and captive adults and juveniles kept by Kauffeld (1943b) readily ate both *S. undulatus*, and *Anolis carolinensis*. In at least some populations, nestling birds may also be important prey. In the Chiracahua Mountains, *C. pricei* seems to be an important predator on nestling yellow-eyed juncos, *Junco phaeonotus* (Cumbert and Sullivan, 1990). Klauber (1972) found a mouse in a wild adult, and Wright and Wright (1957) listed both mice and a "shed skin" as foods of captives. Martin (1974) stated that they also eat invertebrates, but this is debatable.

Crotalus pricei usually retains its hold on a lizard following a strike, but strikes, releases, and then trails mice (Cruz et al., 1987). Lizards are usually struck in the thoracic region (Armstrong and Murphy, 1979). Much tongue flicking occurs while the mouse trail is followed. *C. pricei* never waves its tail to lure lizards, as do some other small pit vipers (Kauffeld, 1943b).

Digestion is rapid after a lizard meal, with defecation usually taking place on the second or third day after feeding, and often again on the fifth day (Kauffeld, 1943b).

Venom and bites

Adult *C. pricei* 402–516 mm long that I measured had fangs 2.0–3.3 mm long. Klauber (1972) reported that the body length/fang length and head length/fang length ratios were 155 and 7.5, respectively.

The venom is probably highly toxic, but yields are low; Klauber (1972) reported a dry yield of only 8 mg per adult, and Minton and Weinstein (1984) recorded a yield of only 4.1 mg, per snake. The LD_{50} (mg/kg) for mice is 0.95 for intravenous and 11.39 for subcutaneous injections, and the minimum lethal dose for a 350-g pigeon is 0.2 mg of dried venom (Klauber, 1972). The venom shows no protease activity (Minton and Weinstein, 1984).

Four humans bitten by *C. pricei* experienced both local and systemic symptoms more serious than expected from such a small rattlesnake (Minton and Weinstein, 1984), so one should not be careless around this species.

Predators and defense

Doubtless the juveniles, at least, occasionally fall victim to birds of prey (*Bubo*, *Buteo*), skunks (*Mephitis*), badgers (*Taxidea*), coyotes (*Canis*), and kingsnakes (*Lampropeltis*), but humans probably kill more than all natural predators combined by overcollection and habitat destruction. Potential threats to *C. pricei* in Arizona include mining, grazing, overcollecting, logging, and recreational or other development (Johnson and Mills, (1982).

This is a shy snake, that will quickly crawl to shelter if possible. Its rattle is soft and at times barely audible above accompanying insect noises. Although usually mild mannered, it will strike if provoked.

Populations

Crotalus pricei pricei is probably the most frequently encountered rattlesnake at higher elevations within its range (Armstrong and Murphy, 1979). Several may occur in a small space of only a few square meters (Lowe et al., 1986). It is quite common in the Chiricahuas of southeastern Arizona, but somewhat less so in the four other mountain habitats it occupies in Arizona (Klauber, 1972).

Remarks

Electrophoretic studies of venom proteins by Foote and MacMahon (1977) indicate that *C. pricei* is most closely related to *C. cerastes* and *C. willardi*; however, a study of its dorsal scale microdermatoglyphics by Stille (1987) shows its nearest relatives to be *C. mitchelli* and *C. cerastes*.

Crotalus ruber (Cope, 1892)
Red diamond rattlesnake
Plates 40, 41.

Recognition

This attractive rattlesnake (maximum length, 162 cm) is brick red, reddish gray, or pinkish brown with a series of 30–40 light-bordered, diamond-shaped marks on its back and 2–7 conspicuous white or gray and black rings on its tail. In some individuals the diamonds may be poorly outlined. Two diagonal light stripes are present on the side of the head. The unmarked venter is white to cream. Usually 29(25–33) rows of keeled scales occur at midbody, ventrals total 179–206, and on the underside of the tail are 15–29 subcaudals; the anal plate is entire. Head scalation consists dorsally of a rostral, 2 small internasals, several canthals (prefrontals are absent), 2 supraoculars, and 4–10 intersupraoculars, and laterally of 2 nasals, 1–2 loreals, 1–2 preoculars, 2 postoculars, several suboculars, 14–16(12–19) supralabials, and 17–18(13–21) infralabials. The prenasal touches the supralabials, the postnasal rarely contacts the upper preocular, and the internasals touch the rostral. The first pair of infralabials is unusually transversely divided. The long hemipenis is bifurcate with a divided sulcus spermaticus and about 62 spines and 50 fringes per lobe (Klauber, 1972). One to three spines are present in the crotch between the lobes. Dentition includes a fang

Crotalus ruber ruber (Barry W. Mansell).

on the maxilla, 3 palatine teeth, 8(6–9) pterygoid teeth, and 9–10(8–11) teeth on the dentary.

Males have 179–203 ventrals, 22–29 subcaudals, and 3–7 dark tail rings; females have 188–206 ventrals, 15–25 subcaudals, and 2–5 dark tail rings.

Sibley (1951) reported the maximum length of a *C. ruber* as 190.5 cm, but this has not been confirmed (Klauber, 1972).

Karyotype and fossil record: Unknown.

Distribution

Crotalus ruber occurs from San Bernardino and Los Angeles counties, California, southward through Baja California, Mexico (except the northeastern portion). It also occurs on several islands in the Gulf of California, and on Isla de Santa Margarita in the Pacific Ocean (Campbell and Lamar, 1989).

Geographic variation

Three subspecies are recognized, but only the nominate race, *Crotalus ruber ruber* (Cope, 1982), the red diamond rattlesnake, occurs in the United States.

Distribution of *Crotalus ruber*.

Confusing species

Four other rattlesnakes are sympatric with *Crotalus ruber* in southern California. *C. atrox* is gray to grayish brown, has dark spots in its diamonds, and does not have the first pair of infralabials divided transversely. *C. viridis helleri* has a darker gray or grayish-brown body, narrower tail rings of which the lighter ones are the same color as the trunk, and more than two internasals touching the rostral scale. *C. mitchelli* has numerous black speckles on its body, indistinct tail rings, and the first nasal scale separated from the rostral by small scales. *C. cerastes* is light colored (gray, pink, or tan) and has a prominent hornlike projection over each eye. One other species of *Crotalus* also lives in southern California: *C. scutulatus* is gray to olive or brown, has the dark rings on the tail narrower than the light rings, and has enlarged scales lying between its supraoculars.

Habitat

The red diamond rattlesnake can be found in suitable areas from near the Pacific Ocean to altitudes of about 1,500 m. It most commonly lives in rocky areas or habitats with thick vegetation (desert scrub, thornscrub, cacti, chaparral, and pine-oak woods). Occasionally, it also ventures into cultivated areas and grasslands. *Crotalus ruber* is most common in the western foothills of the Coast Ranges, but also lives in the dry, rocky inland valleys.

Behavior

The earliest and latest dates of annual appearance reported by Wright and Wright (1957) are 20 February and 9 October, but Klauber (1972) and Armstrong (in Armstrong and Murphy, 1979) have collected *Crotalus ruber* during every month. The greatest amount of activity occurs in the spring months from March to June, when it can be found abroad during the day or at dusk, especially on warm days following rains. In the hottest months it seems almost entirely nocturnal; Klauber (in Wright and Wright, 1957) found active *C. ruber* at 1840, 1955, 2220, 2330, and 0325 hours at air temperatures of 18–30°C. The cloacal temperature of an active snake recorded by Brattstrom (1965) was 24°C. Summer daily retreats include rock crevices, animal burrows, and brush piles. In the winter, cool temperatures bring on lethargy, and these snakes spend much time underground in rock crevices and animal burrows. Several may congregate at such sites, and Klauber (1972) mentioned one such den where 24 were blasted out of a rock crevice during road construction.

Crotalus ruber is an accomplished climber, and has often been seen in above-ground situations. Klauber (1972) found it in a small bush 46 cm high, and on the tops of prickly pear cacti. He also notes cases where it has been seen 75–90 cm off the ground stretched along the tops of buckwheat (*Eriogonum*) bushes, at least 1.8 m high on the limb of a sumac bush, and 1.5 m high on the top branches of a cactus.

The red diamond rattlesnake is not adverse to entering water, and is an accomplished swimmer. Klauber (1972) saw them swimming in reservoirs in San Diego County, California, and was told that they occasionally swim in the Pacific Ocean.

Male *C. ruber* often participate in bouts of dominance combat, and its behavior during these bouts was one of the first sequences recorded in rattlesnakes. During these encounters the two males face each other with one lying on top of the other. The heads and necks are raised and sway from side to side while the tongues are continually flicked in and out. One male rises above the other and bends his head toward that of the other snake until the heads touch. The head of the lower male may be raised only 3–4 cm above his body, but he usually also elevates the anterior portion of his body 30–40 cm. The higher positioned male rises at the same time to maintain his superior position. The two snakes then push their raised bodies and necks against each other. The snakes may separate, then rejoin to repeat the process until one male pushes the other to the ground and attempts to pin it there. At this point the pinned male usually crawls away (Shaw, 1948a).

Reproduction

The size at which sexual maturity is first attained has been inadequately determined for *Crotalus ruber*. Wright and Wright (1957) gave 60 cm as the shortest adult length, but the smallest gravid female examined by Klauber

(1972) was 73.3 cm. These would be one-year-old animals (see below).

Wild *C. ruber* have been seen courting or mating from early March into May; under captive conditions mating may occur at almost any time during the year but most often takes place from February to June. Copulations by captives witnessed by Perkins (in Klauber, 1972) lasted from more than two hours to almost 23 hours. Nudging, examination with the tongue, and spastic jerking may all be displayed by the male while courting, and during coitus slight pulsations of his tail may be evident (pers. obs.). During a lengthy mating the male may change body positions several times, and the attached pair may move about, one crawling and dragging the other along with it.

The ovoviviparous young are born in August and September after an incubation period of 141–190 days. Broods may contain 3–20 young (Klauber, 1972), but 8–10 are most common. Newborn young are grayer than adults, and average 310–315(299–350) mm in total length.

Crotalus ruber is known to have hybridized with *C. viridis helleri* both in the wild and captivity (Klauber, 1972; Armstrong and Murphy, 1979).

Growth and longevity

Klauber (1972) analyzed the growth patterns of 249 *C. r. ruber* from San Diego County, California. Neonates, averaging about 300 mm in total length, appeared in September, had grown to 350 mm by mid-October, and showed further growth to about 390 mm with a button rattle by November. After emerging from hibernation in March, the snakes grew to nearly 500 mm and two rattle segments by mid-April. Toward the end of April, three rattles were most common and some had grown to 600 mm (few were shorter than 450 mm). Snakes were 450–650 mm long in May. The first four rattle strings appeared in June when snakes were 460–700 mm long. In July, *C. ruber* ranged from 500 to 700 mm and some had five rattles. By September, the yearlings had reached 670 mm, more than double the birth lengths. Klauber thought two-year-old snakes probably averaged about 870 mm with seven to eight rattles (sexual dimorphism was evident), and three-year-old males were about 940 mm and females 840 mm. By the next spring an additional 25 mm had been added to the length and the tail had as many as 9–11 rattles. Males exceeded females in length by about 10%.

Crotalus ruber does well in captivity, and one individual lived 14 years and 6 months at the San Diego Zoo (Shaw and Campbell, 1974).

Food and feeding

Crotalus ruber is an active hunter, usually ambushing small mammals, although other animals may also be taken. Of 57 snakes containing prey examined by Klauber (1972), 53 had eaten mammals, three had lizards, and one, a bird. Mammals taken include rabbits (Sylvilagus), ground squirrels (Ammospermophilus, Spermophilus), kangaroo rats (Dipodomys), wood rats (Neotoma), mice (Mus, Peromyscus, Microtus), and the spotted skunk (Spilogale) (Tevis, 1943; Klauber, 1972). Klauber (1972) found the whiptail lizard, Cnemidophorus tigris, in subadults and a "sumac leaf" (probably accidentally ingested) in another, and reported that a captive ate a Crotalus viridis helleri.

Young snakes feed on mice and lizards, adults on larger prey. Most animals eaten are first struck, later followed by odor trail, and then swallowed after they have died. However, carrion is consumed (Cowles and Phelan, 1958). Cunningham (1959) fed a Microtus dead two days to a captive, and Patten and Banta (1980) reported the taking of a road-killed Dipodomys by a wild C. ruber. Although Hammerson (1981) thought that the snake had probably first bitten the kangaroo rat, Patten (1981) provided an additional description of its condition which seems to indicate that it was a roadkill. Zoo captives readily take prekilled white mice (Mus musculus), so why should not wild snakes eat dead rodents?

Venom and bites

This large snake has relatively long fangs; specimens 90–130 cm long measured by Klauber (1939) had fangs 9.5–12.9 mm. The venom is usually injected during the bite, but Klauber (1972) had a C. ruber strike at him when he was beyond range and received a spray of venom or saliva droplets. This snake had previously injured its jaw slightly, and

Klauber thought that the injury may have caused the juice to accumulate on the snake's lower jaw, which was then thrown onto him by the force of the lunge.

The venom is normally colored apricot-yellow (Klauber, 1972). It seems relatively low in toxicity when compared with some other species of Crotalus (Minton, 1956), but the red diamond's large size allows a greater volume of venom to be injected during a strike, and this snake must be considered potentially very dangerous. It is capable of killing a human (Shaw and Campbell, 1974). The lethal venom dose for a human is about 100 mg and the normal total yield 150–350 mg (Dowling, 1975). A typical adult C. ruber secretes 0.72 ml (0.24 g dried) of venom in a strike, and an exceptional one may inject 1.65 ml (0.55 g dried; Amaral, 1928). The average dried venom yield of an adult is 364 mg, and the maximum, 668 mg (Klauber, 1972). The larger the snake, the greater the venom yield: Klauber (1972) reported the following high liquid yields by snake length; 60 cm (0.24 ml), 70 cm (0.43 ml), 80 cm (0.70 ml), 90 cm (1.07 ml), 1 m (1.57 ml), 1.1 m (2.20 ml), and 1.2 m (3.04 ml). The minimum lethal dose for a 350-g pigeon is 0.12–0.40 mg (Klauber, 1972), for 22-g mice 0.11 mg (Macht, 1937), and for 20-g mice 0.08 mg (Githens and Wolff, 1939). The lethal venom dose to cats was decreased only slightly after 26–27 years of storage; a complete neuromuscular block of the diaphragm was produced in 22 minutes (Russell et al., 1960). Proteolytic activity is 6–15 times greater in adult venom than in that from juveniles (MacKessy, 1985).

Case histories and symptoms of human envenomation by C. ruber are presented in Clarke (1961), Lyons (1971), and Klauber (1972); pain, swelling, and discoloration are early symptoms. Lyons (1971) mentioned bloody diarrhea and a dramatic decrease in blood platelets over a four-day period, which then took two weeks to recover. He attributed the loss of platelets to destruction in the blood, as the bone marrow production cells did not seem affected. Clarke (1961) was bitten by an adult male C. r. ruber (132 cm, 2.5 kg) through a sack and his trousers; both fangs were imbedded below the knee. The bite occurred at 1345; a tourniquet was applied by 1348 and a transverse incision made over the fang marks by

1350. The victim was taken to a hospital and there received 30 cc of antivenin between 1420 and 1845. An additional 10 cc was administered at 2215 the next day. Swelling of the leg was great and pain was intense until the third day. Blood pressure was high during the first 24 hours after the bite but was lower thereafter. Body temperature rose slightly. Antibiotics were administered to guard against infection. Clarke was released from the hospital nine days later, and was able to walk almost normally in two weeks.

Predators and defense

Kingsnakes (*Lampropeltis*) and other serpent-eating snakes probably take juveniles and small adults. Birds of prey (*Bubo, Buteo*) and carnivorous mammals, such as coyotes (*Canis*), badgers (*Taxidea*) and skunks (*Mephitis*), are also major predators on the young, but adult *C. ruber* have few enemies other than humans. The automobile and rifle have taken their toll, but habitat destruction probably has caused the greatest loss.

Crotalus ruber is apparently susceptible to the venom of some other rattlesnakes. Shaw and Campbell (1974) told of a pet red diamond rattlesnake dying after being bitten by a new-caught sidewinder, *C. cerastes*, placed in its aquarium.

Crotalus ruber is mild mannered for such a large snake. At times it will lie quietly without even rattling when closely approached. It may sometimes even be caught without attempting to bite, but do not let this fool you. It can bite at the least expected moment, and if cornered or provoked some individuals will put up a spirited fight, coiling, rising up, and striking. However, such behavior seems rare in this snake, and we must agree with Wright and Wright (1959) that "it is a handsome snake which has caused little grief."

Populations

No quantitative data on the population dynamics of *C. ruber* have been published. Locally, it may be a very common snake, especially around a hibernaculum. Of over 12,000 snakes collected or observed from 1923 to 1938 in San Diego County by Klauber (1972), *C. ruber* made up 1.1–10.6% in any zone, but was most numerous in the inland valleys (10.6%), desert foothills (9.1%), foothills (8.3%) and along the coast (6.6%). It constituted 7.8% of all snakes.

Remarks

Gloyd (1940) thought *Crotalus ruber* to be closely related to the other two diamondback rattlesnakes, *C. atox* and *C. adamanteus*, and the Mexican species, *C. exsul* and *C. tortugensis*. Studies on venom proteins by Foote and MacMahon (1977) support its closeness with *atrox* and *adamanteus* (*exsul* and *tortugensis* were not tested).

Crotalus scutulatus (Kennicott, 1861)
Mojave rattlesnake
Plates 42, 43

Recognition:

This species (maximum length, 129 cm) is probably our most dangerous rattlesnake, as its venom is largely neurotoxic. It is greenish gray, olive, yellowish green, or brown with 27–44 dark-gray or brown, light-bordered, diamond-shaped or oval to hexagonal dorsal blotches. The light scales separating these blotches are usually unmarked with dark pigment. A dark, light-bordered stripe extends from the orbit downward to the corner of the mouth, and a pair of dark blotches is usually present on the occiput. Alternating light-gray and dark rings (2–8) cross the base of the tail. The darker rings are narrower than the light ones. The venter is white to cream with only a slight amount of pigment encroaching along the sides of the ventral scutes, and some small dark spots under the tail. Approximately 25(21–29) rows of keeled scales occur at midbody. Beneath are 165–192 ventrals, 15–29 subcaudals, and an undivided anal plate. The rostral is usually higher than wide, and in contact with two small internasals. No prefrontals are present, but 6–21 scales lie between the internasals and intersupraocular scales. Normally only two rows of intersupraoculars lie between the large supraocular scales, but sometimes three are present (those most posterior are smaller). Two nasal scales are present; the prenasal touches the first supralabial, but the postnasal rarely contacts the upper preocular. Usually, 1 loreal, 2 preoculars, 2 postoculars, several suboculars, 14–15(12–18) supralabials and 15–16(12–18) infralabials are present on each side of the head. Lobal adornments on the bifurcate hemipenis consist of approximately 49 spines and 40 fringes per lobe, and 0–2 spines in the crotch; the sulcus spermaticus is divided

(Klauber, 1972). In addition to the maxillary fang, 3 palatine, 7(6–8) pterygoidal, and 9–10 dentary teeth are present.

Males have 165–190 ventrals, 21–29 subcaudals, and 5(3–8) dark tail rings; females have 167–192 ventrals, 15–25 subcaudals, and 3–4(2–6) dark tail rings.

Karyotype

As in other *Crotalus* species, the Mojave rattlesnake has a total of 36 diploid chromosomes: 16 macrochromosomes (including 4 metacentric, 6 submetacentric, and 4 subtelocentric), and 20 microchromosomes. The pair of macrochromosomes which determine sex are either ZW (Z metacentric; W submetacentric) in females or ZZ in males (Baker et al., 1972; Zimmerman and Kilpatrick, 1973).

Fossil record

Pleistocene (Rancholabrean) fossils of *Crotalus scutulatus* have been found in Distrito de Zumpango, México (Brattstrom, 1954a, 1955) and in Deadman Cave, southern Arizona (Mead et al., 1984).

Distribution

Crotalus scutulatus ranges from southwestern Washington County, Utah, Lincoln and Clark counties, Nevada, and Kern, Los Angeles, and San Bernardino counties, California, southeast through Arizona and southward from Trans-Pecos, Texas, and southwestern Hildalgo and Otero counties, New Mexico, to the southern edge of the Mexican Plateau in Puebla and adjacent Veracruz.

Distribution of *Crotalus scutulatus*.

Geographic variation

Two subspecies have been described, but only *Crotalus scutulatus scutulatus* (Kennicott, 1861), the Mojave rattlesnake (described above), occurs in the United States. Several different venom populations occur in Arizona (see Venom and bites).

Confusing species

Crotalus atrox has wider tail rings (three or more scales wide to two or less in *C. scutulatus*), and the posterior white border to the dark eyestripe extends to the corner of the mouth (it ends just above or bends and extends posteriorly before reaching the corner of the mouth in *scutulatus*). *C. viridis* has more than two internasals, and its tail rings are poorly developed.

Habitat

In the United States, the Mojave rattlesnake is primarily a lowland form living in deserts or dry brushy grasslands where the typical vegetation is palo verde, mesquite, creosote, or cacti. It has been found from near sea level to elevations of 2,072 m (in Arizona), and seems to prefer rocky foothills.

Behavior

Very little is known of the habits of the Mojave rattlesnake. It has been observed above ground from 19 March (Woodbury and Hardy, 1947) to November (Gates, 1957), but most activity probably takes place from late April or early May through September. May, July, and August seem to be the months of peak activity. In Chihuahua, *Crotalus scutulatus* is most active in July and August, but particularly in August (Reynolds, 1982). Precipitation is greatest there in July, but most snakes are taken about one month after the peak in rainfall.

Although primarily nocturnal in the summer months (1959–2337 hours; Klauber, in Stebbins, 1954), early morning and late afternoon foraging occurs in the spring and fall, and at higher elevations *C. scutulatus* may be abroad during the day even in summer.

Much time is spent in animal burrows or under rocks, and Wright and Wright (1957) reported that wild honeybees are often found in the same hole (double trouble for the would-be predator or herpetologist). Such burrows help the snake in its thermoregulation by providing a cooler haunt during the hot, summer days and a warmer retreat when the air is cool in the spring and fall. The snakes may hibernate in these if the burrows are deep enough, but if too shallow, they must seek retreats below the frost line. One was found in an Arizona cave on 6 June (Gates, 1957).

Crotalus scutulatus frequently basks in the early morning, often on or along the side of a road. This is particularly true after they have fed the night before.

Brattstrom (1965) recorded four body temperatures of 22.2–34.0°C (mean 30.0°) from wild *C. scutulatus*, and Klauber (1972) found them active at air temperatures of 17–27°C. Pough (1966) reported the average body temperature recorded from 19 *C. scutulatus* was 26.5°C (21.9–29.8), while Brattstrom reported a critical maximum body temperature of 42°C.

A possible incident of climbing by this snake was reported by Boone (1937) who reported that a "yellow Pacific rattlesnake" (thought to have been *C. scutulatus* by Klauber, 1972) climbed 1.5 m into a mesquite tree.

When placed in water, the Mojave rattlesnake is a good swimmer, moving by lateral undulations with its head held high out of the water (Klauber, 1972).

Reproduction

Only a few anecdotal reports have been published on the reproductive biology of *Crotalus scutulatus*. The female sexual cycle is practically unknown, although Minton (1958) did find a gravid female on 4 July.

In Chihuahua, Mexico, male testicular activity peaks in August when the seminiferous tubule diameter is significantly larger than in other months. Testicular length does not vary significantly from month to month; testicular mass does, however, peaking in September. Spermiogenesis occurs from July through Au-

gust, and mature sperm are present in the accessory ducts throughout the summer (Jacob et al., 1987).

No descriptions of natural courtship or mating have been published. During August 1975, Jacob et al. (1987) found a heterosexual pair of *C. scutulatus* lying on a road. Soon a second male emerged onto the highway directly in line with the original pair and less than 2 m behind; however, no copulation took place. A courtship between a male *C. scutulatus* and a female *C. durissus* at the San Diego Zoo took place in October and concluded in a copulation on 9 November (Perkins, 1951). The five young from this union were born 170 days later on 28 April, and later proved fertile (Klauber, 1972). Another captive hybridization between a female *C. scutulatus* and a male *C. viridis* resulted from a May mating, with the 12 young born in October (Klauber, 1972). A June captive mating between a male *C. cerastes* and a female *C. scutulatus* produced four young on 25 November (Powell et al., 1990).

In the wild, parturition takes place within a period extending from July to September (Gloyd, 1937; Gates, 1957; McCoy, 1961; Stebbins, 1985), but most newborn young have been found in mid-August. Litters contain 2 (Stebbins, 1985) to 13 (Klauber, 1972) young (average for 22 litters, 7.8). Neonates are 220–283 mm in total length (average, 256.5 mm).

A natural hybridization between *C. scutulatus* and *C. viridis* from Hudspeth County, Texas, was reported by Murphy and Crabtree (1988), and, in addition to the three captive matings mentioned above that resulted in hybrid young, Cook (1955) noted another successful captive cross between *C. scutulatus* and *C. viridis*. Venom characteristics indicate that *C. scutulatus* and *C. viridis* regularly hybridize in New Mexico (Glenn and Straight, 1990). Jacob (1977) concluded after an intensive investigation that it was highly improbable that *C. scutulatus* and *C. atrox* crossbred in nature, but did relate an occurrence of hybridization between the species in captivity. This is interesting, as Kauffeld (1957) reported antagonistic behavior by *C. atrox* toward *C. scutulatus*. When placed in a bag containing a Mojave rattlesnake, the *C. atrox* seemed extremely frightened and tried vigorously to escape, as if it had met with a dire enemy at close quarters. The Mojave paid it no attention, but the diamondback butted it smartly with its coils in a surprisingly violent display.

Growth and longevity

No growth data is available, but a female Mojave rattlesnake lived 13 years, 19 days in captivity (Bowler, 1977).

Food and feeding

Crotalus scutulatus feeds mainly on mammalian prey. Klauber (1972) found mammal remains in 21 specimens and lizard remains in only two, and Reynolds and Scott (1982) found mammals in 91.7% of the 48 Mojave rattesnakes they examined. In the study by Reynolds and Scott, kangaroo rats (*Dipodomys*) occurred in 39.6% of the stomachs containing food items, pocket mice (*Perognathus*) in 20.8%, white-footed mice (*Peromyscus*) in 16.5%, and ground squirrels (*Spermophilus*) in 10.4%; other mammals taken were the cottontail (*Sylvilagus*), 2.1%, and jackrabbit (*Lepus*), 4.2%. Captives have eaten house mice (*Mus*) and wood rats (*Neotoma*) (Kauffeld, 1943a; Vorhies and Taylor, 1940). Reptile prey includes whiptail lizards (*Cnemidophorus*) and leaf-nosed snakes (*Phyllorhynchus*) (Kauffeld, 1943a; Johnson et al., 1948; Klauber, 1972). Spadefoot toads (*Scaphiopus*) are also taken (Cromwell, 1982), and the Mojave rattler may occasionally eat bird eggs (Boone, 1937). The remains of a centipede and what appeared to be insects have also been found in a Mojave rattlesnake (Klauber, 1972).

Prey is selected on the basis of size; that either too large or too small is rejected, and potential prey that could possibly harm the snake is not accepted (Reynolds and Scott, 1982). Mammals bitten are typically released immediately and allowed to wander off until dying from the venom. The snake follows the scent trail of the animal, and after examining the body with its tongue to determine if it has died, eventually swallows its prey. During trailing the tongue is frequently flicked out, and as the snake nears the stricken mammal, the tongue-flick rate increases. Prey are taken from ambush or by actively foraging; infrared detection, olfaction, and sight all seem important in prey capture.

Venom and bites

For its size, *Crotalus scutulatus* does not have particularly long fangs; seven individuals 81.2–108.5 cm long had fangs only 6.7–8.8 mm long (Klauber, 1939). Long fangs are probably not needed, however, as its neurotoxic venom is extremely virulent.

Venom from the Mojave rattlesnake affects the heart, skeletal muscles, and neuromuscular junctions. The lethal portion of the venom appears to be an acidic protein which has been shown to block transmission of impulses from nerve to muscle (Castilonia et al., 1980). This toxin has been isolated from *C. scutulatus* and termed Mojave toxin. The complete amino acid sequence of the basic subunit of this toxin is now known (Aird et al., 1990), and it is very similar to the basic subunits of related toxins from the venom of South American *C. durissus* and North American *C. viridis concolor*. However, neurolytic Mojave toxin is not the only virulent one present (Weinstein et al., 1985; Martinez et al., 1990). A fibrinogenolytic hemorrhagic toxin has also been isolated from *C. scutulatus* by Martinez et al. (1990). Also, concentrations of protein, and thus presumably of these different toxins, vary between individual snakes (Johnson, Stahnke et al., 1968) so that in some cases smaller *C. scutulatus* may have more potent venom than larger snakes. However, larger snakes with less potent venom can be just as deadly because of their potential of injecting greater amounts of venom per bite. The seriousness of the envenomation is dependent upon both the quantity and quality of the venom injected. Surprisingly, however, protease activity of this venom is relatively low (Mackessy, 1988).

Studies of the venom yield and toxicity in *C. scutulatus* have produced differing results, but all agree that the venom injected in one bite is sufficient to cause the death of its natural prey as well as humans. The venom seems to be ten times as toxic as any other North American crotalid snake (Minton, 1958; Minton and Minton, 1969). Several *C. scutulatus* milked of their venom at the San Diego Zoo produced an average yield in dried venom of 77 mg per fresh adult, with a maximum of 141 mg (Klauber, 1972), and Minton and Minton (1969) reported an average venom yield of 50–90 mg per adult. The minimum lethal dose for a 350-g pigeon is 0.05 mg (Githens and George, 1931). It takes slightly more to kill a mouse: 0.24 mg for a 22-g mouse (Macht, 1937). The LD_{50} for mice is 0.31 mg/kg (Minton, 1958). For humans the story is no better: the estimated dry venom dosage needed to kill an adult human is probably only 10–15 mg (Minton and Minton, 1969).

Electrophoretic studies have revealed variation in the Mojave toxin from the different populations in the Big Bend, Texas, and in southeastern Arizona (Rael et al., 1984). This suggests that several genetically diverse groups occur within *C. s. scutulatus*. Further testing has shown that two distinct venom populations and a zone of intergradation occur in Arizona (Glenn et al., 1983; Glenn and Straight, 1989; Wilkinson et al., 1991). The venom of the population from southeast to northwest Arizona (venom A) contains the toxin "Mojave toxin" and is lacking in hemorrhagic and specific proteolytic activities. The other population (venom B), from central Arizona from east and northeast of Phoenix south to near Tucson, does not contain Mojave toxin but does produce hemorrhagic and proteolytic activities (presumably it contains more of the hemorrhagic toxin discovered by Martinez et al., 1990). This population overlaps that with venom A to the north, west, and south of this region, producing individuals with an intergrade (A + B) venom. The venoms of *C. scutulatus* from regions between the venom A and venom B populations in Arizona were examined for the presence of Mojave toxin by immunochemical assay, lethality by mouse intraperitoneal LD_{50} proteolytic activity, and hemorrhagic activity in mice. Venom protein consituents were analyzed using reverse-phase high-pressure liquid chromotography. Seven venoms contained both the Mojave toxin of venom A and the proteolytic and hemorrhagic activities of venom B. The intraperitoneal LD_{50} values of the A + B venoms were 0.4–2.6 mg/kg, compared to 0.2–0.5 mg/kg for venom A individuals and 2.1–5.3 mg/kg for the venom B individuals. High-pressure liquid chromotography illustrated that the A + B venoms exhibited a combined protein profile of venom A and venom B. These data indicate that an intergrade zone exists between the two venom types which arcs around the western and southern regions of the venom B population.

Within these regions, three major venom types can occur in *C. scutulatus*.

Complete case histories involving human envenomation are rare, but the results of a bite by *C. scutulatus* are usually severe with death occurring in a high percentage of those untreated (Hardy, 1983, 1986; Russell, 1967b). Many of the bites are by captive snakes, and in Arizona most bites by *C. scutulatus* seem to occur in this manner (Hardy, 1983, 1986). Of fatal snake bites recorded in Arizona, those by *C. scutulatus* and *C. atrox* make up the greatest percentage. Another contributing factor in Arizona is that the Mojave rattlesnake is often very common in some heavily populated areas. Dr. Frederick Shannon wrote to Klauber (1972) that in one case of *C. scutulatus* bite the victim developed double vision and difficulty in speech and in swallowing water, classic symptoms of neurotoxic envenomation (a morbid sidenote on this report is that Dr. Shannon himself was later to succumb to the bite of a *C. scutulatus* he was attempting to capture in the wild). Other symptoms of envenomation by the Mojave rattlesnake include local pain and swelling (edema), blood blisters, ecchymosis, necrosis, fragility of veins, cardiopulmonary arrest, elevated heart rate, lowered blood pressure, decrease in blood platelets and fibrinogen levels, increase in fibrinolytic split products, shock, renal failure, drooping eyelids, depression, and diarrhea (Hardy, 1983, 1986; Smith, 1990).

Predators and defense

Nothing has been reported on natural predators of this snake, but humans slaughter many each year, particularly on the highways.

Although *Crotalus scutulatus* usually calms down readily after a short period in captivity, in my experience, some wild individuals are extremely nervous, excitable and aggressive (Klauber, 1972, thought them more peaceful than *C. atrox*, but both are highly reactive). If not allowed to escape, it will coil, rise up, continuously rattle, and strike viciously. Occasionally, this snake will even advance on the disturber. The Mexican subspecies *C. s. salvini* may flatten its neck in a hoodlike display (Glenn and Lawler, 1987) or even its head and trunk (Armstrong and Murphy, 1979). Armstrong and Murphy (1979) thought *C.*

scutulatus one of the most aggressive species they had encountered. Several struck so violently that their entire body became momentarily airborne. It is also common for them to strike through collecting bags, so one must be very careful. Keeping one of these potent snakes as a pet is taking an uncalled-for risk!

A curious instance of defensive behavior was related by Kauffeld (1969) of a Mojave rattler that had apparently learned to associate a certain coat with its owner, who invariably wore the coat when working with this snake. One day he did not wear the coat, and the snake, which had always been exceptionally quiet and phlegmatic, rattled violently and coiled defensively. However, when its owner put on the coat the snake quieted down. The reaction was so consistent in repeated trials that the man was convinced that the coat made the difference.

Populations

No serious studies have been conducted on the relative density or population dynamics of *Crotalus scutulatus*, but it appears to be at least locally very common in some areas of Arizona. It is probably the most common snake around Wickenburg (Gates, 1957). Of 425 rattlesnakes collected in central Arizona, 147 (34.6%) were *C. scutulatus;* it was surpassed in total numbers captured only by *C. atrox* (Klauber, 1972).

Remarks

Gloyd (1940) thought *C. scutulatus* close to the ancestral stock, and somewhat intermediate to the lines of North American rattlesnake evolution that produced the *C. viridis* and *C. atrox* groups (but somewhat closer to *viridis*). Electrophoretic studies of venom proteins by Foote and MacMahon (1977) have not upheld this relationship, as they found it closest to *C. tigris, C. molossus,* and *C. horridus,* but Glenn and Straight (1990) have reported that venom toxins of *C. viridis* from southwestern New Mexico showed more similarity to Mojave toxin of *C. scutulatus* than to similar toxins from *C. v. concolor,* indicating hybridization between the two species in that region.

Campbell and Lamar (1989) were reluctant to use the common name Mojave rattlesnake, since proportionally very little of the species range lies within the Mojave Desert.

With the rapid urban expansion of some cities in Arizona where this snake is common, it will come into more frequent contact with humans, bringing about a potentially dangerous scenario, particularly regarding children and pets. Our poor knowledge of its ecological requirements does not help, and an intensive study of its natural history is needed.

Crotalus tigris (Kennicott, in Baird, 1859)
Tiger rattlesnake
Plates 44, 45

Recognition

This is a small-headed (body length at least 25 times longer than head length), short (maximum length, 90 cm) rattlesnake with a proportionately large rattle. The body color ranges from gray or lavender to pink, yellowish brown, or orange. A series of 35 to 52 faint, irregularly shaped gray, olive, or brown bands crosses the back. No other species of *Crotalus* has crossbands on the anterior portion of the body. The posteriormost bands are often the darkest. The 4 to 10 tail rings are indistinct. The venter is cream to greenish brown with gray, lavender, or brown spots on each scute. Head markings are poorly developed, but a dark cheek stripe is usually present. At midbody are 23(20–28) rows of keeled scales; on the underside are 156–183 ventrals and 16–27 subcaudals; the anal plate is undivided. Enlarged dorsal head scales include a wider-than-high rostral, 2 small internasals, 4 canthals, 2 supraoculars, and 3–8 intersupraoculars. Prefrontals are absent. On each side of the head are 2 nasals, 1–2 loreals, 2–3 preoculars, 1 (occasionally 2) postocular, several suboculars, usually 12–14(11–16) supralabials, and usually 13–15(11–16) infralabials. The

Crotalus tigris (John H. Tashjian, San Diego Zoo).

149

Crotalus tigris (Richard D. Bartlett).

prenasal usually touches the first supralabial, but the postnasal rarely contacts the upper preocular. The shortened hemipenis is bifurcate with a divided sulcus spermaticus. Each lobe contains about 64 spines and 40 fringes, but only 1 or 2 spines occur in the crotch. Dentition includes a maxillary fang, 3 palatine teeth, 7–9 pterygoid teeth, and 9–10 teeth on the dentary.

Males have 156–172 ventrals, 23–27 subcaudals, and 6–10 dark tail rings; females have 164–183 ventrals, 16–27 subcaudals, and 4–7 dark rings on the tail.

Karyotype and fossil record: Unknown.

Distribution

Crotalus tigris occurs in isolated populations from south-central Arizona southward through almost all of Sonora, Mexico, and has also been found on Isla Tiburón in the Gulf of California. Fowlie (1965) thought that the present isolated foothill distribution may have resulted from the displacement of the tiger rattlesnake in the intervening areas by *C. atrox* and *C. cerastes*.

Geographic variation

Although differences in ground color and patterns occur within populations, no subspecies are recognized.

Confusing species

Crotalus mitchelli has several small scales separating the rostral and prenasals, and a large head. *C. viridis* is blotched rather than banded, and has a distinct light stripe on the side of the face. *C. molossus* has a black or dark-brown tail and large prefrontals that touch medially. *C. atrox* and *C. scutulatus* have light, diagonal stripes extending backward and down from in front and in back of the eye. *C. cerastes* has hornlike projections over its eyes.

Habitat

The tiger rattlesnake is found exclusively in rocky canyons, ravines, and hillsides in deserts or mesquite grasslands at elevations to 1465 m. Typical plants in its microhabitat are creosote, various cacti (particularly saguaro and ocotillo), paloverde, and mesquite.

Crotalus tigris (Kennicott, in Baird, 1859)
Tiger rattlesnake
Plates 44, 45

Recognition

This is a small-headed (body length at least 25 times longer than head length), short (maximum length, 90 cm) rattlesnake with a proportionately large rattle. The body color ranges from gray or lavender to pink, yellowish brown, or orange. A series of 35 to 52 faint, irregularly shaped gray, olive, or brown bands crosses the back. No other species of *Crotalus* has crossbands on the anterior portion of the body. The posteriormost bands are often the darkest. The 4 to 10 tail rings are indistinct. The venter is cream to greenish brown with gray, lavender, or brown spots on each scute. Head markings are poorly developed, but a dark cheek stripe is usually present. At midbody are 23(20–28) rows of keeled scales; on the underside are 156–183 ventrals and 16–27 subcaudals; the anal plate is undivided. Enlarged dorsal head scales include a wider-than-high rostral, 2 small internasals, 4 canthals, 2 supraoculars, and 3–8 intersupraoculars. Prefrontals are absent. On each side of the head are 2 nasals, 1–2 loreals, 2–3 preoculars, 1 (occasionally 2) postocular, several suboculars, usually 12–14(11–16) supralabials, and usually 13–15(11–16) infralabials. The

Crotalus tigris (John H. Tashjian, San Diego Zoo).

149

Plate 3 (opposite top). Sonoran coral snake,
Micruroides euryxanthus euryxanthus, Cochise
County, Arizona. Photo by Steve W. Gotte.
Plate 4 (opposite bottom). Sonoran coral snake,
Micruroides euryxanthus euryxanthus. Arizona-
Sonora Desert Museum. Photo by John H.
Tashjian.
Plate 5 (above). Eastern harlequin coral snake,
Micrurus fulvius fulvius. Savannah River Ecology
Laboratory. Photo by Carl H. Ernst.

Plate 6 (above). Eastern harlequin coral snake, *Micrurus fulvius fulvius*, Lee County, Florida. Photo by Richard D. Bartlett.

Plate 7 (opposite top). Texas harlequin coral snake, *Micrurus fulvius tener*. California Academy of Sciences. Photo by John H. Tashjian.

Plate 8 (opposite center). Yellow-bellied sea snake, *Pelamis platurus*. George V. Pickwell collection. Photo by John Sneed.

Plate 9 (opposite bottom). Yellow-bellied sea snake, *Pelamis platurus*. George V. Pickwell collection. Photo by John Sneed.

Plate 10 (above). Southern copperhead,
Agkistrodon contortrix contortrix, Columbus County,
North Carolina. Photo by Carl H. Ernst.
Plate 11 (opposite top). Broad-banded
copperhead, *Agkistrodon contortrix lacticinctus*.
California Academy of Sciences. Photo by John H.
Tashjian.
Plate 12 (opposite bottom). Northern copperhead,
Agkistrodon contortrix mokasen, Kentucky. Photo by
Roger W. Barbour.

Plate 13 (below). Osage copperhead, *Agkistrodon contortrix phaeogaster*, Douglas County, Kansas. University of Kansas collection. Photo by John H. Tashjian.

Plate 14 (opposite top). Trans-Pecos copperhead, *Agkistrodon contortrix pictigaster*, western Texas. Photo by Richard D. Bartlett.

Plate 15 (opposite bottom). Eastern cottonmouth, *Agkistrodon piscivorus piscivorus*, Beaufort County, North Carolina. Photo by Carl H. Ernst.

Plate 16 (opposite). Florida cottonmouth, *Agkistrodon piscivorus conanti*, Duval County, Florida. Photo by Barry W. Mansell.
Plate 17 (below). Western cottonmouth, *Agkistrodon piscivorus leucostoma*, Kentucky. Photo by Roger W. Barbour.

Plate 18 (opposite top). Eastern massasauga, *Sistrurus catenatus catenatus*. Dallas Zoo. Photo by John H. Tashjian.

Plate 19 (opposite center). Desert massasauga, *Sistrurus catenatus edwardsi*. Chuck Hanson collection, Tucson, Arizona. Photo by John H. Tashjian.

Plate 20 (opposite bottom). Western massasauga, *Sistrurus catenatus tergeminus*. Photo by Barry W. Mansell.

Plate 21 (directly below). Carolina pigmy rattlesnake, *Sistrurus miliarius miliarius*, gray phase. Ray Folsom collection, Hermosa Beach, California. Photo by John H. Tashjian.

Plate 22 (bottom). Carolina pigmy rattlesnake, *Sistrurus miliarius miliarius*, red phase, Hyde County, North Carolina. Photo by Richard D. Bartlett.

Plate 23 (opposite top). Dusky pigmy rattlesnake, *Sistrurus miliarius barbouri*, Florida. Ray Folsom collection, Hermosa Beach, California. Photo by John H. Tashjian.

Plate 24 (opposite bottom). Western pigmy rattlesnake, *Sistrurus miliarius streckeri*, Land Between the Lakes, western Kentucky. Photo by Roger W. Barbour.

Plate 25 (below). Eastern diamondback rattlesnake, *Crotalus adamanteus*, Collier County, Florida. Photo by Roger W. Barbour.

Plate 26 (opposite top). Western diamondback rattlesnake, *Crotalus atrox*, Texas. Photo by Roger W. Barbour.

Plate 27 (opposite bottom). Mojave sidewinder, *Crotalus cerastes cerastes*, San Diego County, California. San Diego Zoo. Photo by John H. Tashjian.

Plate 28 (top of page). Southern sidewinder, *Crotalus cerastes cercobombus*. Pet Corral, Red Rock, Arizona. Photo by John H. Tashjian.

Plate 29 (directly above). Colorado Desert sidewinder, *Crotalus cerastes laterorepens*. Staten Island Zoo. Photo by John H. Tashjian.

Plate 30 (left). Timber rattlesnake, *Crotalus horridus horridus*, yellow phase, eastern Kentucky. Photo by Roger W. Barbour.

Plate 31 (right). Timber rattlesnake, *Crotalus horridus horridus*, dark phase, Menifee County, Kentucky. Photo by Roger W. Barbour.

Plate 32 (opposite top). Canebrake rattlesnake, *Crotalus horridus artricaudatus*. San Diego Zoo. Photo by John H. Tashjian.

Plate 33 (opposite center). Mottled rock rattlesnake, *Crotalus lepidus lepidus*, Valverde County, Texas. Photo by Richard D. Bartlett.

Plate 34 (opposite bottom). Banded rock rattlesnake, *Crotalus lepidus klauberi*. Photo by Barry W. Mansell.

Plate 35 (below). Southwestern speckled rattlesnake, *Crotalus mitchelli pyrrhus*. Photo by Barry W. Mansell.

Plate 36 (opposite top). Panamint speckled
rattlesnake, *Crotalus mitchelli stephensi*, Nevada.
Ray Folsom collection, Hermosa Beach, California.
Photo by John H. Tashjian.

Plate 37 (opposite bottom). Northern black-tailed
rattlesnake, *Crotalus molossus molossus*, Cochise
County, Arizona. Photo by Roger W. Barbour.

Plate 38 (below). Northern black-tailed rattlesnake,
Crotalus molossus molossus, western Texas. Photo
by Richard D. Bartlett.

Plate 39 (opposite top). Twin-spotted rattlesnake,
Crotalus pricei pricei. San Diego Zoo. Photo by John
H. Tashjian.
Plate 40 (opposite bottom). Red diamond
rattlesnake, *Crotalus ruber ruber*, San Diego
County, California. Photo by John H. Tashjian.
Plate 41 (above). Red diamond rattlesnake,
Crotalus ruber ruber. Photo by Barry W. Mansell.

Plate 42 (above). Mojave rattlesnake, *Crotalus scutulatus scutulatus*, Clark County, Nevada. Photo by John H. Tashjian.
Plate 43 (opposite top). Mojave rattlesnake, *Crotalus scutulatus scutulatus*. Photo by Richard D. Bartlett.
Plate 44 (opposite bottom). Tiger rattlesnake, *Crotalus tigris*. Photo by Barry W. Mansell.

Plate 45 (below). Tiger rattlesnake, *Crotalus tigris*. Photo by Richard D. Bartlett.

Plate 46 (opposite top). Prairie rattlesnake, *Crotalus viridis viridis*, Colorado. Photo by Roger W. Barbour.

Plate 47 (opposite center). Grand Canyon rattlesnake, *Crotalus viridis abyssus*, Grand Canyon, Arizona. Arizona-Sonora Desert Museum. Photo by John H. Tashjian.

Plate 48 (opposite bottom). Arizona black rattlesnake, *Crotalus viridis cerberus*. Photo by Richard D. Bartlett.

Plate 49 (opposite top). Midget faded rattlesnake, *Crotalus viridis concolor*. Louisiana Purchase Gardens and Zoo, Monroe, Louisiana. Photo by John H. Tashjian.

Plate 50 (opposite bottom). Southern Pacific rattlesnake, *Crotalus viridis helleri*, San Diego County, California. Photo by John H. Tashjian.

Plate 51 (directly below). Great Basin rattlesnake, *Crotalus viridis lutosus*. Photo by Richard D. Bartlett.

Plate 52 (bottom). Hopi rattlesnake, *Crotalus viridis nuntius*. Chuck Hanson collection, Tucson, Arizona. Photo by John H. Tashjian.

Plate 53 (top). Northern Pacific rattlesnake, *Crotalus viridis oreganus*, Calaveras County, California. Photo by John H. Tashjian.
Plate 54 (center). Arizona ridgenose rattlesnake, *Crotalus willardi willardi*. Photo by Richard D. Bartlett.

Plate 55 (bottom). New Mexico ridgenose rattlesnake, *Crotalus willardi obscurus*. San Antonio Zoo. Photo by John H. Tashjian.

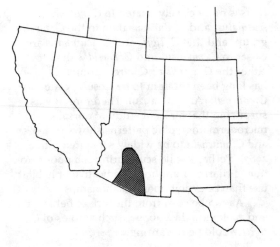

Distribution of *Crotalus tigris*.

Behavior

This species becomes active in late spring or early summer (it emerges after rains), and can be found abroad through September. It begins to hibernate in October or early November, and remains dormant during the winter. It is chiefly nocturnal, beginning daily foraging at dusk and continuing into the night until its environment becomes too cold, but some individuals can occasionally be found basking during the daylight hours, particularly on cool days. Nothing is known of its thermal requirements or foraging area.

Activity is not restricted to the ground, and *C. tigris* has been found in bushes 60 cm above the desert floor (Stebbins, 1954; Klauber, 1972). Despite being an inhabitant of the desert, the tiger rattlesnake swims readily with typical lateral undulations when placed in water (Klauber, 1972).

Reproduction

Compared to other North American rattlesnakes, practically nothing is known of the reproductive biology of *Crotalus tigris*. The shortest known pregnant female was 61.6 cm (Klauber, 1972). Mating probably takes place in April, and Stebbins (1954) reported that two females taken in October contained "eggs." The 210–215 mm young are born alive.

Growth and longevity

The growth rate is unknown. A wild-caught adult survived 15 years, 3 months, and 3 days in captivity (Bowler, 1977).

Food and feeding

Although it probably ambushes much of its prey, as do other rattlesnakes, *Crotalus tigris* is also an active hunter, as evidenced by Armstrong and Murphy (1979) having taken one as it was investigating a wood rat nest.

There is some debate as to whether the prey of the tiger rattlesnake is mostly restricted to lizards because of its small head, or if it is capable of swallowing larger-bodied mammalian prey. Ortenburger and Ortenburger (1926) found one eating a whiptail lizard (*Cnemidophorus*), and Kauffeld (1943a) reported that his captives readily ate lizards. On the other hand, Klauber (1972) found mammalian hair in the intestinal tract of an adult, and the following rodents have been reported as prey: kangaroo rats (*Dipodomys*), pocket mice (*Perognathus*), deer mice (*Peromyscus*), wood rats (*Neotoma*), and pocket gophers (*Thomomys*) (Stebbins, 1954, 1985; Fowlie, 1965). Possibly smaller *C. tigris* rely heavily on lizards for food, while adults depend more on rodents.

Venom and bites

Crotalus tigris is behaviorally unpredictable, sometimes rattling when approached and at other times remaining silent. However, it is generally thought to be inoffensive, which is fortunate as it produces a potent whitish venom with neurotoxic elements (Klauber, 1972). Also, owing to its smaller head and shorter fangs (4.0–4.6 mm; Klauber, 1939), the venom yield per adult is low: 6.4–11 mg dried venom per fresh adult (Klauber, 1972; Minton and Weinstein, 1984; Weinstein and Smith, 1990), 0.18 ml (Amaral, 1928). The LD_{50} (mg/kg) for mice is 0.070 intraperitoneal, 0.056 intravenous, and 0.21 subcutaneous; there seems to be low protease activity and no hemolytic activity (Minton and Weinstein, 1984; Weinstein and Smith, 1990). Weinstein and Smith (1990) isolated four toxins; one comprised about 10% of the total venom protein and had an intraperitoneal LD_{50} of 0.05 mg/kg in mice. This particular toxin showed complete immunoidentity with crotoxin and Mojave toxin, indicating the presence of isoforms of crotoxin and/or Mojave toxin in the venom of *C. tigris*.

Tiger rattlesnake envenomation of humans produces little local reaction and no significant systemic symptoms.

Remarks

Brattstrom (1964) proposed that *Crotalus viridis* was the probable ancestor of *C. tigris*, *C. mitchelli*, *C. cerastes*, and the Mexican *C. enyo*. However, electrophoretic studies of venom proteins by Foote and MacMahon (1977) suggest that *C.*

tigris is more closely related to *C. mitchelli*, *C. scutulatus*, and *C. durissus* than to the *C. viridis* group, and that *C. tigris* and *C. scutulatus* are closer to *C. molossus* and *C. horridus* than to either the *C. viridis* or *C. atrox* groups. *C. tigris* has long been thought to be closely related to *C. mitchelli* (Amaral, 1929). Therefore, it was surprising that a recent study of dorsal scale microdermatoglyphic patterns shows *C. tigris* and *C. mitchelli* to be widely separated (Stille, 1987). Perhaps scale sculpturing is a poor taxonomic character and more indicative of habitat use than of evolutionary relationships.

As can be seen from the lack of behavioral and ecological data above, such studies of *C. tigris* would be rewarding.

Crotalus viridis (Rafinesque, 1818)
Western rattlesnake
Plates 46–53

Recognition:

Crotalus viridis is the most widely distributed rattlesnake in the western United States and Canada, and also the most variable in North America, with nine subspecies. This makes a composite description difficult, so, after reading the description that follows, the descriptions and ranges of the various subspecies presented under Geographic variation should also be examined.

Crotalus viridis (maximum length, 162.5 cm) varies in ground color from gray, olive, greenish gray, greenish brown, brown, yellowish brown, tan, salmon, and reddish to black. Individuals from Colorado and Wyoming are almost lime green immediately after shedding (David Duvall, pers. comm.). The dorsal body pattern consists of a series of 20–50 brown or black, light-bordered blotches, which may become more like crossbands near the tail. The tail has a series of alternating light and dark, poorly emphasized bands hardly more pronounced than the background body color. Two light diagonal stripes occur on the side of the face, one extending backward from in front of the orbit to the supralabials and the other from in back of the eye to in front of and above the corner of the mouth from where it runs backward onto the neck. Narrow, light, transverse stripes may pass across the face in front of the eyes and over the supraocular scales. The underside is gray, cream or white with no dark markings. Body scales lie in 21–29 rows at midbody; on the venter are 158–196 ventrals, 13–31 subcaudals, and an undivided anal plate. This is the only rattlesnake with more than two internasals in contact with the rostral, which is usually higher than wide. Behind the rostral, and touching it, are 3–4(1–8) internasals, followed by at least eight scales before the 4–6(1–9) intersupraoculars (prefrontals are absent). The nasal scale is divided, with the prenasal touching the supralabials in some individuals, and the postnasal seldom contacting the upper preocular scale. Other lateral scalation consists of 1(1–3) loreal, 2 preoculars, several suboculars, 2(3) postoculars, 14–15(10–19) supralabials, and 15–16(11–20) infralabials. Each of the two lobes of the hemipenis is adorned with about 40–92 spines and 26–37 fringes, but no spines lie in the crotch between them; the sulcus spermaticus divides to extend up each lobe (Klauber, 1972). The dental formula consists of the fang on the maxilla, 6–8(10) pterygoidal, 2–3(4) palatines, and 9–10(6–11) dentary teeth.

Males have 158–190 ventrals, 18–31 subcaudals, and 3–15 dark tail bands; females have 164–196 ventrals, 13–26 subcaudals, and 2–11 dark tail bands.

Karyotype

The karyotype consists of 36 chromosomes: 16 macrochromosomes and 20 microchromosomes (Monroe, 1962; Baker et al., 1972; Zimmerman and Kilpatrick, 1973; Porter et al., 1991). Body macrochromosomes include 4 metacentric, 6 submetacentric, and 4 subtelocentric, while the sex-determining macrochromosomes are either ZZ in the male or ZW in the female. Baker et al. (1972) reported the Z is submetacentric and the W subtelocentric, but Zimmerman and Kilpatrick (1973) described these chromosomes as being metacentric and submetacentric, respectively.

The isozyme patterns of a number of gene loci have been examined by Murphy and Crabtree (1985), and the reader is referred to that paper for details.

Fossil record

Numerous fossils of *Crotalus viridis* have been found: Pliocene (Clarendonian), Kansas, Nebraska; Pleistocene—Blancan, Kansas; Irvingtonian, Kansas; Rancholabrean, California, Idaho, Iowa, Kansas, Nevada; Holocene, Nevada (Holman, 1979, 1981; Mead et al., 1989). There is also a questionable Rancholabrean series of fragmented vertebrae from the Lower Grand Canyon, Arizona, that may be either from this species or *C. mitchelli* (Van Devender et al., 1977).

Distribution

The western rattlesnake ranges from southcentral British Columbia, southeastern Al-

berta, and southwestern Saskatchewan southeastward through the United States to extreme western Iowa, Nebraska, and Kansas, and south to northern Baja California, northern Chihuahua, and northwestern Coahuila in Mexico. Elevations occupied range from near sea level to over 3,300 m.

Geographic variation

Nine subspecies are currently recognized, but one of these is confined to a Mexican island. The eight occurring within the range of this book are as follows. *Crotalus viridis viridis* (Rafinesque, 1818), the prairie rattlesnake, ranges generally east of the Rocky Mountains from Alberta, Saskatchewan, Montana, eastern Idaho, and North Dakota south barely to north-

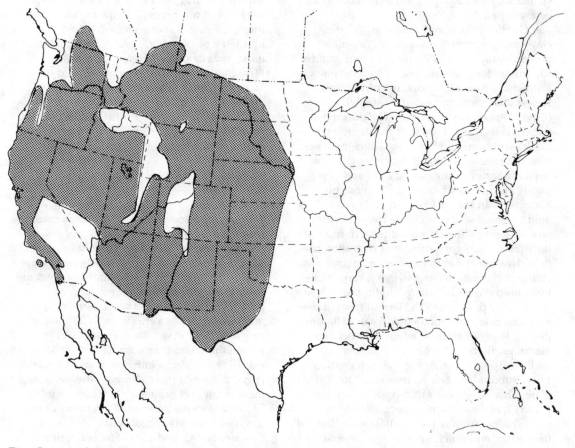

Distribution of *Crotalus viridis*.

Crotalus viridis viridis (John H. Tashjian, Arizona- Sonora Desert Museum).

eastern and southeastern Arizona, New Mexico, western Texas, northern Chihuahua, and northwestern Coahuila. It is greenish gray to brownish gray with oval to quadrangular-shaped, light-bordered, dark blotches and 2 loreals, and it reaches a total length of 145 cm. *C. v. oreganus* Holbrook, 1840, the northern Pacific rattlesnake, ranges west of the Rockies from British Columbia south to San Luis Obispo and Kern counties, California. It is dark gray, olive, yellowish brown, brown, or black, with hexagonal, oval, or almost circular dark blotches having well-defined light borders, and usually only one loreal scale. It grows to 162.5 cm. *C. v. helleri* Meek, 1906, the southern Pacific rattlesnake, is found from southwestern San Luis Obispo and Kern counties, California, south to northern Baja California, also occurs on the islands of Santa Catalina and Coronado del Sur. Its color is similar to that of *C. v. oreganus,* and it also usually has only a single loreal on each side of the head, but its dark blotches are angular or diamond shaped, and it reaches a total length of 135 cm. *C. v. cerberus* (Coues, 1875), the Arizona black rattlesnake, is found in Arizona, where it ranges from Mohave and Yavapai counties southeast to Apache, Graham, and

Pima counties, and in adjacent Grant County, New Mexico. It is dark gray, olive, dark brown, or black with large, poorly defined dark blotches and usually two loreals. Maximum length is 100 cm. *C. v. concolor* Woodbury, 1929, the midget faded rattlesnake, is found in the basins of the Colorado and Green rivers in southwestern Wyoming, eastern Utah, and western Colorado. This small subspecies grows only to about 70 cm, is cream, yellowish brown, or tan with rectangular to oval, only slightly darker (often faint or absent) body blotches and two loreals. *C. v. abyssus* Klauber, 1930, the Grand Canyon rattlesnake, is found only on the floor of the Grand Canyon in Coconino County, Arizona. It is salmon or red with oval, rough-edged blotches, which fade with age, and two loreals; its total length reaches 98 cm. *C. v. lutosus* Klauber, 1930, the Great Basin rattlesnake, ranges between the Rocky and Sierra Nevada mountains from southern Idaho, southwestern Oregon, and northwestern California south through western Utah and Nevada to northwestern Arizona. The ground color is gray, yellowish brown, or tan with elliptical to oval brown or black blotches widely separated by lighter pigment. Two loreals are present, and it has a

Crotalus viridis cerberus (John H. Tashjian, Arizona- Sonora Desert Museum).

body length to 135 cm. *C. v. nuntius* Klauber, 1935, the Hopi rattlesnake, lives in northeastern and north-central Arizona and adjacent extreme northwestern New Mexico. It is another small subspecies, barely reaching 70 cm, and is pink, red, or reddish brown with irregularly oval to rectangular, narrowly light-bordered blotches, and two loreals.

Confusing intergrade individuals may occur in areas where the ranges of two or more of these subspecies meet.

Confusing species

Although several other species of *Crotalus* have facial stripes and a blotched body pattern like *C. viridis,* none normally has more than two internasals touching the rostral scale, and their tail bands are usually well defined. *Sistrurus catenatus* has nine enlarged scales on the top of its head.

Habitat

As it has such a broad geographic range, *C. viridis* can be found in a variety of habitats, including woodlands, scrub areas, prairie grass-lands, shrub-steppes, desert margins, coastal and sand dunes; usually, south-facing rocky outcroppings with deep crevices, or prairie dog (*Cynomys*) towns are located within migratory distance. In Canada, the distribution of this snake seems to be determined by the availability of suitable rock crevice hibernacula, and since such features are usually found only in river valleys, *C. viridis* is restricted to such places (Pendlebury, 1977; Gannon, 1978)

Behavior

Over most of its range *Crotalus viridis* is active only from April through September; however, in southern California some active individuals may be found in every month (Klauber, 1972). Emergence from hibernation is generally earlier in the south (late February to early April) and later in the north (late March to mid-April). The first spring appearance of western rattlesnakes in south-central Wyoming occurs from 23 April to 25 June, with the mean date of emergence about 28 May (Graves and Duvall, 1990). The earliest recorded appearances of active individuals are 11 February (North Dakota; Wheeler, 1947) and 1 March (California; Fitch

Crotalus viridis lutosus (Richard D. Bartlett).

and Twining, 1946). No differences exist between the dates of spring emergences of males and females. Canadian populations enter hibernation earlier (usually in September; Gannon and Secoy, 1985), while *C. viridis* in southern populations may finally become inactive in late November or early December (Howell and Wood, 1957). The pivotal body temperature for both arousal and dormancy is 10°C (Jacob and Painter, 1980), but a temperature of 16°C may be necessary to bring them out of the den (Woodbury, 1951). Increases in ground temperature are probably more responsible for the emergence of *C. viridis* than the air temperature (Jacob and Painter, 1980). Males and females emerge from the hibernacula at about the same date (Fitch, 1949; Graves and Duvall, 1990).

After emergence in the spring, western rattlesnakes usually remain near the hibernaculum for several days to weeks before dispersing to summer feeding areas, and some basking may occur at this time, but not for long. David Duvall (pers. comm.), however, has not observed basking in Wyoming, where the rattlesnakes begin their spring migrations almost immediately after emergence. The summer foraging range is occupied until September, when the return migration to the

hibernaculum begins. From late September through November, the snakes remain near the den, frequently basking, possibly to help in the digestion of a last meal, or to aid the ovaries mature, but feeding less often as the air temperature drops. They enter the den for good in December and, although some may move about within it, remain there until the next spring (Ludlow, 1981, Gannon and Secoy, 1985; Duvall, King et al., 1985).

Typical hibernacula include prairie dog towns and other deep animal burrows (particularly those of the desert tortoise *Gopherus agassizii*), the walls of cisterns or wells, caves, and crevices in south-facing rock outcrops. All of these must extend below the normal frost-line for the region. Within the den, dormant rattlesnakes may lie individually coiled or in bunches. Often other reptiles and amphibians also overwinter in the same den, particularly if it is the only suitable site for some distance. Once the snakes are torpid, warm-blooded animals may move into the den (foxes, badgers, even prey species such as packrats, prairie dogs, and burrowing owls). Body temperatures of hibernating *C. viridis* recorded by Jacob and Painter (1980) dropped to as low as 6°C, and a significant decrease in heart beat occurred after the body temperature fell below

10°C. Body temperatures of hibernating *C. viridis* recorded by Macartney et al. (1989) were 2–7°C. Cooling of the body temperature during hibernation may be necessary for proper reproduction the next year (Tryon, 1985).

Body temperature decreases more slowly when snakes are lying together in groups than when alone (White and Lasiewski, 1971; Graves and Duvall, 1987); physical contact with several other snakes, however, does not help raise the body temperature (Graves and Duvall, 1987). The longer the snake is chilled, the longer the period of time for arousal once temperatures again rise.

Sexton and Marion (1981; Marion and Sexton, 1984), while studying *C. virdis* in an artificial den, reversed the thermal gradient from that normally expected in a wild den, and found that the snakes moved toward and congregated near the den entrance instead of deeper underground. Apparently the snakes sought the warmer portions of a thermal gradient that would have under natural conditions brought them deeper into the den as winter progressed. However, at a natural den in southern British Columbia, Macartney et al. (1989) found the spring thermal gradient to be weak or nonexistent during most of the emergence period. Usually less than a 2–3°C difference existed between the deepest and the shallowest subsurface temperatures.

Crolalus viridis experiences a reduction in body weight during hibernation (Hirth, 1966a). This is probably due to evaporative water loss from the inactive animal, but also involves metabolic reduction of stored fat. Adults may show a weight loss of about 4–9%, while juveniles may lose as much as 20–50% of their body weight (Hirth, 1966a; Klauber, 1972; Parker and Brown 1974). This may be an important factor influencing overwinter survivability of small *C. viridis*, particularly neonates (Charland, 1989, has reported a survivorship rate of only 55% for neonates in their first winter, but that survivorship seemed independent of both weight and condition at birth). Aggregation of rattlesnakes into bunches of several individuals whose bodies are in contact results in reduced water loss (White and Lasiewski, 1971). If given the opportunity *C. viridis* will drink water (Stark, 1984); apparently *C. viridis* must maintain an extracellular fluid volume of about 41.9% of their mass (Smits and Lillywhite, 1985).

Hibernating dens in many parts of the range have been used for centuries (certainly from before the white man reached North America), and theoretically those *C. viridis* now inhabiting them are the descendents of the first snakes to overwinter there. Den fidelity exists (Duvall, King et al., 1985), and marked animals have returned to the same hibernaculum for several years, even having to crawl long migrations to do so.

During the spring and fall, *C. viridis* often prowl in the morning and again in the late afternoon or evening. The hours in between are spent either undercover, coiled in the shade, or basking, depending on the air temperature. During the summer, most activity is nocturnal. Populations occurring at both higher altitudes and latitudes seem to be more diurnally active in the summer. Western rattlesnakes kept in an outside enclosure in Saskatchewan were active from about 0900 to 2000 hours in the spring, 0600 to 2100 hours in the summer, and 0900 to 1900 hours in the fall (Gannon and Secoy, 1985).

Body temperatures recorded from wild, active *C. viridis* have ranged from 9.3°C for one lying extended on the ground on a March morning that was capable of crawling slowly away when disturbed to 37.8°C for one crossing a paved road on a June afternoon (Cunningham, 1966). Most activity, however, seems to take place at body temperatures of 20–35°C (Stebbins, 1954; Brattstrom, 1965; Cunningham, 1966; Hirth and King, 1969; Vitt, 1974). The critical thermal maximum for *C. viridis* is 38°C, and body temperatures of 41–42°C are lethal (Brattstrom, 1965). Klauber (1972) experimented to see how cooler temperatures affected the activity of *C. viridis;* at 15°C and 11°C the snakes were active, but sluggish, at 5°C they could still strike, but for only short distances and very slowly, and at 3.5° one could still crawl slowly about. The frequency of rattling is directly proportional to body temperature (Chadwick and Rahn, 1954).

Crotalus viridis become lighter in color when subjected to body temperatures as high as 34°C (Rahn, 1942a). Their melanophores are in most cases maximally contracted. As their body temperature drops to 30°C, the melanophores disperse and they gradually assume their normal color at a room temperature of 23°C. Animals chilled to about 8°C exhibit maximal darkening. Such pigmental changes may aid in thermoregulation by either absorbing or reflecting infrared radiation from the sun, or by dissipating body heat either more rapidly or more slowly.

Two interesting observations were made by Hirth and King (1969) during their study of body temperatures. Two newborn *C. viridis* collected in the fall had body temperatures of only 16 and 18°C, while the lowest adult temperature recorded was 21°C. Smaller snakes probably heat and cool faster than larger ones. Their second interesting finding was that the mean body temperatures of adult females were 2.1 and 2.0°C higher than those of males in the spring and summer, respectively. This preference for a slightly higher temperature may be correlated with female viviparity.

Several studies have been conducted to determine if differences exist in the thermoregulatory behavior of gravid females and nongravid females. In an indoor study by Gier et al. (1989), radioimplanted female *C. v. oreganus* in both conditions were placed in an enclosure containing a thermal gradient. All of the females regulated their body temperatures by shifting between the warmer and cooler portions of the enclosure, with most periods of inactivity spent near a heat source. Most maintained body temperatures that fluctuated around 29.5°C. There was no significant differences in mean body temperature, standard deviation or range of body temperature between gravid and nongravid snakes, whether before or after birth, but body temperature for gravid females changed significantly after parturition. The prebirth mean of gravid snakes was 30.7°C (26.9–34.6), but the postbirth mean was only 28.5°C (23.3–32.0). Therefore, there was a tendency for those females carrying young to thermoregulate higher, and with less variability, before than after birth. These results suggest that the optimum temperature is higher, and the temperature range narrower, for embryonic development than for the activities of the postpartum female. Also, the body temperature of gravid individuals tended to decrease gradually throughout the duration of the study. Preference for a higher body temperature by the gravid female may be a reaction to an environment generally too cool, whereas preference for a lower body temperature by females may be a reaction to an environment too warm for optimum embryonic development.

A second study using radiotelemetry and female *C. v. oreganus* was conducted in an outdoor enclosure by Charland and Gregory (1990). Both gravid and nongravid females exhibited a triphasic diel pattern of body temperature variation throughout the summer. Body temperature rose rapidly in the morning, remained stable in the afternoon, and declined slowly at night. During the afternoon stable plateau, the mean body temperature of gravid (31.7°C) and nongravid (30.7°C) females did not differ significantly, but temperatures of gravid snakes were significantly less variable. The mean plateau body temperature of gravid females increased across the summer. There was no seasonal change in mean body temperature for nonreproductive females, but the variance in body temperature increased considerably as autumn approached. This increase in variance suggests that nonreproductive females abandon precise thermoregulation as fall approaches, while gravid females, still burdened with developing young, continue to thermoregulate precisely.

Defensive behavior may also be affected by body temperature. Goode and Duvall (1989) designed a quantitative field study to test the prediction that defensive responses of free-ranging *C. v. viridis* would more rapidly escalate to striking as body temperature decreased and that escape would be used as body temperature increased. Twenty-two males, 17 nongravid females and 17 gravid females were approached and grasped at midbody with a pair of foam-padded snake tongs. All were threatened in the same consistent manner. Male and nongravid females did not exhibit a temperature-dependent shift in defensive behavior, but gravid females did. They showed a significant positive regression between body temperature and duration. This may be related to the elevated body temperature usually maintained by females carrying young, and may be in response to their greater vulnerability during the gestation period.

Scarcity of water is a problem in many of the xeric habitats occupied by *C. viridis*. When water is available, such as in puddles after rains, the snake often inbibes to capacity, but rainfall may be little and infrequent, or fall in places where it will not collect in puddles. Aird and Aird (1990) reported a behavior in *C. v. lutosus* that aids in concentrating rainfall for drinking. During a rain storm, an adult female formed a tight concentric coil with her flattened body that allowed rainwater to accumulate to about 5 cm deep in the groove between two coils. After the rain ceased, she drank from this water supply for nearly 30 minutes.

Western rattlesnakes may migrate long distances to and from the hibernaculum in the spring and fall; Duvall (1986) reported migrations of 11 km or more for *C. v. viridis* in Wyoming. These long movements are not usually direct or nonstop, although Duvall, King et al. (1985) reported that a radio-equipped male traveled 245 m and a similarly equipped female 118 m in one to two days during migration; King and Duvall (1990) recorded mean distances moved per movement of 237 m for males and 137 m for females. Instead the snakes seek out good feeding areas where the density of mice (*Peromyscus*) is high, and may spend some time there before crawling onward (Duvall, King et al., 1985), an average of 25.3 days by males and only 6.7 days by females (Duvall, Goode et al., 1990; King and Duvall, 1990). In tests in which caged deer mice (*Peromyscus*) were placed in the paths of migrating snakes, females remained near the cage for about 24 hours and then moved on while males became virtually immobile once the caged mice were found (Duvall, Goode et al., 1990). In contrast, similar behavior by both sexes was observed during laboratory studies by Duvall, Chiszar et al. (1990). Olfaction seems to play the major role in finding these *Peromyscus*-rich areas, and odor trails are detected either by tongue flicking or, possibly, mouth gaping (Graves and Duvall 1983; Duvall, King et al., 1985; Duvall, Chiszar et al., 1990). By traveling along paths of very high angular fixity or straightness, males, more than females, minimize the likelihood of covering the same area twice as they search either for food or mates. Prey odor detection brings about a slowing of male activity. Although males and females differ in their migratory and search behavior, spring migration, at least in Wyoming, is for finding food (Duvall, Goode et al., 1990).

Once the summer feeding area is reached, a more or less permanent activity range is established and, unless prey become scarce, maintained for the rest of the summer. The mean size of these activity ranges in California is 12.1 ha for males and 6.5 ha for females (Fitch and Glading, 1947), but Stark (in Macartney et al., 1988) found these home ranges to be only 2.4 ha for males and 1.8 ha for females (presumably in Alberta).

While both males and females search for small mammal prey in the first half of the brief 3.5-month summer in Wyoming, males but not females search for mates in the second half of the season (King and Duvall, 1990). Females continue to forage for the duration of the season. Consequently, males must search and locate two goals in time and space (food and mates) each season, while females need search for only one (food). Males exhibit significantly greater spatial searching efficiency during spring foraging than do females, allowing males to concentrate foraging activities into the first half of the annual activity season, so the second half can be almost exclusively devoted to mate searching. The mating system of *C. v. viridis* in Wyoming is best described as prolonged mate-searching polygyny.

Of 10 *C. viridis* displaced 50–100 m from their hibernating den, nine returned within 10 days (Hirth, 1966b). Those that returned had been released at all four compass points from the den. Conspecific odor detection may have been a major orientating mechanism in finding the den, but later studies by King et al. (1983) failed to prove this. Instead, field observations of radio-equipped snakes by Duvall, King et al. (1985) seem to indicate the use of fixed-angle, sun-compass orientation in homing to the hibernaculum.

The western rattlesnake sometimes crawls into trees or brush while foraging, or possibly basking. Shaw (1966) found a *C. viridis* about 4.3 m above ground in the crotch of an oak tree, and Owens (in Klauber, 1972) observed one at least 7.6 m above ground dangling from a tree hole. Cunningham (1955) saw them on numerous occasions coiled as high as 1.5 m above ground in thickets. Lakanen (in Klauber, 1972) saw one climb a tree that sloped some 25 or 30°, and an individual that I observed basking in South Dakota had climbed almost 2 m up a slanting rock surface.

Crotalus viridis also swims well, as shown in tests performed by Klauber (1972). Klauber also related several cases of this snake having been seen swimming in nature. Although it probably does not dive when in water, it seems fairly well equipped to do so. Ferguson and Thornton (1984) found that it had a greater oxygen storage capacity, hemoglobin content, and blood volume than the water snake *Nerodia sipedon*. It could remain underwater 30.13 minutes.

Like those of other species of *Crotalus*, male *C. viridis* take part in combat dominance bouts (Klauber, 1972; Thorne, 1977), and the courtship "dance" reported by Gloyd (1947) was undoubtedly male combat.

A possible mutualistic relationship between a juvenile *C. viridis* and red ants (*Formica*) was observed by Duvall, King et al. (1985). The incident occurred on a cool, sunny morning in late June. The snake remained frozen motionless for about five minutes in a curious posture with its body covered with ants. It appeared to have recently shed, and was lying on the ground with three points of its body barely touching the substrate, with its head, tail, cloacal region, and other body parts held as high off the ground as possible. The tail was raised high, resulting in exposure of just a small portion of the anal vent. Possibly the ants were removing ectoparasites or small pieces of unshed skin, which would have been of value to both parties, but Duvall, King et al. noted that other *C. viridis* seemed in discomfort when ants crawled over them, and Duvall (pers. comm.) now believes this was a predatory attack by the ants.

Reproduction

Since there is much variation in the total lengths achieved by the subspecies of *Crotalus viridis*, the total length at which sexual maturity occurs is also variable. Klauber (1972) listed the following lengths of the smallest gravid female of each species he had examined: *abyssus*, 684 mm; *cerberus*, 701 mm; *concolor*, 522 mm; *helleri*, 596 mm; *lutosus*, 557 mm; *nuntius*, 395 mm; *oreganus*, 503 mm; and *viridis*, 890 mm. Female *C. v. lutosus* from Utah mature in 3–4 years at lengths of 564–693 cm (Glissmeyer, 1951).

The length at which maturity is reached may even vary between populations of the same subspecies. In northern Utah, most female *C. v. oreganus* larger than 550 mm snout-vent length (and with five or more rattles) are sexually mature (Diller and Wallace, 1984), while in British Columbia females of this same race mature at about 650 mm snout-vent length in five to seven years and produce their first litters at 700–670 mm at probable ages of seven to nine years (Macartney and Gregory, 1988; Macartney et al., 1990).

Male *C. v. oreganus* mature in northern Idaho at snout-vent lengths of at least 520 mm (with four or more rattles), but the mean snout-vent length of mature snakes, 706 mm, is significantly greater than that of mature females, 659 mm (Diller and Wallace, 1984). In southern British Columbia, males reach maturity in three to four years at snout-vent lengths of about 535 mm (Macartney et al., 1990).

Once maturity is reached, males are capable of breeding annually, so the potential lifetime reproductive output is greater for males than for females, a polygynous system (see discussion in Duvall et al., 1991). In New Mexico, the spermatogenic cycle begins early (Aldridge, 1979a). Most males collected in mid-April have a Sertoli syncytium present, and some have dividing spermatogonia in the syncytium. Recrudescence and spermatogenesis continue through May, and by early June mature sperm are present in the seminiferous tubules. Peak sperm production occurs from mid-June through September, and spermiation apparently takes place between mid-June and mid-October. By October and November the walls of the seminiferous tubules contain fewer layers of cells and in some the syncytium is beginning to form. Sertoli syncytium with spermatogonia is present during hibernation. Tubule diameter is smallest in the spring, but increases as recrudescence occurs until it is about twice the diameter in mid-July (this trend may not be true, but instead an artifact of histological preparation). Temperature plays a major role in initiating spermatogenesis. In western Idaho, sperm may be present in the vas deferens in every month (Diller and Wallace, 1984).

Aldridge (1979a) studied the effects of varying photoperiods (10-hour photoperiod, no photoperiod, or gradually increasing photoperiod) and ambient temperatures (8°C,

22°C) or variable temperatures of 19.5–31.5°C on sperm production in New Mexican *C. viridis*. Snakes kept at 22°C with or without a photoperiod had a Sertoli syncytium and spermatogonia present, but those kept with a photoperiod had significantly greater seminiferous tubule diameters. Six of eight males kept at variable temperatures had sperm form in the tubules. No males maintained at 8°C, with or without a photoperiod, initiated spermatogenesis.

The female cycle seems biennial, with approximately 50–70% of the females breeding in any given year, but females in some southern populations may reproduce annually (Diller and Wallace, 1984; Fitch, 1985a), and King and Duvall (1990) have presented evidence for a triennial female reproductive cycle in Wyoming. This three-year cycle probably consists of two successive seasons of feeding and follicle development with a peak of sexual receptivity and mating occurring in the second season associated with maximum follicle development, and a third season in which ovulation takes place in spring and fertilization is accomplished with stored sperm, and pregnancy in summer is followed by birth of live young in autumn.

The ovarian cycle of *C. viridis* has been studied by several investigators in various parts of the range (Rahn, 1942b; Fitch, 1949; Glissmeyer, 1951; Aldridge, 1979b; Gannon and Secoy, 1984; Diller and Wallace 1984; Macartney and Gregory, 1988). These studies basically agree in their descriptions of the female reproductive cycle, except for minor differences, and slight differences in timing probably brought about by latitudinal effects. Therefore, the description by Aldridge (1979b) for New Mexican snakes is presented here as the model. The vitellogenesis cycle is divided into two periods. The first state (primary vitellogenesis) occurs in follicles from microscopic size to 4–6 mm in length. Follicles 4–6 mm in length are present in the ovary at any time of the year. They remain at this size, owing to a lack of yolk deposition, until initiation of the second stage of vitellogenesis (secondary vitellogenesis). Yolk deposition begins in summer and follicles enlarge to 15–20 mm long by hibernation. No apparent follicular growth occurs during the winter. Yolk deposition resumes in the spring, and ovulation occurs in May to early June. Re-

productive females produce more body fat than females not reproducing that year (Diller and Wallace, 1984).

In California, mating in *C. v. helleri* and *C. v. oreganus* occurs in the spring (Fitch and Glading, 1947; Fitch, 1949; Klauber, 1972), but elsewhere, particularly in the north, most copulations have been observed during the period from mid-July to early September (Wright and Wright, 1957; Duvall, King et al., 1985; Macartney and Gregory, 1988; Graves and Duvall, 1990; King and Duvall, 1990; Duvall, pers. comm.). Males probably find and identify females by olfactory cues, and such ability is probably innate (Scudder et al., 1988). Odors seem enhanced by female ecdysis (Klauber, 1972; Macartney and Gregory, 1988). Courted females have usually completed shedding less than 48 hours previously (Macartney and Gregory, 1988).

A detailed description of courtship and mating in *C. v. oreganus* was published by Hayes (1986); an edited version retaining most of the terminology developed by Gillingham (1979) follows. In most courting *C. viridis*, Phase I consists of chases (female flees, male pursues) or chase-mounts (male maintains contact during pursuit). Phase II begins when the male exhibits tail-search copulatory attempts, which are tactile attempts by the male's tail to locate the female's cloaca. During the mating observed by Hayes (1986), the female usually remained coiled and quiescent. Only occasionally did she flee and the male give chase. However, chases often interrupted rather than preceded Phase II behavior. Phase I behavior was therefore distinct from Phase II only in the initial absence (2.6, 7.4 min) of tail-search copulatory attempts. When chases interrupted Phase II behavior, tail-search copulatory attempts were quickly resumed once contact was reestablished. Most of the male's courtship efforts consisted of Phase II events. After three or four tongue flicks, forward jerking or flexions of the entire body (which audibly shook the rattle) developed. During this flexion cycle, rapid tongue flicks were directed toward the dorsal surface of the female (all regions), often making contact. Tongue flicks generally occurred between flexions. The flexion cycle ceased when the male initiated tail-search copulatory attempts. During this tail-search cycle, the male generally withdrew his head and anterior trunk from the female as both tongue flicks and flexions ceased. Outstretched positions may be necessary for copulation, as coiling by the female seemed to hinder courtship. A rest cycle, marked by the absence of any male movement, then followed until flexions began again and the entire three-part sequence was repeated. The average durations of the flexion cycle, tail-search cycle, and rest cycle were 35.1 seconds, 7.2 seconds and 25.5 seconds, respectively. The entire sequence took an average of 67.9 seconds. The number of forward jerks during the flexion cycle averaged 62.4. The rate of forward jerks (or the flexion rate) during uninterrupted cycles was nearly constant (1.78/second), so the flexion rate was probably not a function of courtship duration or sexual arousal of the male. The duration of the flexion cycle, number of forward jerks, and duration of the rest cycle were all correlated with the sequence duration. Neither the flexion rate nor tongue-flick rate was observed to exceed 2.0/second. Rapid lateral tail whipping by the female often occurred in response to tactile movements of the male. The duration of tail whips ranged from 1.14 to 7.35 seconds (mean, 2.13). In 12 instances unusual pulsations of the tail terminus consisting of sine waves directed posteriorly were observed immediately following tail whipping. Forward-jerking flexions and tongue flicking by the male were interrupted immediately subsequent to female tail whipping (this suggests that the male responded to what may be a visual-tactile display indicating the degree of female receptivity).

Tail whipping by females did not reduce copulatory success in *C. atrox*, but tail waves accompanied by slight cloacal extrusion and possibly by pheromonal discharge stimulated increased tongue flicking by males (Gillingham et al., 1983). The tail-terminus pulsations seen by Hayes (1986) may likewise be associated with pheromonal emissions from the cloaca, but this was not observed. Copulation (Phase III) did not take place during these observations.

The rate of tongue flicking by *C. v. oreganus* appeared to increase with excitement. It is generally believed that flexions stimulate both the male and female, but empirical evidence is lacking. The role of the tongue, however, is clearer; a functional male vomeronasal system is necessary for courtship. At 0400 hours (in darkness) on 13 August 1984, courtship was observed by Hayes (1986) for the last time under red light when the female was permanently reintroduced into the pen. Some rattlesnakes are known to continue copulation (for up to 28 hours) well into darkness, but the report by Hayes is the first to document initiation of rattlesnake courtship in darkness. The failures of several extensive field studies to observe or report courtship activities in *C. viridis* suggest that sexual encounters may normally occur at night. During warm weather, rattlesnakes are typically active nocturnally at which time a male would likely encounter and follow the sex-pheromone trail of a breeding female.

During a courtship and mating sequence of *C. v. viridis* witnessed by Duvall, King et al. (1985), the male pressed his chin along the entire length of the female's body, possibly to calm her. The male trailed the receptive female for a day before the courtship and mating sequence (which lasted about 90 minutes; 10 minutes of chin pressing, about 10 minutes to evert, position, and insert the hemipenis, and 70 minutes of actual coitus). David Duvall writes that he and his coinvestigators in Wyoming have now witnessed 12 copulations that have all lasted about 90 minutes.

Although most mating activity observed in the field has involved only a pair of *C. viridis*, aggregations of up to eight snakes may occur (Macartney and Gregory, 1988). Mating activity in British Columbia coincided with the peak of the male spermatogenic activity (Macartney and Gregory, 1988). As mating in most populations takes place in late summer when ovarian follicles are small, and ovulation does not occur until the next spring, viable sperm must be stored overwinter in the oviducts.

Some gravid females may immigrate from the hibernaculum in the spring, but most remain relatively close and form maternity colonies either there or in secondary nearby dens. Gravid females can often be found basking near the maternity den. Little if any feeding may occur during the period of pregnancy, and the females must depend on the fat bodies they have built up to carry them through the gestation period. After the birth of the young, the emaciated females apparently must double their weight before mating again, and those that cannot do so in a single summer must delay mating for another year or more (Macartney and Gregory, 1988; Charland and Gregory, 1989). Females using maternity colonies give birth sooner than those that migrate away from the den, and their young may have a better chance to find suitable overwintering sites before the beginning of winter (Duvall, 1986).

The young are born during the period August to early October after a gestation period of about 110 days (Diller and Wallace, 1984). The average number of young in 516 clutches listed in published accounts is 9.9 (1–25; 5,110 young). Litter size generally is proportional to female body length, so the shorter subspecies usually produce less than 10 young per litter, while larger races may produced more than this (Klauber, 1972). Northern populations often produce smaller litters (MacCartney and Gregory, 1988). Younger females may also produce smaller litters; Hammerson (1982) reported that while female *C. v. viridis* average only six young per litter, older females often give birth to 15–16 young. At birth, total length averages about 248(180–305) mm and weight about 15(13–16) g. However, the size of the young is correlated with female length, and females of the short subspecies *nuntius* may produce young only 190–200 mm long. The first few hours after birth are often spent close to the mother.

Crotalus viridis exhibits much ontogenetic variation. The patterns are usually vivid in juveniles, but fade with age, so that the series of lateral blotches may almost totally disappear in adults. Some juveniles may have yellowish or orange pigment on the tail.

A study by Crabtree and Murphy (1984) to determine if multiple inseminations occur in *C. viridis* that could produce litter mates of different parentage was inconclusive.

Crotalus viridis occasionally mates with *C. ruber* (Klauber, 1972; Armstrong and Murphy, 1979) and *C. scutulatus* (Murphy and Crabtree, 1988; Glenn and Straight, 1990) in nature. Captive western rattlesnakes have also hybridized with these same two species (Cook, 1955; Klauber, 1972).

Growth and longevity

In any population, the growth rate of individual snakes is quite variable, depending on for-

aging success and climatic conditions, making it difficult to present an overall growth calculation for *C. viridis*. It is known, however, that growth slows with age and length in both sexes. In Utah, 400-mm male *C. v. lutosus* grew an average of 207 mm (51%) between captures, while males 500 mm long grew only 75 mm (15%), those 600 mm long grew 56 mm (9%), those 700 mm grew 38 mm (5%), 800-mm males grew 19 mm (2%), and males 900 mm long increased only 16 mm (1%) (Heyrand and Call, 1951). Females had slower growth rates. Those 400 mm long grew 134 mm (33%) between captures, 500-mm females grew 71 mm (15%), those 600 mm long increased 41 mm (7%), 700-mm females grew 27 mm (4%), and those about 790 mm grew only 12 mm (1.5%).

Newborn *C. v. helleri* average 275–280 mm in length, but most young snakes found in late September and October are 300–340 mm long (Klauber, 1972). Some of this growth may be due to absorption of the egg yolk. Most young *C. v. helleri* enter their first hibernation at a length of about 350 mm. By mid-April of the next year they have grown to an average of about 380 mm (310–460). In May they average 400 mm (310–500), in June 430 mm (310–580), in July 450–500 mm, and by September the now one-year-old snakes are about 540 mm long. They enter the second hibernation at an average length of 600 mm. By their second September males usually exceed 800 mm, but females are only about 720 mm.

Crotalus v. oreganus from British Columbia have a highly variable annual growth rate in juveniles that decreases in both magnitude and variability with increasing body length (Macartney et al., 1990). Males have slightly greater growth rates than females. The mean snout-vent lengths of males for the first three years are 285, 358, and 453 mm, respectively; females average 282 and 352 mm snout-vent length during years 1 and 2. The average annual growth for adult males of several size classes is: < 750 mm, 45 mm; < 850 mm, 33 mm; < 950 mm, 22 mm; < 1,050 mm, 17 mm; and only 5 mm for those longer than 1,049 mm. The average annual growth for adult females is: < 750 mm, nongravid 28 mm, gravid 13 mm; < 850 mm, nongravid 17 mm; gravid 4 mm; and < 950 mm, nongravid 14 mm, gravid almost 9 mm. The nonfeeding by gravid females is clearly reflected in these figures.

In Saskatchewan, *C. v. viridis* are 195–229 mm at birth, while those in their second summer are 350—595 mm long (Gannon and Secoy, 1984). The average growth for the first year is 337 mm, and 212 mm the second year. In a northern California population of *C. v. oreganus*, the snakes grow an average of only 220 mm their first year (Fitch, 1949). In Utah, neonate *C. v. lutosus* have an average increase of 127 mm their first year.

The longevity record for the species is by a *C. v. viridis* kept 27 years, 9 months in captivity in Kansas that was estimated to have been two years old when caught (Bailey et al., 1989). Fitch (1949) thought some may survive about 16–20 years in the wild.

Food and feeding

Crotalus viridis is the most widespread and variable rattlesnake in western North America. Consequently, its several subspecies occur within the ranges of numerous potential prey species, and this snake seems to have the most variable diet of any rattlesnake in North America. Warm-blooded prey (rodents, lagomorphs, and birds) seem to be preferred, but lizards are also frequently taken. Probably small reptiles make up the greatest food bulk of juveniles in most subspecies, and possibly are a major food source for the two smallest subspecies, *concolor* and *nuntius*. However, Graves (1991) caught a neonate 24-cm *C. v. viridis* that had eaten a 12.1-g (77% of the snake's weight) *Peromyscus maniculatus*. The only ontogenetic shift in prey of *C. v. oreganus* found by Wallace and Diller (1990) was an increase in mammalian prey size with snake body length; snakes less than one year took shrews, but as the snakes grew, they captured more mice and voles, and finally prey as large as cottontail rabbits. Apparently, prey selection is limited by the size that can easily be swallowed, and prey size increases as the snakes grow.

The following list of prey species has been taken either from the literature presented below or is based on personal observations: mammals—shrews (*Sorex*), shrew moles (*Neurotrichus*), pikas (*Ochotona*), cottontail rabbits (*Sylvilagus*), juvenile jackrabbits and snowshoe hares (*Lepus*), pocket gophers (*Geomys*, *Thomomys*), kangaroo rats (*Dipodomys*, *Microdipodops*), pocket mice (*Perognathus*), juvenile

muskrats (*Ondatra*), chipmunks (*Eutamias*), juvenile yellow-bellied marmots (*Marmota*), ground squirrels (*Ammospermophilus*, *Spermophilus*), juvenile prairie dogs (*Cynomys*), juvenile tree squirrels (*Sciurus*, *Tamiasciurus*), wood rats (*Neotoma*), various mice (*Mus*, *Clethrionomys*, *Microtus*, *Peromyscus*, *Reithrodontomys*), and brown rats (*Rattus*); birds (including eggs and nestlings)—juvenile pheasants (*Phasianus*), quail (*Lophortyx*, *Oreortyx*), juvenile domestic chickens (*Gallus*), juvenile turkeys (*Meleagris*), grouse (*Dendragapus*), juvenile burrowing owls (*Speotyto*), woodpeckers (*Picoides*), mourning doves (*Zenaida*), bushtit (*Psaltriparus*), horned larks (*Eremophila*), mockingbirds (*Mimus*), western meadowlarks (*Sternella*), Brewer's blackbirds (*Euphagus*), bluebirds (*Sialia*), warblers (*Dendroica*), starlings (*Sturnus*), towhees (*Pipilo*), junco (*Junco*), lark buntings (*Calamospiza*), lark sparrows (*Chondestes*), song sparrows (*Melospiza*), Savannah sparrows (*Passerculus*), vesper sparrows (*Pooecetes*), and Gambel's sparrow (*Zonotrichia*); lizards—race runners (*Cnemidophorus*), alligator (*Gerrhonotus*), earless (*Holbrookia*), skinks (*Eumeces*), fence (*Sceloporus*), horned (*Phrynosoma*), and side-blotched (*Uta*); snakes—juvenile western rattlesnakes (*Crotalus viridis*, both in wild and captivity), kingsnakes (*Lampropeltis*, captivity), and leaf-nosed snakes (*Phyllorhynchus*, captivity); spadefoots (*Scaphiopus*); leopard frogs (*Rana*); fish-trout and salmon; and insects—mormon crickets, grasshoppers, beetles, and hymenopterans (Fitch and Twining, 1946; Fitch, 1949; Hamilton, 1950; Mosimann and Rabb, 1952; Stebbins, 1954, 1985, Wright and Wright, 1957; Cunningham, 1959; Fowlie, 1965; Bullock, 1971; Powers, 1972; Klauber, 1972; Banta, 1974; Conant, 1975; Young and Miller, 1980; Hammerson, 1982; Lillywhite, 1982; Duvall, King et al., 1985, Duvall, 1986; Genter, 1984; Gannon and Secoy, 1984; Jaksic and Greene, 1984; Diller and Johnson, 1988; MaCartney, 1989; Brown, 1990; Wallace and Diller, 1990).

Not all of the above may be taken as live prey, for carrion is sometimes consumed (Cunningham, 1959), and I once came upon a 875-mm *C. viridis* swallowing a road-killed thirteen-lined ground squirrel (*Spermophilus tridecemlineatus*) on a road in North Dakota.

Stomach and scatological examinations of California *C. v. oreganus* by Fitch and Twining

(1946) showed the California ground squirrel (*Spermophilus beecheyi*) to be the chief prey, followed by kangaroo rats (*Dipodomys*), cottontails (*Sylvilagus*), white-footed mice (*Peromyscus*), pocket mice (*Perognathus*) and pocket gophers (*Thomomys*). Mammals predominated, but two birds, four lizards, and a spadefoot toad were identified among the remains. Juvenile ground squirrels (*Spermophilus*) constituted 81% of ingested biomass by *C. viridis* from southwestern Idaho (Diller and Johnson, 1988), and in northern Idaho, voles (*Microtus*), deer mice (*Peromyscus*) and cottontail rabbits (*Sylvilagus*) accounted for nearly 92% of the biomass injested and 80% of the total number of prey taken (Wallace and Diller, 1990). In a study on the extent of snake predation on lizards in California, Jaksic and Greene (1984) found only 26 lizards (9%) among 285 prey items in *C. viridis*.

First-year *C. viridis* have greater consumption rates (expressed as percent of body mass) than do older age classes (Diller and Johnson, 1988). The production efficiency for all *C. viridis* (proportion of prey mass consumed that is used for growth) is 28%. The annual predation rate by the rattlesnakes was 14% of the juvenile ground squirrel population and 5–11% of the population of juvenile cottontails. Feeding usually takes place from mid-April to early October in most populations. In British Columbia, *C. viridis* feeds most often from June to August, and rodents make up 91%, shrews 5%, and birds 4% of the prey consumed (Macartney, 1989). Newborn snakes and small juveniles prey on the smallest mammals, whereas adults eat larger prey of a greater diversity.

In most populations, gravid females usually do not feed during gestation or after parturition, but Wallace and Diller (1990) reported that in northern Idaho, reproductive females feed from the spring to early August, and again in the fall after the birth of the young. They thought feeding during the breeding year may explain how some females can reproduce in consecutive years.

Crotalus viridis uses two predatory strategies: either that of an actively searching forager or that of a sit and wait ambusher (Diller, 1990). The first is used when prey are scattered, or often during migration to or from the den (Duvall, King et al., 1985). The second is used more often (particularly by males) when colonies of rodents have been identified and the snake can lie near the openings of burrows or rodent runs. The snake is capable of only short bouts of maximal activity before exhaustion (Ruben, 1983), so the latter strategy is more efficient from an energy standpoint. During maximum activity, *C. viridis* has only moderate powers of anaerobiosis and limited powers of aerobiosis. This low endurance might be associated with exercise-related bone dissolution, resulting in hypercalcemia which may function to further debilitate the already poorly developed blood-buffering capacity of the snake (Ruben, 1983).

Prey may be detected by vision, infrared emissions (particularly from birds and mammals), or olfaction. Movement of the prey seems to be the primary visual component which brings on further exploratory behavior (Scudder and Chiszar, 1977), principally a rise in the rate of tongue flicking (Chiszar, Taylor et al., 1981), that provides additional olfactory information. There is also an integration of visual and infrared information in the bimodal neurons of the optic tectum which receive input from both the retina and pit organ (Newman and Hartline, 1981). This is extremely important during crepuscular or nocturnal hunting. However, a visual component may not be necessary for an accurate strike; a congenitally blind *C. v. oreganus* not only could successfully strike mice, but accurately directed the strikes to the more vulnerable anterior region (Kardong and MacKessy, 1991).

The pit organ may not be the only means of detecting prey body heat. Chiszar, Dickman et al. (1986) found that *C. viridis* who had been anesthetized so that the trigeminal nerve could not mediate electrophysiological responses of the pit organs to thermal stimulation still exhibited behavioral responses to thermal cues. They believed that either an auxiliary infrared-sensitive system (nociceptors) or the common temperature sense could be responsible. Intraoral thermal stimulation elicits response from the superficial maxillary branch of the trigeminal nerve (Dickman et al., 1987). These responses to oral heat stimulation are independent of any responses associated with thermal stimulation of the loreal pit organs. Histological preparations of tissues from the upper lip, palate, and fang sheath reveal dense ramifying neurons in the epidermal layers of the fang sheaths that are morphologically similar to the infrared-sensitive neurons in pit organ membranes. Visual and thermal cues are sufficient to bring on a strike response in *C. viridis*. When free-ranging *C. v. viridis* were presented warm models of deer mice (*Peromyscus*) devoid of mice odors, the snakes readily struck at the models (Hayes and Duvall, 1991). In every case the core temperatures of the models (mean 32.4°C, 21.5–38.5) were 1.5–4.5°C warmer than the background air temperatures. Since no odor identification could be made from the models, this indicates that olfactory cues may be unnecessary for the release of predatory strikes.

Olfaction, however, also plays a major role in prey detection, as indicated by the spring migratory patterns (especially of males; Duvall, Goode et al., 1990) and the modified behavior of *C. viridis* with vomeronasal organs closed by sutures (Graves and Duvall, 1985a). Snakes rendered avomic by this procedure have lower tongue-flick rates when presented prey odor stimuli. Normal western rattlesnakes may mouth gap and actually shake their heads in two or three rapid horizontal jerks to help bring odors to the vomernasal organ (Graves and Duvall, 1983, 1985b). These results are in direct contrast to the conclusion by Chiszar, Scudder and Knight (1976) that, based on tongue-flick-rate experiments, the vomeronasal system of *C. viridis* is not activated by olfactory cues.

Perhaps prey-chemical preferences must be learned. Chiszar and Radcliffe (1977) reported the absence of these in neonate *C. viridis*, but they have been demonstrated in numerous tests on adults. The tests on neonates involved responses (or lack of) to five nonprey organisms (cricket, garter snake, fish, salamander, worm) and two prey types (lizard, mouse). No significant difference in tongue-flick rates was detected for either group. However, later tests using the odor of fish mucus did bring on greater numbers of tongue flicks than did mouse odor (Chiszar, Scudder and Smith, 1979), unless the mouse was first struck and subsequently tasted. The significance of the response to fish odors is unclear. Tests by Chiszar, Radcliffe, Smith et al. (1981) on food-deprived *C. viridis* resulted in hunger-related increases in tongue-flick rates in response to mouse odors.

When prey wander into striking range, *C. viridis* usually bites it only once, after which the prey may walk freely off while the venom digests it (Klauber, 1972). Lizards and small rodents are usually held, but large rodents are released (Radliffe et al., 1980; Chiszar, Radcliffe, Byers et al., 1986). Tests on striking behavior of *C. v. oreganus* by Kardong (1986a,b) indicate that the larger the snake, the more likely it is to hold its prey, rather than quickly release it. Snakes have a ready reserve of venom sufficient to envenomate up to four mice in close succession without loss of killing effectiveness, but when presented several mice in succession, the later mice are often struck repeatedly or held in the mouth rather than released. This behavior probably results from the depletion of and reduced access to venom reserves (Hayes et al., 1991).

The head or chest are most often struck, and envenomation of these areas most quickly leads to death. Small prey may simply be swallowed alive without envenomation by larger *C. viridis*. Retention of prey following a bite increases the severity of the envenomation, and a poor strike is usually rapidly followed by a second or third. However, holding prey also increases the chance of a retaliatory bite with possible injury to the snake.

Once prey has been struck, and possibly tasted at that time, a chemical search image is created resulting in an increase in the rate of tongue flicking (O'Connell et al., 1981; Chiszar, Radcliffe, O'Connell et al., 1982; Radcliffe et al., 1986; Cruz et al., 1987; Melcer and Chiszar, 1989a,b; Melcer et al., 1990; Chiszar, Hobika et al., 1991; Chiszar, Radcliffe et al., 1991; Furry et al., 1991; Robinson and Kardong, 1991). Post-strike head orientation and tongue flicking may facilitate location of the trail left by the dying prey. No prey-trailing behavior is elicited unless the prey has been envenomated (Golan et al., 1982; Diller, 1990). Western rattlesnakes locate the departure trail of a bitten mouse before they locate the entry trail of an unstruck mouse (Lee et al., 1988). *C. viridus* can even distinguish between the trails of individual struck sibling mice (Furry et al., 1991). Carcasses of struck mice, even if concealed, are located before those of unenvenomated mice (Chiszar, Nelson et al., 1988). It is apparently not the odor of venom that attracts *C. viridis* to the proper prey trail.

Robinson and Kardong (1991) demonstrated that venomoid snakes in which the venom duct was ligated to prevent venom release still show a preference for trails of struck mice over those of nonstruck mice.

The rate of tongue flicking increases as the dead prey is approached (the rate of tongue flicking is higher after predatory than after defensive strikes; O'Connell et al., 1982) and may be greater during a nocturnal strike sequence (O'Connell et al., 1983). Once the odor trail of the injured prey is located, the snake holds his head to within 2 cm of the odor cue, and 75% of the tongue flicks are directed to the trail (Golan et al., 1982), but most searches only last up to about 15 minutes. If environmental odors interfere with the searching, the flick rate is increased (Scudder et al., 1983). Adult mice envenomated by adult *C. viridis* may crawl more than 670 cm in an open field before dying (Estep et al., 1981), so the snake may have to work hard to reach its meal. Wild house mice are capable of traveling significantly farther than laboratory-reared house mice after being bitten (Hayes and Galusha, 1984).

Chiszar et al. (1991) reported that the tongue-flick rate is higher after warm models of prey are struck than after the same snakes struck identical models at ambient temperature. They thought this indicates that the snake's decision to deliver a predatory strike is not automatically followed by a full expression of searching behavior. The snake must make a second decision, probably using additional information, before initiating search behavior.

Envenomated mice are preferred over nonenvenomated mice (Duvall et al., 1978; Chiszar, Duvall et al., 1980), and odors from the anterior end of the mouse help the snake quickly find the head for easier swallowing (Duvall, et al., 1980). These odors from the nasal-oral tissues of the mouse may also produce a swallowing modal action pattern in *C. viridis*.

Venom and bites

Fang length is positively correlated with body length in venomous snakes, so the largest subspecies of *C. viridis* have the greatest mean fang lengths, while the smallest subspecies have shortest mean fang lengths. Klauber (1939) measured the fangs of six subspecies. Presented below are his results; following the subspecific name is the number of specimens examined in parentheses, their body length range, and their fang length range, respectively: *viridis* (590), 700–1000 mm, 6.3–8.4 mm; *abyssus* (5), 656–981 mm, 5.3–8.4 mm; *concolor* (5), 522–735 mm, 4.1–5.2 mm; *lutosus* (5), 793–1,173 mm, 6.0–8.8 mm; *nuntius* (5), 458–540 mm, 3.8–4.8 mm; *oreganus* (4 localities, 22), 547–1,140 mm, 4.1–9.6 mm.

The venom of *C. viridis* is predominately hemorrhagic in its effects, but that of some subspecies also contains neurotoxic peptide elements (Russell, 1983). The overall toxicity of the venom is slightly higher than that of the larger species *C. atrox,* and this potency coupled with the high irritability of many individuals makes *C. viridis* a very dangerous snake. Fortunately, bites are relatively infrequent in occurrence.

The chemistry of the venom has been described by Schaeffer et al. (1972a,b), Tu (1977), Cameron and Tu (1977), Russell (1983), Aird et al. (1988, 1991) and Soto et al. (1989), and the reader is referred to those publications for detailed information beyond the scope of this book. Two color shades of venom occur, white (colorless) or yellow; they differ biochemically and may be produced in different venom glands of the same snake. White venom has fewer lower molecular weight components and is less toxic (Johnson et al., 1987). Juveniles have colorless venom, but qualitatively it is similar to the yellow venom of adults (Fiero et al., 1972).

Hemorrhagic, neurologic, and proteolytic activities may all result during the development of a single bite. Hemorrhagic activity may occur in as short a time as 18 minutes (Soto et al., 1989). Neurologic symptoms may be paralytic in nature, but not as severe as those resulting from bites by *C. scutulatus* or the South American rattlesnake *C. durissus* and its close relatives, although the venom of some New Mexican *C. v. viridis* contains the Mojave toxin usually associated with *C. scutulatus* (Glenn and Straight, 1990), and the venom from *C. v. concolor* has a similar amino acid sequence to the Mojave toxin in the basic subunit of its related toxins (Aird et al., 1990).

Venom of *C. viridis* produces a precipitous fall in systemic arterial pressure and flow, a rise in systemic venous pressure, pulmonary artery pressure, and cisternal pressure along with changes in respiration and the electrocardiogram and electroencephalogram (Russell and Michaelis, 1960; Schaeffer et al., 1973). A peptide in the venom can cause cardiovascular failure, probably by an increase in vascular permeability to protein and erythrocytes (Schaeffer et al., 1979). Injected rats experience venom shock characterized by hypotension, lactacidemia, hemoconcentration, hypoproteinemia, and death (Schaeffer et al., 1979). There is a drop in the peripheral platelet count in rats, but the venom does not seem to damage the precursors in the bone marrow (Wingert et al., 1981). Symptoms recorded by Hutchison (1929) resulting from human envenomation include swelling, pain, weakness, giddiness, breathing difficulty, hemorrhage, weak pulse, heart failure, nausea and vomiting, secondary gangrene infection, ecchymosis, paralysis, unconsciousness or stupor, nervousness, and excitability. Russell (1960) reported that the bite produces pronounced changes in consciousness, profuse weakness, sweating, respiratory difficulties, and tingling or numbness over the bitten area and over the tongue, mouth, and scalp. Death in untreated cases has occurred within 18 hours to five days, and within 18–27 hours in several poorly treated cases (Hutchison, 1929). Venom of the subspecies *concolor* is 10–30 times more toxic than that of any other subspecies (Glenn and Straight, 1977). Over (1928) and Klauber (1972) present case histories of bites.

Hemorrhagic intensity varies between populations of *C. v. lutosus*. Venoms from snakes occupying northern ranges in Utah have high hemorrhagic ability, while those of individuals from southern Utah and northern Arizona have a lower hemorrhagic capability (Adame et al., 1990).

Potential venom yield increases with body length (Klauber, 1972). Adult *C. viridis* produce average dried venom yields of 75–160 mg (Minton and Minton, 1969), and Klauber (1972) reported that the yield in dried venom per fresh adult for several subspecies varies from 22 to 112 mg, with maximum yields of 72–390 mg. The average amount of dry venom se-

creted at one time is 65–90 mg (Amaral, 1928). The lethal venom dose for a human adult is 70–160 mg for *C. v. helleri* (Minton and Minton, 1969). Minimum lethal doses for some other animals are: pigeons, 0.08–0.10 mg (Githens and George, 1931); 20-g mice, 0.025–0.07 mg (Githens and Wolff, 1939); and 22-g mice, 0.045–0.14 mg (Macht, 1937).

Hayes et al. (1991) studied the quantity of venom expended while striking by *C. v. viridis*. Enzyme-linked immunosorbent assays of skin washings and whole-animal homogenates were used to measure quantities of venom (dry weight) released. The mean mass of venom expended in a single predatory bite was 14 mg; approximately 89% of the venom was injected into the tissues (12 mg) while the rest was left on the skin surface (1.5 mg). Thirty-two percent of the available venom (44 mg) was released in a single bite, which was more than previously reported for other viperid species.

Grazing horses and cattle are sometimes bitten, but usually survive. Martin (1930) reported that his dog was struck seven or eight times by *C. v. oreganus* over a period of a few years and always apparently fully recovered, but Schaeffer et al. (1973) calculated the LD_{50} for dogs was 0.05 mg/kg of body weight.

Crotalus viridis is not immune to the venom of its species, and fatal bites have occurred between individuals in captivity, particularly between those of different subspecies (Sanders, 1951; La Rivers, 1973).

Venom of *C. viridis* is remarkably stable. That of a snake does not vary electrophoretically from the time of its emergence in the spring until hibernation in the fall (Gregory-Dwyer et al., 1986; Johnson, 1987); however, there may be considerable variation between individuals. Dried venom of this species retains most of its toxicity for mice and cats even after storage for 26–27 years (Russell et al., 1960).

Predators and defense

As *Crotalus viridis* is so widespread and often common, more observations of predation upon it have been published than for any other rattlesnake. The following animals are known predators: marmots (*Marmota*), skunks (*Mephitis*), badgers (*Taxidea*), domestic cats (*Felis*), bobcats

(*Lynx*), domestic dogs and coyotes (*Canis*), foxes (*Vulpes*), domestic hogs (*Sus*), golden eagles (*Aquila*), red-tailed hawks (*Buteo*), Cooper's hawks (*Accipiter*), great horned owls (*Bubo*), wild turkeys (*Meleagris*), roadrunners (*Geococcyx*), kingsnakes (*Lampropeltis*), whipsnakes (*Masticophis*) and racers (*Coluber*) (Fitch, 1949; Klauber, 1972; Toweill, 1982; Duvall, King et al., 1985; Duvall, 1986). Fitch (1949) also saw a California jay (*Aphelocoma californica*) warily circle a small western rattlesnake coiled on a lawn, and thought the bird could have overpowered the snake if it had attacked. *C. viridis* sometimes eats its own kind, both in nature and captivity (Bullock, 1971; Klauber, 1972; Powers, 1972; Banta, 1974; Lillywhite, 1982), and a *C. ruber* swallowed one at the San Diego Zoo (Klauber, 1972).

Fitch (1949) reported that red-tailed hawks (*Buteo jamaicensis*) and coyotes (*Canis latrans*) were the worst natural predators on his study population of *C. v. oreganus* in California. Of 4,110 total vertebrate prey items recorded from the hawks, 74 were rattlesnakes, and of 1,924 vertebrate prey items of the coyote, 17 were *C. viridis*. In each case, this amounts to less than 1% of the total vertebrates taken by these predators. In Wyoming, golden eagles (*Aquila*) and hawks (*Buteo*) seem to be the leading predators (David Duvall, pers. comm.).

Accidents involving other wild animals often prove fatal. Deer, pronghorn, goats, horses, and cattle often trample *C. viridis* (Klauber, 1972), and Metter (1963) reported finding a *C. viridis* with porcupine quills protruding from its head.

Despite its many natural enemies, *C. viridis* has the most to fear from humans who usually kill it on sight. Many are slaughtered on our highways each year, and habitat destruction has taken a great toll in some states. These rattlesnakes have even been poisoned during rodent control (Campbell, 1953) and from apparently eating prey from a solid waste storage and disposal area (irradiation with cesium; Arthur and Janke, 1986). Formerly large populations overwintered in specific dens, but many of these have been systematically eradicated with gas, bullets, and explosives (Martin, 1930; Klauber, 1972; Brown and Parker, 1982) so that now *C. viridis* is in danger of totally disappearing from areas where it was once the most common snake.

When disturbed *C. viridus* is a determined, spirited fighter. I will never forget my first encounter with one. Surprisingly this did not occur in the wild, but instead on the stage of a college auditorium. As an undergraduate interested in herpetology, I had volunteered to help the late Carl Kauffeld set up a show he was to give at the college. As we unpacked his props and animals, he came to a box that he remarked held "hot stuff" and told me to stand back. When I did, he removed an adult prairie rattlesnake from the box and placed it on the demonstration table. The snake immediately went into a defensive coil with its head and anterior portion of the body raised high, and began to rattle continuously. I did not have to be told not to approach it, for that snake was the meanest I had seen to that point, quite in contrast to the eastern copperheads and timber rattlesnakes with which I was already familiar. The snake remained coiled, not moving from the table, as we worked around it (needless to say, out of striking range). Whenever I walked around the table, the snake rotated its coils so that it always faced me, as if it had taken an instant dislike to me. This behavior, with the constant rattling, continued for more than 30 minutes, and ceased only when Kauffeld placed the snake in a compartment in one of the props. I came to the show that day just to see if it would "perform" again. That little rattler did not disappoint me, bursting balloons and acting continually agitated throughout its part of the show. I assume this is the same irritable *C. viridis* to which Kauffeld referred in his 1969 book.

The defensive posture described above is similar to that of *C. atrox*, and while *C. viridis* is a shorter snake and does not rise as high, it certainly gets your attention. I have since observed little difference in the behavior of wild *C. viridis* in the Dakotas and Wyoming from that of the captive described above. If the snake is approached too close it will strike viciously, and on one occasion I had an 810-mm adult advance toward me each time it struck. It is said that some subspecies are more mild mannered than others (Douglas, 1966; Young and Miller, 1980), but I cannot verify this as all those I have encountered have had bad tempers.

Duvall, King et al. (1985) studied the defensive sequences of more than 200 *C. viridis* in Wyoming. Usually it will first rely on its cryptic coloration and pattern to escape detection, especially when sufficient cover is present. On realizing it has been discovered, it normally tries to crawl directly away from the threat. If further hassled, the snake will then either coil and rise high or stretch out its body so that the head and upper third of the body are positioned, or cocked, for a potential bite while the snake backs away with the posterior two-thirds of its body. If this does not help, the snake may simply hide its head under the central (usually widest) portion of the body (I have seen this occur only once). Finally, if further disturbed, it will strike. It seems to be the movement of the enemy rather than its physical contour that directs the attention of the snake (Scudder and Chiszar, 1977).

Gravid females seem to rely more on procrypsis and escape, but when disturbed, because of their normally elevated body temperatures, they are more reactive than either males or nongravid females (Goode and Duvall, 1989). Postpartum females essentially follow the above warning sequence; perhaps in some cases this may involve defense of the newborn young (Graves, 1989).

The snake is also capable of projecting fluid from its anal glands in a spray of fine droplets for a distance of about a meter (Fitch, 1949; pers. obs.). Studies by Graves and Duvall (1988) seem to indicate that an alarm pheromone may be released from the cloacal glands by defensively aroused C. viridis. Individuals exposed to odors from alarmed, sham-operated conspecifics with intact cloacal glands exhibited increased heart rates when subsequently presented with a threatening stimulus. In contrast, the heart rates of snakes exposed to a control odor, or to odors of alarmed conspecifics with their cloacal glands surgically removed, remained near normal.

Sound frequencies of the rattle range from mean lows of 3.28–3.34 kHz to mean highs of 10.40–10.62 kHz; the mean width of the sound band is 7.06–7.33 kHz (Fenton and Licht, 1990).

Lillywhite (1974) has noted an interesting case of procrypsis in C. viridis in recently burned chaparral. The dark (often black) coloration of the rattlesnake renders it cryptically colored during post-fire years, as they are easily confused with charred stalks of wood plants scattered within the burned area.

Populations

If the habitat is proper, C. virdis may be the most common local snake, particularly in the Dakotas and adjacent Montana and Wyoming, but in many areas populations have decreased or disappeared owing to direct human predation at hibernacula. Generally the largest populations existing today occur in isolated areas.

The longest studied den is in northwestern Utah (Woodbury, 1951; Hirth and King, 1968; Parker and Brown, 1973, 1974; Brown and Parker, 1982). Rattlesnakes using this hibernaculum were studied for 10 years by Woodbury (1951) and his students, and more than 900 different snakes were marked from 1939 to 1950. The population size was estimated to be 769 snakes in 1940, but by 1950 had dwindled to only 235. During three consecutive years, 1964–1966, Hirth and King (1968) collected only 58, 53, and 55 C. viridis, respectively, at the den (an annual biomass of about 20 kg), and by 1969–1973, the population had dropped to only about 12–17 individuals (Parker and Brown, 1974). Another Utah den contained only 69 adults in 1949 after an estimated 300 C. viridis, had been killed there in 1937 (Woodbury and Hansen, 1950). Fitch (1949) reported a density of approximately 2.9/ha at his California site (males 1.3/ha, females 1.6/ha). The population at a den in Saskatchewan was estimated to be 149 (Gannon and Secoy, 1984), and at one in Wyoming, 42 adults (Duvall, King et al., 1985). Overwintering mortality is high, particularly among juveniles. Gannon and Secoy (1984) reported that the proportion of the young-of-the-year decreased from 39% of the population to only 12.7% from the fall of 1976 to the spring of 1977. Fitch (1985b) calculated a mortality rate of 60% in first-year young and 50% in each subsequent year for a population in Kansas. He thought few live to be older than eight years in the wild; however, most populations are composed predominately of adults with snout-vent lengths greater that 500 mm.

The sex ratio of a typical population does not vary significantly from 1:1 (Fitch, 1949; Julian, 1951; Gannon and Secoy, 1984; Diller and Wallace, 1984; Duvall, et al., 1985; King and Duvall, 1990), but the juvenile to adult ratio may vary from 1:1 or 1:2 (Diller and Wallace, 1984; Macartney, 1985) to 2:1 (Fitch, 1949).

Remarks

Gloyd (1940), based on body pattern, thought *C. viridis* most closely related to *C. scutulatus*. However, Brattstrom (1964) proposed that it was morphologically closest to *C. mitchelli, C. tigris, C. cerastes* and *C. atrox*. Klauber (1972) agreed, for the most part, with both Gloyd and Brattstrom. Recent electrophoretic studies of *Crotalus* venom by Foote and MacMahon (1977) place *C. viridis* with *C. atrox* and *C. ruber*.

Comparison of venom proteins of *C. v. abyssus* with those of the subspecies *concolor*, *lutosus* and *nuntius* shows *abyssus* to be more closely related to *lutosus* than the other subspecies (Young et al., 1980).

The western rattlesnake seems to be in-volved in several mimicry systems. Gopher snakes (*Pituophis melanoleucus*) share aspects of coloration, pattern, and defensive behavior with sympatric subspecies of *C. viridis* in what seems to be a case of Batesian mimicry (Kardong, 1980; Sweet, 1985). Another possible case of mimicry involving *C. viridis* is that of the hiss of the burrowing owl (*Speotyto cunicularia*) and the sound of the snake's rattle (Rowe et al., 1986).

No discussion of *C. viridis* is complete without mention of the involvement of the sub-species *nuntius* in the Hopi Indian snake dance. During the dance the snakes are carried in the mouths and hands of the participants (for a detailed description and discussion, see Klauber, 1972).

Crotalus willardi (Meek, 1906)
Ridgenose rattlesnake
Plates 54, 55

Recognition

This small (maximum length, 67 cm) montane serpent was the last species of rattlesnake to be described from the United States. It is gray to brown or reddish brown with a series of 18–26 white to cream-colored, dark-bordered crossbars, and only 1–3 anterior rings on the tail. Three longitudinal rows of small, indistinct dark spots occur along the sides. Individuals from Arizona have two white longitudinal stripes on the side of the face; one extending diagonally backward from the prenasal scale below the eye to the corner of the mouth, and a second extending backward along the supra- and infralabials. A median white, vertical stripe extends downward from the rostral and mental scales. Snakes from New Mexico lack the rostral-mental stripe and have only faded longitudinal stripes on the side of the face. The pink, cream, or tan venter is mottled with dark brown or black. At midbody are 25–27 rows of keeled body scales. Beneath are 144–159 ventrals, 21–35 subcaudals, and an undivided anal plate. The rostral is higher than wide, anteriorly pointed, and slightly upturned. It combines with the recurved outer edges of the 2 internasals and 2/2 canthals to produce a distinctly raised snout. As many as 60 small scales (usually 20–40) occupy the intercanthal

Crotalus willardi willardi (John H. Tashjian, Arizona- Sonora Desert Museum).

to parietal areas (no prefrontal scales are present), and there are 6–9 intersupraoculars; the 2 supraoculars are reduced in size. The nasals are followed by 2(1–3) loreals, 2–3 preoculars (the postnasal does not touch the upper preocular, but the prenasal contacts the supralabials), and 3–4 postoculars; 13–14(12–17) supralabials and infralabials are present. The hemipenis is bilobed with a divided sulcus spermaticus, and about 56 short, heavy spines and 16 fringes on each lobe; 1–2 spines lie in the crotch between the lobes (Klauber, 1972). The reduction of the lobal spines to reticulations is quite sudden (Harris and Simmons, 1976). The dental formula is 1 maxillary (the fang), 6(5–7) pterygoidal, 1–2 palatine, and 8 dentary teeth.

Males have 144–154 ventrals, 25–35 subcaudals, and tails 9.1–11.5% of the entire body length; females have 149–159 ventrals, 21–32 subcaudals, and tails only 7.9–9.8% the length of the entire body.

Karyotype and fossil record: Unknown.

Distribution

Most of the geographic range of *Crotalus willardi* lies within Mexico, as it is known from only a few localities in south-central Arizona and extreme southwestern New Mexico. In Mexico, it occurs from north-central and northwestern Sonora and eastern Chihuahua southward through west-central Durango to southwestern Zacatecas. Being an upland species, its range lies between 1,600 and 2,750 m elevation.

Geographic variation

Five subspecies have been described, but only two live in the United States. *Crotalus willardi willardi* Meek, 1906, the Arizona ridgenose rattlesnake, ranges from the Huachuca, Patagonia, and Santa Rita mountains in southeastern Arizona southward into the Sierra de los Ajos, Sierra Azul, and Sierra de Cananea of northern Sonora, Mexico. It has well-developed white stripes on the sides of the face and vertically on the rostral and mental scales, a brownish to reddish-brown back, and little dark spotting on the head. *C. w. obscurus* Harris and Simmons, 1976, the New Mexico ridgenose rattlesnake, lives only in the Animas and Peloncillo mountains of New Mexico, and the Sierra de San Luis of extreme northeastern Sonora and western Chihuahua, Mexico. It lacks the vertical white stripe on the rostral and mental scales, and the lateral facial strips are faded or absent. Its back may be either gray or brownish, and the head is heavily marked with dark spots.

The New Mexican population when first discovered was thought to represent the Mexican subspecies *C. w. silus* (Bogert and Degenhardt, 1961), but reevaluation by Harris and Simmons (1976) showed it to be an undescribed form which they named *C. w. obscurus*.

Confusing species

The raised rostrum and pattern of white facial stripes set this species apart from all other rattlesnakes in the United States. The western hognose snake, *Heterodon nasicus,* also has an upturned snout, but lacks the loreal pit, elliptical pupil, and tail rattle of *C. willardi.*

Habitat

This is a montane woodland species associated most often with pine-oak or scrub oak forests where it lives along streams about shaded rock outcrops with crevices, in rock piles, or in old stumps. Usually a thick mat of leaf litter or

Distribution of *Crotalus willardi.*

pine needles covers the ground. This microhabitat is more humid and less open than those occupied by its sympatric relatives, *C. pricei* and *C. lepidus*.

Behavior

There is an appalling lack of knowledge of the natural behavior of *Crotalus willardi*. The annual activity period in Arizona extends from April or early May to mid-November (Johnson, 1983; Lowe et al., 1986). The colder months are spent undercover, probably deep within rock crevices or within rotting stumps or logs.

Crotalus willardi is most active during the day, particularly in the spring and fall, but some nocturnal foraging may take place in the hottest summer months. Daily activity reaches its peak on warm, humid mornings, but some will emerge late in the afternoon, especially after rains. Toward fall, when daytime temperatures are dropping, the activity period shifts to the afternoon. Active *C. willardi* have been found when air temperatures varied from 23.9°C (Kauffeld 1943a) to between 30 and 32°C (Wright and Wright, 1957; Brattstrom, 1965), but most foraging and basking takes place when air temperatures are 24–29°C (Armstrong and Murphy, 1979).

Reproduction

The total body length at which maturity is attained by either sex has not been determined. Female *C. w. willardi* 481 and 460 cm long have given birth (Klauber, 1972; Quinn, 1977), and a 548-mm male *C. w. willardi* has courted and successfully mated (Tryon, 1978).

The male and female sexual cycles are also unknown. However, as females of other species of *Crotalus* experience a biennial cycle and Martin (1975c, 1976) has reported an approximate 13-month gestation period in captivity, Tryon (1978) thought the female cycle to be at least biennial.

Courtship activity and copulation have occurred in captive *C. willardi* on 29 January, 17–19 April, 16, 19 and 22 June, 15 and 28 July, in early August, and on 8–9 September (Martin, 1975b, 1976; Tryon, 1978; Armstrong and Murphy, 1979), and Lowe et al. (1986) observed mating in the wild in July. A pair of *C. w. willardi* observed by Armstrong and Murphy (1979) were joined from 0800 to 1700 hours, and Tryon (1978) mentioned matings in this subspecies that lasted 15–24 hours, but which may have been even longer, as entire sequences were not witnessed. Coitus often follows soon after female ecdysis.

The best description of courtship and mating behavior in this species was reported by Tryon (1978) for a pair of *C. w. willardi:* The pair was separated during hibernation, and the female was introduced into the male's enclosure periodically throughout the warm months. Upon introduction, the male immediately began courting with rapid, longitudinal chin rubbing (1/sec) and tongue flicks (2–3/sec) against the female dorsally and dorsolaterally. If unreceptive, she reacted with an immediate, rapid slapping of the tail from side to side and an attempt to escape, or she assumed a tight coil. The female indicated receptiveness by a lack of tail slapping and slowly crawled about the enclosure, which intensified the male's courtship attempts. Once the male assumed a position of loose coils on and parallel to the female's body, she raised her tail slightly and ceased forward momentum. The male then looped his tail under hers and attempted to align the cloacae with several anterior-posterior strokes. If intromission was not accomplished the sequence of chin rubbing, tail tightening, and cloacal search began again. Once intromission was accomplished, most movement subsided in both snakes for approximately 1–2 min, whereupon chin rubbing and tongue flicking was periodically initiated by the male. Entire body movement was reduced during actual coitus. On several occasions, a rapid head-bobbing movement was observed in the female during coitus. This, in turn, accelerated the chin-rubbing behavior in the male. Copulation was effected 19 June 1975 and 22 June 1976. The female had shed several hours prior to the initiation of each coitus. Nonreceptivity by the female was observed until actual ecdysis occurred.

A captive male *C. w. amabilis* observed during courtship by Guese (in Armstrong and Murphy, 1979) directed head bobbing and tongue flicking across a coiled, resting female's back (3–5/5 seconds). He moved his uplifted tail in both a horizontal and vertical plane with an undulating motion. The sequence lasted five minutes. A captive male *C. w. silus* began twitching (1/sec) after discovering a female, and tested the entire dorsal surface of her body with his tongue (2 flicks/second). He rubbed along her back for approximately five minutes by holding his head at a 30° angle and sliding the area of the mental scale forward for 1 cm (Armstrong and Murphy, 1979).

Litters of *C. w. willardi* contain two (Stebbins, 1985) to nine (Klauber, 1972) young, but average about four to six young. The only reported litter of *C. w. obscurus* had nine young (Martin, 1976). There seems to be a positive correlation between the number of young produced and female body size. Twenty-four neonate *C. w. willardi* averaged 186.7 mm (165–214) in total body length, and 6.0 g (5.4–7.0) in weight. The nine young *C. w. obscurus* averaged 205 mm (200–212) in length and 7.4 g (6.8–8.0) in weight (Martin, 1976). Captive and wild birth dates range from 1 August to 10 September.

Young *C. w. willardi* are brownish at birth, and may or may not have yellowish tails (Martin 1975a,c; Quinn, 1977; Tryon, 1978). Young from the only known brood of *C. w. obscurus* were dark brown with distinctly blackish tails (Martin, 1976).

An apparent natural hybridization between *C. willardi* and *C. lepidus* has occurred in the Peloncillo Mountains of New Mexico (Campbell, Brodie et al., 1989).

Growth and longevity

The growth rate has not been studied. However, *Crotalus willardi* may have a long life span; a captive female lived for 21 years, 3 months, and 24 days (Bowler, 1977).

Food and feeding

Crotalus willardi captures its prey either by striking it from ambush or by actively hunting for it. When rodents are bitten, they are usually released at once, and later trailed by olfactory cues. Lizards are usually struck in the body region and are retained until dead.

This small rattlesnake has a varied diet. Wild *C. w. willardi* eat mice (*Peromyscus*), lizards (*Sceloporus*), birds (rufous-crowned sparrows, *Aimophila ruficeps*), scorpions, and centipedes (Klauber, 1949b, 1972; Fowlie, 1965; Martin,

1975b, Parker and Stotz, 1977). In captivity, this subspecies has consumed white laboratory mice (*Mus*), lizards (*Urosaurus*), snakes (*Hypsiglena, Trimorphodon*), and centipedes (*Scolopendra*) (Kauffeld, 1943b; Vorhies, 1948; Woodin, 1953; Manion, 1968; Armstrong and Murphy, 1979; Johnson, 1983; Lowe et al., 1986). *C. w. obscurus* has taken white mice (*Mus*) and green anoles (*Anolis*) in captivity (Bogert and Degenhardt, 1961; Martin, 1976).

Juvenile *C. willardi* are probably more dependant on lizards as prey than are adults.

Venom and bites

Klauber (1939) reported fang lengths of 5.3–6.0 mm for *C. willardi*. The total volume of venom available for injection by *C. willardi* is rather small; Klauber (1972) extracted only a total of 3.7 mg dried venom from one *C. w. willardi*. The venom is also relatively weak; the minimum lethal dose for a 20-g mouse is 0.24 mg (Githens and Wolff, 1939), and for a 350-g pigeon, only 0.1 mg (Githens and George, 1931). More recently, Minton and Weinstein (1984) reported a venom yield of 3.1 mg for a *C. w. willardi*, and that the intravenous LD_{50} (mg/kg) for 20–25-g mice was 1.61. There was insufficient venom to determine a subcutaneous LD_{50}, but a dose of 0.33 mg was lethal and produced extensive subcutaneous hemorrhage. The venom possesses a moderate amount of protease activity.

Russell (in Minton and Weinstein, 1984) treated a human bite by *C. willardi* that showed only minimal local signs of envenomation.

Predators and defense

Natural predators of *C. willardi* are unknown, but probably include ophiophagous snakes, birds of prey, and carnivorous mammals. However, human activities are more destructive than those of natural predators, and over-collection of *C. w. obscurus* in New Mexico has severely decimated the Animas population to the point that it should now be considered endangered.

This is a relatively secretive snake of mild disposition. Most lie still without rattling, or may try to crawl away, occasionally shaking their tails as they go. Seldom do they coil and strike at an intruder, but they do have the unpleasant habit of turning and trying to bite the hand holding their neck, so one must be alert to avoid such an accident. Usually all rattling ceases once they are placed in a collecting bag or container.

Populations

In the United States and Mexico, the populations of C. willardi are not continuously distributed, but instead occur in isolated pockets on separate mountain ranges. This has led to evolution of much variation (five subspecies), but also could lead to extinction of some of the smaller isolated populations if human interference or natural conditions become intolerable. Fortunately, the normal microhabitat is one not often visited by humans, and the major populations of C. w. willardi in Arizona are situated in federally protected lands. At some sites in Arizona it may be the most common snake, and Johnson (1983) does not believe the ridgenose rattlesnake is currently threatened in Arizona. As many as six have been found in two hours in one canyon. Based on the number of specimen records, it seems most common in the Huachuca Mountains, followed by the Santa Rita and Patagonia mountains, in that order (Johnson and Mills, 1982).

Klauber (1936) reported a sex ratio for C. w. willardi of 1.15 males for each female, and Quinn (1977) noted a ratio of 1.5:1, but Tryon (1978) found two females to one male in broods of C. w. willardi.

Remarks

Electrophoretic study of rattlesnake venom proteins has showed that C. willardi is most closely related to C. pricei, C. cerastes, C. triseriatus, C. lepidus, and the Mexican Sistrurus ravus (Foote and MacMahon, 1977). Brattstrom (1964) had previously suggested that C. willardi is close to C. lepidus, C. pricei, and C. triseriatus on morphological grounds (C. willardi and C. lepidus are known to interbreed; Campbell, Brodie et al., 1989).

Dodd (in Seigel, 1987) has identified the race C. w. obscurus as at least threatened, principally through overcollection for the pet trade. Someday soon we may have to manage the populations of C. w. obscurus to ensure their survival, so it seems essential that thorough ecological study of both subspecies of C. willardi be conducted in the United States.

Bibliography

Adame, B. L., J. G. Soto, D. J. Secraw, J. C. Perez, J. L. Glenn, and R. C. Straight. 1990. Regional variation of biochemical characterics and antigeneity in Great Basin rattlesnake (*Crotalus viridis lutosus*) venom. Comp. Biochem. Physiol. 97B:95–101.

Adler, K. K. 1960. On a brood of *Sistrurus* from northern Indiana. Herpetologica 16:38.

Ahrenfeldt, R. H. 1955. Two British anatomical studies on American reptiles (1650–1750) II. Edward Tyson: comparative anatomy of the timber rattlesnake. Herpetologica 11:49–69.

Aird, S. D., and M. E. Aird. 1990. Rain-collecting behavior in a Great Basin rattlesnake (*Crotalus viridis lutosus*). Bull. Chicago Herpetol. Soc. 25:217.

Aird, S. D., and N. J. da Silva, Jr. 1991. Comparative enzymatic composition of Brazilian coral snake (*Micrurus*) venoms. Comp. Biochem. Physiol. 99B:287–294.

Aird, S. D., W. G. Kruggel, and I. I. Kaiser. 1990. Amino acid sequence of the basic subunit of Mojave toxin from the venom of the Mojave rattlesnake (*Crotalus scutulatus*). Toxicon 28:669–673.

———. 1991. Multiple myotoxin sequences from the venom of a single prairie rattlesnake (*Crotalus viridis viridus*). Toxin 29:265–268.

Aird, S. D., C. S. Seebart, and I. I. Kaiser. 1988. Preliminary fractionation and characterization of the venom of the Great Basin rattlesnake (*Crotalus viridis lutosus*). Herpetologica 44:71–85.

Aird, S. D., L. J. Thirkhill, C. S. Seebart, and I. I. Kaiser. 1989. Venoms and morphology of western diamondback/Mojave rattlesnake hybrids. J. Herpetol. 23:131–141.

Aldridge, R. D. 1975. Environmental control of spermatogenesis in the rattlesnake *Crotalus viridis*. Copeia 1975:493–496.

———. 1979a. Seasonal spermatogenesis in sympatric *Crotalus viridis* and *Arizona elegans* (Reptilia, Serpentes) in New Mexico. J. Herpetol. 13:187–192.

———. 1979b. Female reproductive cycles of the snakes *Arizona elegans* and *Crotalus viridis*. Herpetologica 35:256–261.

Allen, E. R. and W. T. Neill. 1950a. The eastern diamondback rattlesnake. Florida Wildlife 4(2):10–11.

———. 1950b. The pigmy rattlesnake. Florida Wildlife 5(4):10–11.

———. 1950c. The cane-brake rattlesnake. Florida Wildlife 5(6):18–19, 35.

Allen, E. R., and D. Swindell. 1948. Cottonmouth moccasin of Florida. Herpetologica 4(suppl. 1):1–16.

Allen, M. J. 1933. Report on a collection of amphibians and reptiles from Sonora, Mexico, with the description of a new lizard. Occ. Pap. Mus. Zool. Univ. Michigan (259):1–15.

Allen, W. B., Jr. 1955. Some notes on reptiles. Herpetologica 11:228.

———. 1956. The effects of a massasauga bite. Herpetologica 12:151.

Amaral, A. do. 1927. The anti-snake-bite campaign in Texas and in the subtropical United States. Bull. Antivenin Inst. Amer. 1:77–85.

———. 1928. Studies on snake venoms. I. Amounts of venom secreted by Nearctic pit vipers. Bull. Antivenin Inst. Amer. 1:103–104.

———. 1929. Studies of Nearctic Ophidia. III. Notes on *Crotalus tigris* Kennicott, 1859. Bull. Antivenin Inst. Amer. 2:82–85.

Anderson, B. B., and F. H. Emmerson. 1970. The rattlesnake *Crotalus atrox* in southern Nevada. Great Basin Natur. 30:107.

Anderson, P. K. 1965. The reptiles of Missouri. Univ. Missouri Press, Columbia. 330 pp.

Antoinio, F. B., and J. B. Barker. 1983. An inventory of phenotypic aberrancies in the eastern diamondback rattlesnake (*Crotalus adamanteus*). Herp. Review 14:108–110.

Anon. 1989. Venerable rattler dies—Kansas snake held unofficial record for longevity. Kansas Herpetol. Soc. Newletter (78):8.

Appleby, L. G. 1981. Rattlesnakes. Wildlife (London) 23(1):42–44.

Armstrong, B. L., and J. B. Murphy. 1979. The natural history of Mexican rattlesnakes. Univ. Kansas Mus. Natur. Hist. Spec. Publ. (5):1–88.

Arnberger, L. P. 1948. Gila monster swallows quail eggs whole. Herpetologica 4:209–210.

Arno, S. F. 1969. Interpreting the rattlesnake. Natl. Parks Mag. 43(267):15–17.

Arrington, O. N. 1930. Notes on the two poisonous lizards with special reference to *Heloderma suspectum*. Bull. Antivenin Inst. Amer. 4:29–35.

Arthur, W. J. III, and D. H. Janke. 1986. Radionuclide concentrations in wildlife occurring at a solid radioactive waste disposal area. Northwest Sci. 60:154–159.

Ashton, R. E., Jr., and P. S. Ashton. 1981. Handbook of reptiles and amphibians of Florida. Part 1. The snakes. Windward Publ., Inc., Miami, Florida. 176 pp.

Atkinson, D. A., and M. G. Netting. 1927. The distribution and habits of the Massasauga. Bull. Antivenin Inst. Amer. 1:40–44.

Auffenberg, W. 1963. The fossil snakes of Florida. Tulane Stud. Zool. 10:131–216.

Axtell, R. W. 1959. Amphibians and reptiles of the Black Gap Wildlife Management Area Brewster County, Texas. Southwest. Natur. 4:88–109.

Babb, R. D. 1989. *Micruroides euryxanthus* (Arizona Coral Snake). Size. Herp. Review 2:53.

Babcock, H. L. 1929. The snakes of New England. Natur. Hist. Guide, Boston Soc. Natur. Hist. (1):1–30.

Bailey, R. M. 1942. An intergeneric hybrid rattlesnake. Amer. Natur. 76:376–385.

Bailey, V., M. R. Terman, and R. Wall. 1989. Noteworthy longevity in *Crotalus viridus viridus* (Rafinesque). Trans. Kansas Acad. Sci. 92:116–117.

Baird, S. F. 1859. Reptiles of the boundary. *In* Report of the United States and Mexican Boundary Survey, U.S. 34th Congress 1st Session, Exec. Doc. 108, vol. 2, pt. 2, pp. 1–35.

Baird, S. F., and C. Girard. 1853. Catalogue of North American reptiles in the museum of the Smithsonian Institution. Part I. Serpents. Smithsonian Institution, Washington, D.C. 172 pp.

Baker, R. J., J. J. Bull, and G. A. Mengden. 1971. Chromosomes of *Elaphe subocularis* (Reptilia:Serpentes), with the description of an in vivo technique for preparation of snake chromosomes. Experientia 27:1228–1229.

Baker, R. J., G. A. Mengden, and J. J. Bull. 1972. Karyotypic studies of thirty-eight species of North American snakes. Copeia 1972:257–265.

Banta, B. H. 1965. A distributional check list of the Recent reptiles inhabiting the state of Nevada. Occ. Pap. Biol. Soc. Nevada (5):1–8.

———. 1974. A pre-Columbian record of cannibalism in the rattlesnake. Bull. Maryland Herpetol. Soc. 10:56.

Barbour, R. W. 1950. The reptiles of Big Black Mountain, Harlan County, Kentucky. Copeia 1950:100–107.

———. 1956a. A study of the cottonmouth, *Ancistrodon piscivorus leucostoma*, in Kentucky. Trans. Kentucky Acad. Sci. 17:33–41.

———. 1956b. Poisonous snakes of Kentucky. Kentucky Happy Hunting Grounds. 12(1):18–19, 32.

———. 1962. An aggregation of copperheads, *Agkistrodon contortrix*. Copeia 1962:640.

———. 1971. Amphibians and reptiles of Kentucky. Univ. Press Kentucky, Lexington, Kentucky. 334 pp.

Barnes, M. 1968. Venomous sea snakes (Hydrophiidae). *In* Bücherl, W., et al. (eds.), Venomous animals and their venoms, vol. 1, pp. 285–308. Academic Press, New York.

Barrett, S. L., and J. A. Humphrey. 1986. Agonistic interactions between *Gopherus agassizii* (Testudinidae) and *Heloderma suspectum* (Helodermatidae). Southwest. Natur. 31:261–263.

Bartlett, R. D. 1988. In search of reptiles and amphibians. E. J. Brill, Leiden. 363 pp.

Barton, A. J. 1950. Replacement fangs in newborn timber rattlesnakes. Copeia 1950:235–236.

Baxter, G. 1977. Rattlesnake fact and fiction. Wyoming Wild Life 41(6):12–13, 15.

Baxter, G., and H. Rahn. 1941. Rattlesnakes of Wyoming. Wyoming Wild Life 6(4):1–6.

Baxter, G. T., and M. D. Stone. 1980. Amphibians and reptiles of Wyoming. Wyoming Game Fish Dept. Bull. (16):1–137.

Beavers, R. A. 1976. Food habits of the western diamondback rattlesnake, *Crotalus atrox*, in Texas. Southwest. Natur. 20:503–515.

Beck, D. D. 1985. *Heloderma suspectum cinctum* (Banded Gila Monster). Pattern/coloration. Herp. Review 16:53.

————. 1990. Ecology and behavior of the Gila monster in southwestern Utah. J. Herpetol. 24:54–68.

Bertke, E. M., D. D. Watt, and T. Tu. 1966. Electrophoretic patterns of venoms from species of Crotalidae and Elapidae snakes. Toxicon 4:73–76.

Best, I. B. 1978. Field sparrow reproductive success and nesting ecology. Auk 95:9–22.

Best, T. L., and H. C. James. 1984. Rattlesnakes (genus *Crotalus*) of the Pedro Armendariz Lava Field, New Mexico. Copeia 1984:213–215.

Billing, W. M. 1930. The action of the toxin of *Crotalus adamanteus* on blood clotting. J. Pharm. Exp. Ther. 38:173–196.

Birchard, G. F., C. P. Black, G. W. Schuett, and V. Black. 1983. Foetal–maternal blood respiratory properties of an ovoviviparous snake the cottonmouth, *Agkistrodon piscivorus*. J. Exp. Biol. 108:247–255.

Blair, W. F. 1954. Mammals of the Mesquite Plains Biotic District in Texas and Oklahoma, and speciaion in the central grasslands. Texas J. Sci. 6:235–264.

Blem, C. R. 1981. Reproduction of the eastern cottonmouth *Agkistrodon piscivorus piscivorus* (Serpentes:Viperidae) at the northern edge of its range. Brimleyana (5):117–128.

————. 1982. Biennial reproduction in snakes: an alternative hypothesis. Copeia 1982:961–963.

————. 1987. Development of combat rituals in captive cottonmouths. J. Herpetol. 21:64–65.

Bogert, C. M. 1942. Field note on the copulation of *Crotalus atrox* in California. Copeia 1942:262.

————. 1943. Dentitional phenomena in cobras and other elapids with notes on adaptive modifications of fangs. Bull. Amer. Mus. Natur. Hist. 81:285–360.

————. 1960. The influence of sound on the behavior of amphibians and reptiles. *In* Lanyon, W. E., and W. N. Tavolga (eds.), Animal sounds and communication, pp. 137–320. Amer. Inst. Biol. Sci. Publ. (7):1–443.

Bogert, C. M., and W. G. Degenhardt. 1961. An addition to the fauna of the United States, the Chihuahuan ridge-nosed rattlesnake in New Mexico. Amer. Mus. Novitates (2064):1–15.

Bogert, C. M., and R. Martín del Campo. 1956. The Gila monster and its allies. The relationships, habits, and behavior of the lizards of the family Helodermatidae. Bull. Amer. Mus. Natur. Hist. 109:1–238.

Bonilla, C. A., M. R. Faith, and S. A. Minton. 1973. L-amino acid oxidase, phosphodiesterase, total protein and other properties of juvenile timber rattlesnakes (*Crotalus h. horridus*) venom at different stages of growth. Toxicon 11:301–303.

Bonilla, C. A., W. Seifert, and N. Horner. 1971. Comparative biochemistry of *Sistrurus miliarius barbouri* and *Sistruruus catenatus tergeminus* venoms. *In* Bücherl, W., and E. E. Buckley (eds.), Venomous animals and their venoms, vol. 2, Venomous vertebrates, pp. 203–209. Academic Press, New York.

Boone, A. R. 1937. Snake hunter catches rattlers for fun. Pop. Sci. Monthly 131(4):54–55.

Boquet, P. 1948. Venins de serpents et antivenins. Collections de l'Institut Pasteur, [Paris]. 157 pp.

Bothner, R. C. 1973. Temperatures of *Agkistrodon p. piscivorous* and *Lampropeltis g. getulus* in Georgia. HISS News-J. 1:24–25.

———. 1974. Some observations on the feeding habits of the cottonmouth in southeastern Georgia. J. Herpetol. 8:257–258.

Boulenger, G. A. 1893–1896. Catalogue of the snakes in the British Museum (Natural History). 3 volumes. London.

Bowler, J. K. 1977. Longevity of reptiles and amphibians in North American collections. Soc. Stud. Amphib. Rept. Misc. Publ. Herpetol. Circ. (6):1–32.

Boyer, D. A. 1933. A case report on the potency of the bite of a young copperhead. Copeia 1933:97.

Boyer, D. R. 1957. Sexual dimorphism in a population of the western diamond-backed rattlesnake. Herpetologica 13:213–217.

Branch, W. R. 1979. The venomous snakes of southern Africa. Part 2. Elapidae and Hydrophidae. The Snake 11:199–225.

———. 1982. Hemipeneal morphology of platynotan lizards. J. Herpetol. 16:16–38.

Brattstrom, B. H. 1953. Records of Pleistocene reptiles and amphibians from Florida. Quart. J. Florida Acad. Sci. 16:243–248.

———. 1954a. The fossil pit-vipers (Reptilia:Crotalidae) of North America. Trans. San Diego Soc. Natur. Hist. 12:31–46.

———. 1954b. Amphibians and reptiles from Gypsum Cave, Nevada. Bull. South. California Acad. Sci. 53:8–12.

———. 1955. Records of some Pliocene and Pleistocene reptiles and amphibians from Mexico. Bull. South. California Acad. Sci. 54:1–4.

———. 1964. Evolution of the pit vipers. Trans. San Diego Soc. Natur. Hist. 13:185–268.

———. 1965. Body temperatures of reptiles. Amer. Midl. Natur. 73: 376–422.

Breckenridge, W. J. 1944. Reptiles and amphibians of Minnesota. Univ. Minnesota Press, Minneapolis. 202 pp.

Breen, A. R. 1984. Rhode Island's declining rattlers. Massachusetts Audubon 23:866–868.

Breidenbach, C. H. 1990. Thermal cues influence strikes in pitless vipers. J. Herpetol. 24:448–450.

Breisch, A. R. 1984. Just hanging in there: the eastern massasauga in danger of extinction. Conservationist, New York 39(3):35.

Brennan, G. A. 1924. A case of death from *Heloderma* bite. Copeia (129):45.

Brimley, C. S. 1941–1942. The amphibians and reptiles of North Carolina: the snakes. Carolina Tips 4–5 (19–26).

Brock, O. G. 1981. Predatory behavior of eastern diamondback rattlesnakes (*Crotalus adamanteus*): field enclosure and Y-maze laboratory studies, emphasizing prey trailing behaviors. Diss. Abstr. Int. B. 41:2510.

Brown, B. C. 1950. An annotated check list of the reptiles and amphibians of Texas. Baylor Univ. Press, Waco, Texas. 259 pp.

Brown, C. W., and C. H. Ernst. 1986. A study of variation in eastern timber rattlesnakes, *Crotalus horridus* Linnae (Serpentes: Viperidae). Brimleyana (12):57–74.

Brown, D. G. 1990. Observation of a prairie rattlesnake (*Crotalus viridis viridis*) consuming neonatal cottontail rabbits (*Sylvilagus nuttalli*), with defense of the young cottontails by adult conspecifics. Bull. Chicago Herpetol. Soc. 25:24–26.

Brown, T. W. 1970. Autecology of the sidewinder (*Crotalus cerastes*) at Kelso Dunes, Mojave Desert California. Unpubl. Ph.D. Diss., Univ. California, Los Angeles.

Brown, W. H., and C. H. Lowe, Jr. 1955. Technique for obtaining maximum yields of fresh labial gland secretions from the lizard *Heloderma suspectum*. Copeia 1955:63.

Brown, W. S. 1982. Overwintering body temperatures of timber rattlesnakes (*Crotalus horridus*) in northeastern New York. J. Herpetol. 16:145–150.

———. 1987. Hidden life of the timber rattler. Natl. Geogr. Mag. 172:128–138.

———. 1991. Female reproductive ecology in a northern population of the timber rattlesnake, *Crotalus horridus*. Herpetologica 47:101–115.

Brown, W. S., and F. M. MacLean. 1983. Conspecific scent-trailing by newborn timber rattlesnakes, *Crotalus horridus*. Herpetologica 39:430–436.

Brown, W. S., and W. H. Martin. 1990. Geographic variation in female reproductive ecology of the timber rattlesnake, *Crotalus horridus*. Catesbeiana 10:48.

Brown, W. S., and W. S. Parker. 1982. Niche dimensions and resource partitioning in a Great Basin Desert snake community. *In* Scott, N. J., Jr. (ed.), Herpetological communities, pp. 59–81. U.S. Fish Wildl. Serv., Wildl. Res. Rept. 13.

Brown, W. S., D. W. Pyle, K. R. Greene, and J. B. Friedlaender. 1982. Movements and temperature relationships of timber rattlesnakes (*Crotalus horridus*) in northeastern New York. J. Herpetol. 16:151–161.

Brugger, K. E. 1989. Red-tailed hawk dies with coral snake in talons. Copeia 1989:508–510.

Brush, S. W., and G. W. Ferguson. 1986. Predation of lark sparrow eggs by a massasauga rattlesnake. Southwest. Natur. 31:260–261.

Buijs, H. 1988. Ein Texaanse ratelslang (*Crotalus atrox*) bijt een soortgenoot. Lacerta 47:72–73.

Bullock, R. E. 1971. Cannibalism in captive rattlesnakes. Great Basin Natur. 31:49–50.

Bullock, T. H., and R. B. Cowles. 1952. Physiology of an infrared receptor: the facial pit of pit vipers. Science 115:541–543.

Bullock, T. H., and F. P. J. Diecke. 1956. Properties of an infra-red sense organ in the facial pit of pit vipers. J. Physiol. 134:47–87.

Bullock, T. H., and W. Fox. 1957. The anatomy of the infra-red sense organ in the facial pit vipers. Quart. J. Micro. Sci. 98:219–234.

Burger, J. W. 1934. The hibernation habits of the rattlesnake of the New Jersey Pine barrens. Copeia 1934:142.

Burkett, R. D. 1966. Natural history of cottonmouth moccasin, *Agkistrodon piscivorus* (Reptilia). Univ. Kansas Publ. Mus. Natur. Hist. 17:435–491.

Burns, B. 1969. Oral sensory papillae in sea snakes. Copeia 1969:617–619.

Burns, B., and G. V. Pickwell. 1972. Cephalic glands in sea snakes (*Pelamis, Hydrophis* and *Laticauda*). Copeia 1972:547–559.

Bushey, C. L. 1985. Man's effect upon a colony of *Sistrurus c. catenatus* (Raf.) in northeastern Illinois (1834–1975). Bull. Chicago Herpetol. Soc. 20:1–12.

Cadle, J. E., and V. M. Sarich. 1981. An immunological assessment of the phylogenetic position of New World coral snakes. J. Zoology, London 195:157–167.

Cale, W. G., Jr., and J. W. Gibbons. 1972. Relationships between body size, size of the fat bodies, and total lipid content in the canebrake rattlesnake

(*Crotalus horridus*) and the black racer (*Coluber constrictor*). Herpetologica 28:51–53.

Calmonte, A. 1974. Die Felsenklapperschlange, *Crotalus lepidus lepidus* (Kennicott, 1861). Aquar. Aqua Terra 8:460–462.

———. 1978. Die Schwarzschwanz-klapperschlange in der Freiheit und im Terrarium beobachtet. Aquar. Aqua Terra 12:221–223.

Cameron, D. L., and A. T. Tu. 1977. Characterization of myotoxin a from the venom of a prairie rattlesnake (*Crotalus viridis viridis*). Biochemistry 16:2546–1553.

Campbell, B. 1934. Report on a collection of reptiles and amphibians made in Arizona during the summer of 1933. Occ. Pap. Mus. Zool. Univ. Michigan (289):1–10.

Campbell, H. 1950. Rattlesnakes tangled in wire. Herpetologica 6:44.

———. 1953. Probable strychnine poisoning in a rattlesnake. Herpetologica 8:184.

———. 1958. An unusually long rattle string in *Crotalus atrox* (Serpentes: Crotalidae). Southwest. Natur. 3:233.

———. 1977. The coral snake: a New Mexico treasure. New Mexico Wildlife 22(4):2–5, 26–29.

Campbell, J. A. 1973. A captive hatching of *Micrurus fulvius tenere* (Serpentes, Elapidae). J. Herpetol. 7:312–315.

Campbell, J. A., E. D. Brodie, Jr., D. G. Barker, and A. H. Price. 1989. An apparent natural hybrid rattlesnake and *Crotalus willardi* (Viperidae) from the Peloncillo Mountains of southwestern New Mexico. Herpetologica 45:344–349.

Campbell, J. A., D. R. Formanowicz, Jr., and E. D. Brodie, Jr. 1989. Potential impact of rattlesnake roundups on natural populations. Texas J. Sci. 41:301–317.

Campbell, J. A., and W. W. Lamar. 1989. The venomous reptiles of Latin America. Comstock Publ. Assoc., Cornell Univ. Press, Ithaca, New York. 425 pp.

Campbell, J. A., and D. H. Whitmore, Jr. 1989. A comparison of the skin keratin biochemistry in vipers with comments on its systematic value. Herpetologica 45:242–249.

Carl, G. C. 1960. The reptiles of British Columbia. Handb. British Columbia Prov. Mus. (3):1–65.

Carpenter, C. C. 1958. Reproduction, young, eggs and food of Oklahoma snakes. Herpetologica 14:113–115.

———. 1960. A large brood of western pigmy rattlesnakes. Herpetologica 14:113–115.

———. 1986. An inventory of combat rituals in snakes. Smithsonian Herpetol. Inform. Serv. (69):1–18.

Carpenter, C. C., and G. W. Ferguson. 1977. Variation and evolution of stereotyped behavior in reptiles. *In* Gans, C. and D. W. Tinkle (eds.), Biology of the Reptilia, vol. 7, pp. 335–554. Academic Press, London.

Carpenter, C. C., and J. C. Gillingham. 1975. Postural response of kingsnakes by crotaline snakes. Herpetologica 31:293–302.

Carpenter, C. C., J. C. Gillingham, and J. B. Murphy. 1976. The combat ritual of the rock rattlesnake (*Crotalus lepidus*). Copeia 1976:764–780.

Carr, A. F., Jr. 1940. A contribution to the herpetology of Florida. Univ. Florida Publ. Biol. Ser. 3(1):1–118.

Carr, A. F., Jr., and M. H. Carr. 1942. Notes on the courtship of the cottonmouth moccasin. Proc. New England Zool. Club. 20:1–6.

Carr, A. F., Jr., and C. J. Goin. 1955. Guide to the reptiles, amphibians, and freshwater fishes of Florida. Univ. Florida Press, Gainesville. 341 pp.

Castilonia, R. R., T. R. Pattabhiraman, and F. E. Russell. 1980. Neuromuscular blocking effects of Mojave rattlesnake (*Crotalus scutulatus scutulatus*) venom. Proc. West. Pharmacol. Soc. 23:103–106.

Chadwick, L. E., and H. Rahn. 1954. Temperature dependence of rattling frequency in the rattlesnake, *Crotalus v. viridis*. Science 119:442–443.

Chamberlain, E. B. 1935. Notes on the pygmy rattlesnake *Sisturus miliarius* Linnaeus, in South Carolina. Copeia 1935:146–147.

Chance, B. 1970. A note on the feeding habits of *Micrurus fulvius fulvius*. Bull. Maryland Herpetol. Soc. 6:56.

Chapman, B. R., and S. D. Casto. 1972. Additional vertebrate prey of the loggerhead shrike. Wilson Bull. 84:496–497.

Charland, M. B. 1989. Size and winter survivorship in neonatal western rattlesnakes (*Crotalus viridis*). Can. J. Zool. 67:1620–1625.

Charland, M. B., and P. T. Gregory. 1989. Feeding rate and weight gain in postpartum rattlesnakes: do animals that eat more always grow more? Copeia 1989:211–214.

———. 1990. The influence of female reproductive status on thermoregulation in a viviparous snake, *Crotalus viridis*. Copeia 1990:1089–1098.

Chenowith, W. L. 1948. Birth and behavior of young copperheads. Herpetologica 4:162.

Chermock, R. L. 1952. A key to the amphibians and reptiles of Alabama. Univ. Alabama Mus. Pap. (33):1–88.

Chiasson, R. B. 1982. The apical pits of *Agkistrodon* (Reptilia: Serpentes). J. Arizona-Nevada Acad. Sci. 16:69–73.

Chiasson, R B., D. L. Bentley, and C. H. Lowe. 1989. Scale morphology in *Agkistrodon* and closely related crotaline genera. Herpetologica 45:430–438.

Chiodini, R. J., J. P. Sundberg, and J. A. Czikowsky. 1982. Gross anatomy of snakes. Veterin. Med. Sm. Anim. Clin. 77:413–419.

Chiszar, D., C. Andren, F. Nilson, B. O'Connell, J. S. Mestas, Jr., H. M. Smith, and C. W. Radcliffe. 1982. Strike-induced chemosensory searching in Old World vipers and New World pit vipers. Anim. Learn. Behav. 10:121–125.

Chiszar, D., C. A. Castro, H. M. Smith, and C. Guyon. 1986. A behavioral method for assessing utilization of thermal cues by snakes during feeding episodes, with a comparison of crotaline and viperine species. Ann. Zool., Acad. Zool. India 24:123–131.

Chiszar, D., D. Dickman, and J. Colton. 1986. Sensitivity to thermal stimulation in prairie rattlesnakes (*Crotalus viridis*) after bilateral anesthetization of the facial pits. Behav. Neurol. Biol. 45:143–149.

Chiszar, D., D. Duvall, and K. Scudder. 1980. Simultaneous and successive discriminations between envenomated and nonenvonomated mice by rattlesnakes (*Crotalus durissus* and *C. viridis*). Behav. Neurol. Biol. 29:518–521.

Chiszar, D., G. Hobica, H. M. Smith, and J. Vidaurri. 1991. Envenomation and acquisition of chemical information by prairie rattlesnakes. Prairie Natur. 23:69–72.

Chiszar, D., P. Nelson, and H. M. Smith. 1988. Analysis of the behavioral sequence emitted by rattlesnakes during feeding episodes. III. Strike-induced chemosensory searching and location of rodent carcasses. Bull. Maryland Herpetol. Soc. 24:99–108.

Chiszar, D., B. O'Connell, R. Greenlee, B. Demeter, T. Walsh, J. Chiszar, K. Moran, and H. M. Smith. 1985. Duration of strike-induced chemosensory searching in long-term captive rattlesnakes at National Zoo, Audubon Zoo, and San Diego Zoo. Zoo Biology 4:291–294.

Chiszar, D., and C. W. Radcliffe. 1977. Absence of prey-chemical preference in newborn rattlesnakes (*Crotalus cerastes, C. enyo,* and *C. viridis*). Behav. Biol. 21:146–150.

Chiszar, D., C. W. Radcliffe, T. Byers, and R. Stoops. 1986. Prey capture behavior in nine species of venomous snakes. Psychol. Rec. 36:433–438.

Chiszar, D., C. Radcliffe, R. Boyd, A. Radcliffe, H. Yun, H. M. Smith, T. Boyer, B. Atkins, and F. Feiler. 1986. Trailing behavior in cottonmouths (*Agkistrodon piscivorus*). J. Herpetol. 20:269–272.

Chiszar, D., C. Radcliffe, and F. Feiler. 1986. Trailing behavior in banded rock rattlesnakes (*Crotalus lepidus klauberi*) and prairie rattlesnakes (*C. viridis viridis*). J. Comp. Psychol. 100:368–371.

Chiszar, D., C. W. Radcliffe, B. O'Connell, and H. M. Smith. 1982. Analysis of the behavioral sequence emitted by rattlesnakes during feeding episodes. II. Duration of strike-induced chemosensory searching in rattlesnakes (*Crotalus viridis, C. enyo*). Behav. Neurol. Biol. 34: 261–270.

Chiszar, D., C. W. Radcliffe, R. Overstreet, T. Poole, and T. Byers. 1985. Duration of strike-induced chemosensory searching in cottonmouths (*Agkistrodon piscivorus*) and a test of the hypothesis that striking prey creates a specific search image. Can. J. Zool. 63:1057–1061.

Chiszar, D., C. W. Radcliffe, and K. M. Scudder. 1977. Analysis of the behavioral sequence emitted by rattlesnakes during feeding episodes. I. Striking and chemosensory searching. Behav. Biol. 21:418–425.

Chiszar, D., C. W. Radcliffe, K. M. Scudder, and D. Duvall. 1983. Strike-induced chemosensory searching by rattlesnakes: the role of envenomation-related chemical cues in the post-strike environment. *In* Müller-Schwarze, D., and R. M. Silverstein (eds.), Chemical signals in vertebrates, III, pp. 1–24. Plenum Press, New York.

Chiszar, D., C. W. Radcliffe, H. M. Smith, and H. Baskinski. 1981. Effect of prolonged food deprivation on response to prey odors by rattlesnakes. Herpetologica 37:237–243.

Chiszar, D., C. W. Radcliffe, H. M. Smith, and P. Langer. 1991. Strike-induced chemosensory searching: do rattlesnakes make one decision or two? Bull. Maryland Herpetol. Soc. 27:90–94.

Chiszar, D., and K. M. Scudder. 1980. Chemosensory searching by rattlesnakes during predatory episodes. *In* Müller-Schwarze, D., and R. M. Silverstein (eds.), Chemical signals: vertebrates and aquatic invertebrates, pp. 125–129. Plenum Press, New York.

Chiszar, D., K. Scudder, and L. Knight. 1976. Rate of tongue flicking by garter snakes (*Thamnophis radix haydeni*) and rattlesnakes (*Crotalus v. viridis, Sistrurus catenatus tergeminus,* and *Sistrurus catenatus edwardsii*) during prolonged exposure to food odors. Behav. Biol. 18:273–283.

Chizar, D., K. M. Scudder, L. Knight, and H. M. Smith. 1978. Exploratory behavior in prairie rattlesnakes (*Crotalus viridis*) and water moccasins (*Agkistrodon piscivorus*). Psychol. Rec. 28:363–368.

Chiszar, D., K. M. Scudder, and H. M. Smith. 1979. Chemosensory investigation of fish mucus odor by rattlesnakes. Bull. Maryland Herpetol. Soc. 15:31–36.

Chiszar, D., K. Scudder, H. M. Smith, and C. W. Radcliffe. 1976. Observations of courtship behavior in the western massasauga (*Sistrurus catenatus tergeminus*). Herpetologica 32:337–338.

Chiszar, D., L. Simonsen, C. Radcliffe, and H. M. Smith. 1979. Rate of tongue flicking by cottonmouths (*Agkistrodon piscivorus*) during prolonged exposure to various odors, and strike induced chemosensory searching by the cantil (*Agkistrodon bilineatus*). Trans. Kansas Acad. Sci. 82:49–54.

Chiszar, D., K. Stimac, and T. Boyer. 1983. Effect of mouse odor on visually-induced and strike-induced chemosensory searching in prairie rattlesnakes (*Crotalus viridis*). Chem. Senses 7:301–308.

Chiszar, D., S. V. Taylor, C. W. Radcliffe, H. M. Smith, and B. O'Connell. 1981. Effects of chemical and visual stimuli upon chemosensory searching by garter snakes and rattlesnakes. J. Herpetol. 15:415–423.

Christman, S. P. 1975. The status of the extinct rattlesnake *Crotalus giganteus*. Copeia 1975:43–47.

———. 1980. Patterns of geographic variation in Forida snakes. Bull. Florida St. Mus. Biol. Sci. 25:157–256.

Clark, D. R., Jr. 1963. Variation and sexual dimorphism in a brood of the western pigmy rattlesnake (*Sistrurus*). Copeia 1963:157–159.

Clark, R. F. 1949. Snakes of the hill parishes of Louisiana. J. Tennessee Acad. Sci. 24:244–261.

Clark, W. W., and E. Schultz. 1980. Rattlesnake shaker muscle: 2. Fine structure. Tissue Cell 12:335–351.

Clarke, G. K. 1961. Report on a bite by a red diamond rattlesnake, *Crotalus ruber ruber*. Copeia 1961:418–422.

Clarkson, R. W., and James C. deVos, Jr. 1986. The bullfrog, *Rana catesbeiana* Shaw, in the Lower Colorado River, Arizona–California. J. Herpetol. 20:42–49.

Cochran, D. A. 1954. Our snake friends and foes. Natl. Geogr. Mag. 106:333–364.

Cochran, D. M., and C. J. Goin. 1970. The new field book of reptiles and amphibians. Putnam, New York. 359 pp.

Cohen, A. C., and B. C. Myres. 1970. A function of the horns (suprocular scales) in the sidewinder rattlesnake, *Crotalus cerastes*, with comments on other horned snakes. Copeia 1970:574–575.

Collins, J. T. 1964. A preliminary review of the snakes of Kentucky. J. Ohio Herpetol. Soc. 4:69–77.

———. 1974. Amphibians and reptiles in Kansas. Univ. Kansas Publ. Mus. Natur. Hist. Publ. Ed. Ser. (1):1–183.

———. 1991. Viewpoint: a new taxonomic arrangement for some North American amphibians and reptiles. Herp. Review 22:42–43.

Collins, J. T., R. Conant, J. E. Huheey, J. L. Knight, E. M. Rundquist, and H. M. Smith. 1982. Standard common and current scientific names for North American amphibians and reptiles. 2d ed. Soc. Stud. Amphib. Rept. Herpetol. Circ. (12):1–28.

Collins, J. T., and J. L. Knight. 1980. *Crotalus horridus*. Catalog. Amer. Amphib. Rept. 253:1–2.

Collins, R. F., and C. C. Carpenter. 1970. Organ position–ventral scute relationship in the water moccasin (*Agkistrodon piscivorus leucostoma*),

with notes on food habits and distribution. Proc. Oklahoma Acad. Sci.
498:15–18.

Conant, R. 1929. Notes on a water moccasin in captivity (*Agkistrodon
piscivorus*) (female). Bull. Antivenin Inst. Amer. 3:61–64.

———. 1933. Three generations of cottonmouths, *Agkistrodon piscivorus*
(Lacépède). Copeia 193:43.

———. 1945. An annotated checklist of the amphibians and reptiles of the
Del-Mar-Va Peninsula. Soc. Natur. Hist. Delaware, Wilmington. 8 pp.

———. 1951. The reptiles of Ohio. 2d ed. Notre Dame Press, Notre Dame,
Indiana. 284 pp.

———. 1955. Notes on three Texas reptiles, including an addition to the
fauna of the state. Amer. Mus. Novitates (1726):1–6.

———. 1957. Reptiles and amphibians of the northeastern states, 3d ed.
Zool. Soc. Philadelphia. 40 pp.

———. 1969. Some rambling notes on rattlesnakes. Arch. Environ. Health
19:768–769.

———. 1975. A field guide to reptiles and amphibians of eastern and
central North America, 2d ed. Houghton Mifflin Co., Boston. 429 pp.

———. 1986. Phylogeny and zoogeography of the genus *Agkistrodon* in
North America. *In* Z. Rocek (ed.), Proc. European Herpetol. Meet.,
Prague, 1985, pp. 89–92. Charles Univ., Prague.

Conant, R., and W. Bridges. 1939. What snake is that? Appleton Century,
New York. 163 pp.

Conant, R., and J. T. Collins. 1991. A field guide to reptiles and amphibians: eastern and central North America. Houghton Mifflin Co., Boston.
450 pp.

Cook, F. A. 1954. Snakes of Mississippi. Mississippi Game Fish Comm.,
Jackson. 40 pp.

Cook, F. R. 1966. A guide to the amphibians and reptiles of Saskatchewan.
Saskatchewan Mus. Natur. Hist. Popular Ser. (13):1–40.

———. 1984. Introduction to Canadian amphibians and reptiles. Natl.
Mus. Canada, Ottawa. 200 pp.

Cook, S. F., Jr. 1955. Rattlesnake hybrids: *Crotalus viridis* X *Crotalus
scutulatus*. Copeia 1955:139–141.

Cooper, J. E., and F. Groves. 1959. The rattlesnake, *Crotalus horridus*, in the
Maryland Piedmont. Herpetologica 15:33–34.

Cooper, R. H., and J. C. List. 1979. Further information on the health and
longevity of the Gila monster (*Heloderma suspectum* Cope). Proc. Indiana
Acad. Sci. 88:434–435.

Cooper, W. E., Jr. 1989. Prey odor discrimination in the varanoid lizards
Heloderma suspectum and *Varanus exanthematicus*. Ethology 81:250–258.

Coote, J. 1981. The California mountain kingsnake (*Lampropeltis zonata*) and
the coral snake mimic problem. Herptile 6:17–19.

Cope, E. D. 1861. Contributions to the ophiology of Lower California,
Mexico and Central America. Proc. Acad. Natur. Sci. Philadelphia
13:292–306.

———. 1867. On the Reptilia and Batrachia of the Sonoran Province of the
Nearctic region. Proc. Acad. Natur. Sci. Philadelphia 18:300–314.

———. 1869. Remarks on *Heloderma suspectum*. Proc. Acad. Natur. Sci.
Philadelphia 21:4–5.

———. 1892. A critical review of the characters and variations of the
snakes of North America. Proc. U.S. Natl. Mus. 14:589–694.

————. 1900. The crocodilians, lizards, and snakes of North America. Report U.S. Natl. Mus. 1898:153–1294.

Cottam, C., W. C. Glazener, and G. G. Raun. 1959. Notes on food of moccasins and rattlesnakes from the Welder Wildlife Refuge, Sinton, Texas. Contrib. Welder Wildlife Found. 45:1–12.

Coues, E. 1875. Synopsis of the reptiles and batrachians of Arizona. Washington, One hundredth meridian surveys (Wheeler Report) 5:585–663.

Cowles, R. B. 1938. Unusual defense postures assumed by rattlesnakes. Copeia 1938:13–16.

————. 1941. Winter activities of desert reptiles. Ecology 22:125–140.

————. 1945. Some of the activities of the sidewinder. Copeia 1945:220–222.

————. 1953. The sidewinder: master of desert travel. Pacific Discovery 6(2):12–15.

Cowles, R. B., and C. M. Bogert. 1944. A preliminary study of the thermal requirements of desert reptiles. Bull. Amer. Mus. Natur. Hist. 83: 261–296.

Cowles, R. B., and R. L. Phelan. 1958. Olfaction in rattlesnakes. Copeia 1958:77–83.

Crabtree, C. B., and R. W. Murphy. 1984. Analysis of maternal-offspring allozymes in *Crotalus viridis*. J. Herpetol. 18:75–80.

Crimmins, M. L. 1927. Prevalence of poisonous snakes in the El Paso and San Antonio Districts in Texas. Bull. Antivenin Inst. Amer. 1:23–24.

————. 1931. Rattlesnakes and their enemies in the Southwest. Bull. Antivenin Inst. Amer. 5:46–47.

Cromwell, W. R. 1982. Underground desert toads. Pacific Discovery 35:10–17.

Cross, J. K., and M. S. Rand. 1979. Climbing activity in wild-ranging Gila monsters, *Heloderma suspectum* (Helodermatidae). Southwest. Natur. 24:703–705.

Cruz, E., S. Gibson, K. Kandler, G. Sanchez, and D. Chiszar. 1987. Strike-induced chemosensory searching in rattlesnakes: a rodent specialist (*Crotalus viridis*) differs from a lizard specialist (*Crotalus pricei*). Bull. Psychonomic Soc. 25:136–138.

Culotta, W. A., and G. V. Pickwell. 1991. The venomous sea snakes: a comprehensive bibliography. Krieger Publ. Co., Melbourne, Florida.

Cumbert, T. C., and K. A. Sullivan. 1990. Predation on yellow-eyed junco nestlings by twin-spotted rattlesnakes. Southwest. Natur. 35:367–368.

Cunningham, J. D. 1955. Arboreal habits of certain reptiles and amphibians in southern California. Herpetologica 11:217–220.

————. 1959. Reproduction and food of some California snakes. Herpetologica 15:17–19.

————. 1966. Field observations on the thermal relations of rattlesnakes. Southwest. Natur. 11:140–141.

Curran, C. H. 1935. Rattlesnakes. Natur. Hist. 36:331–340.

Curran, C. H., and C. Kauffeld. 1937. Snakes and their ways. Harper & Bros., New York. 285 pp.

Curtis, L. 1949. The snakes of Dallas County, Texas. Field & Lab. 17:1–13.

————. 1952. Cannibalism in the Texas coral snake. Herpetologica 8:27.

DaLie, D. A. 1953. Poisonous snakes of America. J. Forestry 51:243–248.

Dalrymple, G. H., F. S. Bernardino, Jr., T. M. Steiner, and R. J. Nodell. 1991. Patterns of species diversity of snake community assemblages, with data on two Everglades snake assemblages. Copeia 1991:517–521.

Dalrymple, G. H., T. M. Steiner, R. J. Nodell, and F. S. Bernardino, Jr.

1991. Seasonal activity of the snakes of Long Pine Key, Everglades National Park. Copeia 1991:294–302.

Darlington, P. J., Jr. 1957. Zoogeography: the geographical distribution of animals. John Wiley & Sons, Inc., New York, 675 pp.

Daudin, F. M. 1801–1803. Histoire naturelle, général et particulière, des reptiles. 7 vols. F. Dufart, Paris.

Davenport, J. W. 1943. Field book of snakes of Bexar County, Texas and vicinity. White Mem. Mus., San Antonio, Texas. 132 pp.

Davis, D. D. 1936. Courtship and mating behavior in snakes. Field Mus. Natur. Hist. Zool. Ser. 20:257–290.

Davis, R. A. 1980. Vipers among us. Cincinnati Mus. Natur. Hist. Quart. 17(2):8–12.

Dean, B. 1938. Note on the sea-snakes, *Pelamis platurus* (Linnaeus). Science 88:144–145.

de Cock Buning, T. 1983. Thermal sensitivity as a specialization for prey capture and feeding in snakes. Amer. Zool. 23:363–375.

———. 1984. A theoretical approach to the heat sensitive pit organs of snakes. J. Theor. Biol. 111:509–529.

———. 1985. Qualitative and quantitative explanation of the forms of heat sensitive organs in snakes. Acta Biotheor. 34:193–205.

DeGraaf, R. M., and D. D. Rudis. 1981. Forest habitat for reptiles and amphibians of the Northeast. Forest Service, U.S. Dept. Agric., Washington, D.C. 239 pp.

———. 1983. Amphibians and reptiles of New England: habitats and natural history. Univ. Massachusetts Press, Amherst. 112 pp.

Demeter, B. J. 1986. Combat behavior in the Gila monster (*Heloderma suspectum cinctum*). Herp. Review 17:9–11.

de Wit, C. A. 1982. Yield of venom from the Osage copperhead, *Agkistrodon contortrix phaeogaster*. Toxicon 20:525–527.

Dickinson, W. E. 1949. Field guide to the lizards and snakes of Wisconsin. Milwaukee Publ. Mus., Pop. Sci. Handb. Ser. (2):1–70.

Dickman, J. D., J. S. Colton, D. Chiszar, and C. A. Colton. 1987. Trigeminal responses to thermal stimulation of the oral cavity in rattlesnakes (*Crotalus viridis*) before and after bilateral anesthetization of the facial pit organs. Brain Res. 400:365–370.

Diener, R. A. 1961. Notes on a bite of the broadbanded copperhead, *Ancistrodon contortrix laticinctus* Gloyd and Conant. Herpetologica 17:143–144.

Diller, L. V. 1990. A field observation on the feeding behavior of *Crotalus viridis lutosus*. J. Herpetol. 24:95–97.

Diller, L. V., and D. R. Johnson. 1988. Food habits, consumption rates, and predation rates of western rattlesnakes and gopher snakes in southwestern Idaho. Herpetologica 44:228–233.

Diller, L. V., and R. L. Wallace. 1984. Reproductive biology of the northern Pacific rattlesnake (*Crotalus viridis oreganus*) in northern Idaho. Herpetologica 40:182–193.

Ditmars, R. L. 1931a. The reptile book. Doubleday Doran & Co., Garden City, New York. 472 pp.

———. 1931b. Snakes of the World. McMillan, New York. 207 pp.

———. 1936. The reptiles of North America. Doubleday, Doran & Co., Garden City, New York. 476 pp.

———. 1939. A field book of North American snakes. Doubleday, Doran & Co., New York. 305 pp.

Dixon, J. R. 1956. The mottled rock rattlesnake, *Crotalus lepidus lepidus,* in Edwards County, Texas. Copeia 1956:126–127.

Dobie, J. F. 1965. Rattlesnakes. Little, Brown & Co., Boston, Massachusetts. 201 pp.

Dodge, C. H., and G. E. Folk, Jr. 1960. A case of rattlesnake poisoning in Iowa with a description of early symptoms. Proc. Iowa Acad. Sci. 67:622–624.

Dolley, J. S. 1939. An anomalous pregnancy in the copperhead. Copeia 1939:170.

Douglas, C. L. 1966. Amphibians and reptiles of Mesa Verde National Park, Colorado. Univ. Kansas Publ. Mus. Natur. Hist. 15:711–744.

Dowling, H. G. 1951a. A proposed method of expressing scale reductions in snakes. Copeia 1951:131–134.

———. 1951b. A proposed standard system of counting ventrals in snakes. British J. Herpetol. 1:97–99.

———. 1957. A review of the amphibians and reptiles of Arkansas. Occ. Pap. Univ. Arkansas Mus. (3):1–51.

———. 1959. Classification of the Serpentes: a critical review. Copeia 1959:38–52.

———(ed.). 1975. Yearbook of herpetology. HISS, New York, New York. 256 pp.

Dowling, H. G., and J. M. Savage. 1960. A guide to the snake hemipenis: a survey of basic structure and systematic characteristics. Zoologica 45:17–28.

Drda, W. J. 1968. A study of snakes wintering in a small cave. J. Herpetol. 1:64–70.

Duellman, W. E. 1950. A case of *Heloderma* poisoning. Copeia 1950:151.

Duellman, W. E., and A. Schwartz. 1958. Amphibians and reptiles of southern Florida. Bull. Florida St. Mus. Biol. Sci. 3:181–324.

Dullemeijer, P. 1959. A comparative functional-anatomical study of the heads of some Viperidae. Morphol. Jarhb. 99:881–985.

———. 1961. Some remarks on the feeding behavior of rattlesnakes. Proc. Acad. Wet. Amst. 64C:383–396.

———. 1969. Growth and size of the eye in viperid snakes. Netherlands J. Zool. 19:249–276.

Dullemeijer, P., and G. D. E. Povel. 1972. The construction for feeding in rattlesnakes. Zool. Mededel. 47:561–578.

Dundee, H. A., and W. L. Burger, Jr. 1948. A denning aggregation of the western cottonmouth. Natur. Hist. Misc. (21):1–2.

Dundee, H. A., and D. A. Rossman. 1989. Amphibians and reptiles of Louisiana. Louisiana St. Univ. Press, Baton Rouge. 300 pp.

Dunkle, D. H., and H. M. Smith. 1937. Notes on some Mexican ophidians. Occ. Pap. Mus. Zool. Univ. Michigan (363):1–15.

Dunson, M. K., and W. A. Dunson. 1975. The relation between plasma Na concentration and salt gland Na-K ATPase content in the diamondback terrapin and the yellow-bellied sea snake. J. Comp. Physiol. 101:89–97.

Dunson, W. A. 1968. Salt gland secretion in the pelagic sea snake *Pelamis.* Amer. J. Physiol. 215:1512–1517.

———. 1971. The sea snakes are coming. Natur. Hist. 80:52–60.

———(ed.). 1975. The biology of sea snakes. Univ. Park Press, Baltimore, Maryland. 530 pp.

Dunson, W. A., and G. W. Ehlert. 1971. Effects of temperature, salinity, and surface water flow on distribution of the sea snake *Pelamis*. Limnol. Oceanogr. 16:845–853.

Dunson, W. A., and J. Freda. 1985. Water permeability of the skin of the amphibious snake, *Agkistrodon piscivorus*. J. Herpetol. 19:93–98.

Dunson, W. A., R. K. Packer, and M. K. Dunson. 1971. Sea snakes: an unusual salt gland under the tongue. Science 173:437–441.

Dunson, W. A., and G. D. Robinson. 1976. Sea snake skin: permeable to water but not to sodium. J. Comp. Physiol. 108:303–311.

Dunson, W. A., and G. D. Stokes. 1983. Asymmetrical diffusion of sodium and water through the skin of sea snakes. Physiol. Zool. 56:106–111.

Durkin, J. P., G. V. Pickwell, J. T. Trotter, and W. T. Shier. 1981. Phospholipase A-2 electrophoretic variants in reptile venoms. Toxicon 19:535–546.

Duvall, D. 1986. Shake, rattle, and roll. Natur. Hist. 95:66–73.

Duvall, D., S. J. Arnold, and G. W. Schuett. 1991. Pit viper mating systems: ecological potential, sexual selection, and microevolution. *In* J. A. Campbell and E. D. Brodie, Jr. (eds.), Biology of pit vipers. Cornell Univ. Press, Ithaca, New York. In press.

Duvall, D., D. Chiszar, W. K. Hayes, J. K. Leonhardt, and M. J. Goode. 1990. Chemical and behavioral ecology of foraging in prairie rattlesnakes (*Crotalus viridis viridis*). J. Chem. Ecol. 16:87–101.

Duvall, D., D. Chiszar, and J. Trupiano. 1978. Preference for envenomated rodent prey by rattlesnakes. Bull. Psychononic Soc. 11:7–8.

Duvall, D., M. J. Goode, W. K. Hayes, J. K. Leonhardt, and D. G. Brown. 1990. Prairie rattlesnake vernal migration: field experimental analyses and survival value. Natl. Geogr. Res. 6:457–469.

Duvall, D., K. Gutzwiller, and M. King. 1985. Reconstructing the rattlesnake. BBC Wildlife 3:80–82.

Duvall, D., M. B. King, and K. J. Gutzwiller. 1985. Behavioral ecology and ethology of the prairie rattlesnake. Natl. Geogr. Res. 1:80–111.

Duvall, D., M. King and R. Miller. 1983. Rattler. Wyoming Wild Life 47(10):26–30.

Duvall, D., K. M. Scudder, and D. Chiszar. 1980. Rattlesnake predatory behavior: mediation of prey discrimination and release of swallowing cues arising from envenomated mice. Anim. Behav. 28:674–683.

Dyr, J. E., B. Hessel, J. Suttnar, F. Kornalik, and B. Blomback. 1989. Fibrinopeptide-releasing enzymes in the venom from the southern copperhead snake (*Agkistrodon contortrix contortrix*). Toxicon 27:359–373.

Dyr, J. E., J. Suttnar, J. Simak, H. Fortova, and F. Kornalik. 1990. The action of a fibrin-promoting enzyme from the venom of *Agkistrodon contortrix contortrix* on rat fibrinogen and plasma. Toxicon 28:1364–1367.

Edgren, R A., Jr. 1948. Notes on a litter of young timber rattlesnakes. Copeia 1948:132.

Edmund, A. G. 1960. Tooth replacement phenomena in the lower vertebrates. Royal Ontario Mus. Life Sci. Contrib. 52:1–190.

Ehrlich, S. P. 1928. A case report of severe snake-bite poisoning. Bull. Antivenin. Inst. Amer. 2:65–66.

Emmerson, F. H. 1982. Western diamonback rattlesnake in southern Nevada: a correction and comments. Great Basin Natur. 42:350.

Engelhardt, G. P. 1914. Notes on the Gila monster. Copeia (7):1–2.

———. 1932. Notes on poisonous snakes in Texas. Copeia 1932:37–38.

Engelman, W., and F. J. Obst. 1981. Snakes: biology, behavior and relationships to man. Exeter Books, New York. 222 pp.

Ernst, C. H. 1964. A study of sexual dimorphism in American *Agkistrodon*
fang lengths. Herpetologica 20:214.

———. 1965. Fang length comparisons of American *Agkistrodon*. Trans.
Kentucky Acad. Sci. 26:12–18.

———. 1982. A study of the fangs of snakes belonging to the *Agkistrodon–*
complex. J. Herpetol. 16:72–80.

Ernst, C. H., and R. W. Barbour. 1989. Snakes of eastern North America.
George Mason Univ. Press, Fairfax, Virginia. 282 pp.

Essex, H. E. 1932. The physiologic action of the venom of the water moccasin (*Agkistrodon piscivorus*). Bull. Antivenin Inst. Amer. 5:81.

Estep, K., T. Poole, C. W. Radcliffe, B. O'Connell, and D. Chiszar. 1981.
Distance traveled by mice after envenomation by a rattlesnake (C.
viridis). Bull. Psychonomic Soc. 18:108–110.

Evans, G. M. 1987. The coral snake question. Herptile 12:105–107, 110.

Evans, P. D., and H. K. Gloyd. 1948. The subspecies of the massasauga,
Sistrurus catenatus, in Missouri. Bull. Chicago Acad. Sci. 8:
225–232.

Falck, E. G. J. 1940. Food of an eastern rock rattlesnake in captivity. Copeia
1940:135.

Fenton, M. B., and L. E. Licht. 1990. Why rattlesnake? J. Herpetol.
24:274–279.

Ferguson, J. H., and R. M. Thornton. 1984. Oxygen storage capacity and
tolerance of submergence of a non-aquatic reptile and an aquatic reptile.
Comp. Biochem. Physiol. 77A:183–187.

Fidler, H. K., R. D. Glasgow, and E. B. Carmichael. 1938. Pathologic
changes produced by subcutaneous injection of rattlesnake (*Crotalus*)
venom into *Macaca mulatta* monkeys. Proc. Soc. Exp. Biol. Med. 38:892–
894.

Fiero, M. K., M. W. Seifert, T. J. Weaver, and C. A. Bonilla. 1972. Comparative study of juvenile and adult prairie rattlesnake (*Crotalus viridis viridis*)
venoms. Toxicon 10:81–82.

Finneran, L. C. 1953. Aggregation behavior of the female copperhead,
Agkistrodon contortrix mokeson, during gestation. Copeia 1953:61–62.

Fischman, H. K., J. Mitra, and H. Dowling. 1972. Chromosome characteristics of 13 species in the Order Serpentes. Mammal Chromosomes Newsl.
13:72–73.

Fishbeck, D. W., and J. C. Underhill. 1959. A check list of the amphibians
and reptiles of South Dakota. Proc. South Dakota Acad. Sci. 38:107–113.

Fitch, H. S. 1949. Study of snake populations in central California. Amer.
Midl. Natur. 41:513–579.

———. 1956. Temperature responses in free-living amphibians and reptiles
in northeastern Kansas. Univ. Kansas Publ. Mus. Natur. Hist. 8:417–476.

———. 1958. Home ranges, territories, and seasonal movements of vertebrates of the Natural History Reservation. Univ. Kansas Publ. Mus.
Natur. Hist. 11:63–326.

———. 1960a. Criteria for determining sex and breeding maturity in
snakes. Herpetologica 16:49–51.

———. 1960b. Autecology of the copperhead. Univ. Kansas Publ. Mus.
Natur. Hist. 13:85–288.

———. 1970. Reproductive cycles in lizards and snakes. Univ. Kansas
Mus. Natur. Hist., Misc. Publ. (52):1–247.

——. 1981. Sexual size differences in reptiles. Univ. Kansas Mus. Natur. Hist. Misc. Publ. (70):1–72.

——. 1982. Resources of a snake community in prairie-woodland habitat of northeastern Kansas. *In* Scott, N. J., Jr. (ed.), Herpetological communities, pp. 83–97. U.S. Fish Wildl. Serv., Wildl. Res. Rep. 13.

——. 1985a. Variation in clutch and litter size in New World reptiles. Univ. Kansas Mus. Natur. Hist., Misc. Publ. (76):1–76.

——. 1985b. Observation on rattle size and demography of prairie rattlesnakes (*Crotalus viridis*) and timber rattlesnakes (*Crotalus horridus*) in Kansas. Occ. Pap. Mus. Natur. Hist. Univ. Kansas (118):1–11.

Fitch, H. S., and J. T. Collins. 1985. Intergradation of Osage and broad-banded copperheads in Kansas. Trans. Kansas Acad. Sci. 38:135–137.

Fitch, H. S., and B. Glading. 1947. A field study of a rattlesnake population. California Fish & Game 33:103–123.

Fitch, H. S., and H. W. Shirer. 1971. A radiotelemetric study of spatial relationships in some common snakes. Copeia 1971:118–128.

Fitch, H. S., and H. Twining. 1946. Feeding habits of the Pacific rattlesnake. Copeia 1946:64–71.

Fix, J. D., and S. A. Minton, Jr. 1976. Venom extraction and yields from the North American coral snake, *Micrurus fulvius*. Toxicon 14:143–145.

Fleet, R. R., and J. C. Kroll. 1978. Litter size and parturition behavior in *Sistrurus miliarius streckeri*. Herp. Review 9:11.

Foote, R., and J. A. MacMahon. 1977. Electrophoretic studies on rattlesnake (*Crotalus* and *Sistrurus*) venom: taxonomic implications. J. Biochem. Physiol. 57B:235–241.

Ford, N. B., V. A. Cobb, and W. W. Lamar. 1990. Reproductive data on snakes from northeastern Texas. Texas J. Sci. 42:355–368.

Ford, N. B., V. A. Cobb, and J. Stout. 1991. Species diversity and seasonal abundance of snakes in a mixed pine-hardwood forest of eastern Texas. Southwest. Natur. 36:171–177.

Forsyth, B. J., C. D. Baker, T. Wiles, and C. Weilbaker. 1985. Cottonmouth, *Agkistrodon piscivorus*, records from the Blue River and Potato Run in Harrison County, Indiana (Ohio River Drainage, U.S.A.). Proc. Indiana Acad. Sci. 94:633–634.

Fouquette, M. J., and H. L. Lindsay, Jr. 1955. An ecological survey of reptiles in parts of northwestern Texas. Texas J. Sci. 7:402–421.

Fowlie, J. A. 1965. The snakes of Arizona. Azul Quinta Press, Fallbrook, California. 164 pp.

Fox, W. 1956. Seminal receptacles of snakes. Anat. Rec. 124:519–540.

Freda, J. 1977. Fighting a losing battle. The story of a timber rattlesnake. HERP: Bull. New York Herpetol. Soc. 13:35–38.

Froom, B. 1972. The snakes of Canada. McCelland and Stewart, Toronto. 128 pp.

Frost, D. R., and J. T. Collins. 1988. Nomenclatural notes on reptiles of the United States. Herp. Review 19:73–74.

Funderburg, J. B. 1968. Eastern diamondback rattlesnake feeding on carrion. J. Herpetol. 2:161–162.

Funk, R. S. 1964a. On the food of *Crotalus m. molossus*. Herpetologica 20:134.

——. 1964b. Birth of a brood of western cottonmouths, *Agkistrodon piscivorus leucostoma*. Trans. Kansas Acad. Sci. 67:199.

——. 1964c. On the reproduction of *Micruroides euryxanthus* (Kennicott). Copeia 1964:219.

————. 1965. Food of *Crotalus cerastes laterorepens* in Yuma County, Arizona. Herpetologica 21:15–17.

————. 1966. Notes about *Heloderma suspectum* along the western extremity of its range. Herpetologica 22:254–258.

Furry, K., T. Swain, and D. Chiszar. 1991. Strike-induced chemosensory searching and trail following by prairie rattlesnakes (*Crotalus viridis*) preying upon deer mice (*Peromyscus maniculatus*): chemical discrimination among individual mice. Herpetologica 47:69–78.

Galligan, J. H., and W. A. Dunson. 1979. Biology and status of timber rattlesnake (*Crotalus horridus*) populations in Pennsylvania. Biol. Conserv. 15:13–58.

Gannon, V. 1978. Factors limiting the distribution of the prairie rattlesnake. Blue Jay 36:142–144.

Gannon, V., and D. M. Secoy. 1984. Growth and reproductive rates of a northern population of the prairie rattlesnake, *Crotalus v. viridis*. J. Herpetol. 18:13–19.

————. 1985. Seasonal and daily activity patterns in a Canadian population of the prairie rattlesnake, *Crotalus viridis viridis*. Can. J. Zool. 63:86–91.

Gans, C., and W. B. Elliott. 1968. Snake venoms: production, injection, action. Adv. Oral Biol. 3:45–81.

Garrett, J. M., and D. G. Barker. 1987. A field guide to reptiles and amphibians of Texas. Texas Monthly Press, Austin. 225 pp.

Garton, J. S., and R. W. Dimmick. 1969. Food habits of the copperhead in middle Tennessee. J. Tennessee Acad. Sci. 44:113–117.

Gates, G. O. 1956a. A record length for the Arizona coral snake. Herpetologica 12:155.

————. 1956b. Mating habits of the Gila monster. Herpetologica 12:184.

————. 1957. A study of the herpetofauna in the vicinity of Wickenburg, Maricopa County, Arizona. Trans. Kansas Acad. Sci. 60:403–418.

Gehlbach, F. R. 1956. Annotated records of southwestern amphibians and reptiles. Trans. Kansas Acad. Sci. 59:364–372.

————. 1965. Herpetology of the Zuni Mountains Region, northwestern New Mexico. Proc. U.S. Natl. Mus. 116:243–332.

————. 1972. Coral snake mimicry reconsidered: the strategy of self-mimicry. Forma et Functio 5:311–320.

Genter, D. L. 1984. *Crotalus viridis* (Prairie Rattlesnake). Food. Herp. Review 15:49–50.

Gibbons, J. W. 1972. Reproduction, growth, and sexual dimorphism in the canebrake rattlesnake (*Crotalus horridus atricaudatus*). Copeia 1972:222–226.

————. 1977. Snakes of the Savannah River Plant with information about snakebite prevention and treatment. ERDA's Savannah River Nat. Environ. Res. Park. SRO-NERP-1. 26 pp.

Gibbons, W., R. R. Haynes, and J. L. Thomas. 1990. Poisonous plants and venomous animals of Alabama and adjoining states. Univ. Alabama Press, Tuscaloosa. 348 pp.

Gier, P. J., R. L. Wallace, and R. L. Ingermann. 1989. Influence of pregnancy on behavioral thermoregulation in the northern Pacific rattlesnake *Crotalus viridis oreganus*. J. Exp. Biol. 145:465–469.

Gillingham, J. C. 1979. Reproductive behavior of the rat snakes of eastern North America, genus *Elaphe*. Copeia 1979:319–331.

Gillingham, J. C., and R. E. Baker. 1981. Evidence for scavenging behavior in the western diamondback rattlesnake (*Crotalus atrox*). Z. Tierpsychol. 55:217–227.

Gillingham, J. C., C. C. Carpenter, and J. B. Murphy. 1983. Courtship, male combat and dominance in the western rattlesnake, *Crotalus atrox*. J. Herpetol. 17:265–270.

Gillingham, J. C., and D. L. Clark. 1981. An analysis of prey-searching behavior in the western diamondback rattlesnake, *Crotalus atrox*. Behav. Neurol. Biol. 32:235–240.

Gilmore, C. W. 1938. Fossil snakes of North America. Geol. Soc. Amer. Spec. Publ. (9):1–96.

Githens, T. S., and I. D. George. 1931. Comparative studies of the venoms of certain rattlesnakes. Bull. Antivenin Inst. Amer. 5:31–35.

Githens, T. S., and N. O. Wolff. 1939. The polyvalency of crotalidic antivenins. J. Immunol. 37:33–51.

Glenn, J. L., and H. E. Lawler. 1987. *Crotalus scutulatus salvini* (Huamantlan Rattlesnake). Behavior. Herp. Review 18:15–16.

Glenn, J. L., and R. Straight. 1977. The midget faded rattlesnake (*Crotalus viridis concolor*) venom: lethal toxicity and individual variability. Toxicon 15:129–133.

———. 1989. Intergradation of two different venom populations of the Mojave rattlesnake (*Crotalus scutulatus scutulatus*) in Arizona. Toxicon 27:411–418.

———. 1990. Venom characteristics as an indicator of hybridization between *Crotalus viridis viridis* and *Crotalus scutulatus scutulatus* in New Mexico. Toxicon 28:857–862.

Glenn, J. L., R. C. Straight, M. C. Wolfe, and D. L. Hardy. 1983. Geographical variation in *Crotalus scutulatus scutulatus* (Mojave rattlesnake) venom properties. Toxicon 21:119–130.

Glissmeyer, H. R. 1951. Egg production of the Great Basin rattlesnake. Herpetologica 7:24–27.

Gloyd, H. K. 1928. The amphibians and reptiles of Franklin County, Kansas. Trans. Kansas Acad. Sci. 31:115–141.

———. 1934. Studies on the breeding habits and young of the copperhead, *Agkistrodon mokasen* Beauvois. Pap. Michigan Acad. Sci., Arts, Lett. 19:587–604.

———. 1935. The subspecies of *Sistrurus miliarius*. Occ. Pap. Mus. Zool. Univ. Michigan (322):1–7.

———. 1936. The subspecies of *Crotalus lepidus*. Occ. Pap. Mus. Zool. Univ. Michigan (337):1–5.

———. 1937. A herpetological consideration of faunal areas in southern Arizona. Bull. Chicago Acad. Sci. 5:79–136.

———. 1938. A case of poisoning from the bite of a black coral snake. Herpetologica 1:121–124.

———. 1940. The rattlesnakes, genera *Sistrurus* and *Crotalus*. Chicago Acad. Sci. Spec. Publ. (4):1–266.

———. 1947. Notes on the courtship and mating behavior of certain snakes. Natur. Hist. Misc. (12):1–4.

———. 1969. Two additional subspecies of North American snakes, genus *Agkistrodon*. Proc. Biol. Soc. Washington 82:219–232.

Gloyd, H. K., and R. Conant. 1934. The broad-banded copperhead: a new subspecies of *Agkistrodon mokasen*. Occ. Pap. Mus. Zool. Univ. Michigan (283):1–5.

————. 1943. A synopsis of the American forms of *Agkistrodon* (copper-heads and moccasins). Bull. Chicago Acad. Sci. 7:147–170.

————. 1990. Snakes of the *Agkistrodon* complex: a monographic review. Soc. Stud. Amphib. Rept., Contr. Herpetol. (6):1–614.

Goin, C. J., O. B. Goin, and G. R. Zug. 1978. Introduction to herpetology. 3d ed. Freeman, San Francisco. 378 pp.

Golan, L., C. Radcliffe, T. Miller, B. O'Connell, and D. Chiszar. 1982. Trailing behavior in prairie rattlesnakes (*Crotalus viridis*). J. Herpetol. 16:287–293.

Golay, P. 1985. Checklist and keys to the terrestrial proteroglyphs of the world (Serpentes: Elapidae-Hydrophiidae). Elapsoïdea fondation culturelle, Geneva, Switzerland. 91 pp.

Goode, M. J., and D. Duvall. 1989. Body temperature and defensive behaviour of free-ranging prairie rattlesnakes, *Crotalus viridid viridis*. Anim. Behav. 38:360–362.

Goodman, J. D. 1958. Material ingested by the cottonmouth, *Agkistrodon piscivorus*, at Reelfoot Lake, Tennessee. Copeia 1958:149.

Graham, G. L. 1977. The karyotype of the Texas coral snake, *Micrurus fulvius tenere*. Herpetologica 33:345–348.

Graham, J. B. 1974a. Aquatic respiration in the sea snake *Pelamis platurus*. Respir. Physiol. 21:1–7.

————. 1974b. Body temperatures of the sea snake *Pelamis platurus*. Copeia 1974:531–533.

Graham, J. B., J. H. Gee, J. Motta, and I. Rubinoff. 1987. Subsurface buoyance regulation by the sea snake *Pelamis platurus*. Physiol. Zool. 60(2):251–161.

Graham, J. B., J. H. Gee, and F. S. Robison. 1975. Hydrostatic and gas exchange functions of the lung of the sa snake *Pelamis platurus*. Comp. Biochem. Physiol. 50A:477–482.

Graham, J. B., W. R. Lowell, I. Rubinoff, and J. Motta. 1987. Surface and subsurface swimming of the sea snake *Pelamis platurus*. J. Exp. Biol. 127:27–44.

Graham, J. B., I. Rubinoff, and M. K. Hecht. 1971. Temperature physiology of the sea snake *Pelamis platurus*: an index of its colonization potential in the Atlantic Ocean. Proc. Natl. Acad. Sci. 68:1360–1363.

Grant, M. L., and L. J. Henderson. 1957. A case of Gila monster poisoning with a summary of some previous accounts. Proc. Iowa Acad. Sci. 64:686–697.

Graves, B. M. 1989. Defensive behavior of female prairie rattlesnakes (*Crotalus viridus*) changes after parturition. Copeia 1989:793–794.

————. 1991. Consumption of an adult mouse by a free-ranging neonate prairie rattlesnake. Southwest. Natur. 36:143.

Graves, B. M., and D. Duvall. 1983. Occurrence and function of prairie rattlesnake mouth gaping in a non-feeding context. J. Exp. Zool 227:471–474.

————. 1985a. Avomic prairie rattlesnakes (*Crotalus viridis*) fail to attack rodent prey. Z. Tierpsychol. 67:161–166.

————. 1985b. Mouth gaping and head shaking by prairie rattlesnakes are associated with vomeronasal organ olfaction. Copeia 1985:496–497.

————. 1987. An experimental study of aggregation and thermoregulation in prairie rattlesnakes (*Crotalus viridis viridis*). Herpetologica 43:259–264.

————. 1988. Evidence of an alarm pheromone from the cloacal sacs of prairie rattlesnakes. Southwest. Natur. 33:339–345.

————. 1990. Spring emergence patterns of wandering garter snakes and prairie rattlesnakes in Wyoming. J. Herpetol. 24:351–356.

Graves, B. M., M. B. King, and D. Duvall. 1986. Natural history of prairie rattlesnake (*Crotalus viridis viridis*) in Wyoming. Herptile 11:5–10.

Greding, E. J., Jr. 1964. Food of *Ancistrodon c. contortrix* in Houston and Trinity counties, Texas. Southwest. Natur. 9:105.

Green, N. B., and T. K. Pauley. 1987. Amphibians and reptiles in West Virginia. Univ. Pittsburgh Press, Pittsburgh. 241 pp.

Greene, H. W. 1973. Defensive tail display by snakes and amphisbaenians. J. Herpetol. 7:143–161.

————. 1976. Scale overlap, a directional sign stimulus for prey ingestion by ophiophagous snakes. Z. Tierpsychol. 41:113–120.

————. 1983. Dietary correlates of the origin and radiation of snakes. Amer. Zool. 23:431–441.

————. 1984. Feeding behavior and diet of the eastern coral snake, *Micrurus fulvius*. Univ. Kansas Mus. Natur. Hist. Spec. Publ. (10):147–162.

————. 1988. Antipredator mechanisms in reptiles. *In* Gans, C., and R. B. Huey (eds.), Biology of the Reptilia, vol. 16, Defense and life history, pp. 1–152. Alan R. Liss, Inc. New York.

————. 1990. A sound defense of the rattlesnake. Pacific Discovery 43(4):10–19.

Greene, H. W., and R. W. McDiarmid. 1981. Coral snake mimicry: does it occur? Science 213:1207–1212.

Greene, H. W., and G. V. Oliver, Jr. 1965. Notes on the natural history of the western massasauga. Herpetologica 21:225–228.

Greene, H. W., and W. F. Pyburn. 1973. Comments on aposematism and mimicry among coral snakes. Biologist 55:144–148.

Gregory, P. T., and R. W. Campbell. 1984. The reptiles of British Columbia. Handb. British Columbia Prov. Mus. (44):1–103.

Gregory-Dwyer, V. M., N. B. Egen, A. B. Bosisio, P. G. Righetti, and F. E. Russell. 1986. An isoelectric focusing study of seasonal variation in rattle-snake venom proteins. Toxicon 24:995–1000.

Grobman, A. B. 1978. An alternative solution to the coral snake mimic problem (Reptilia, Serpentes, Elapidae). J. Herpetol. 12:1–11.

Groombridge, B. 1986. Comments on the *M. pterygoideus glandulae* of crotaline snakes (Reptilia: Viperidae). Herpetologica 42:449–457.

Groves, J. D. 1977. Aquatic behavior in the northern copperhead, *Agkistrodon contortrix mokasen*. Bull. Maryland Herpetol. Soc. 13:114–115.

Guidry, E. V. 1953. Herpetological notes from southeastern Texas. Herpetologica 9:49–56.

Guilday, J. E. 1962. The Pleistocene local fauna of the Natural Chimneys, Augusta County, Virginia. Ann. Carnegie Mus. 36:87–122.

Gumbart, T. C., and K. A. Sullivan. 1990. Predation on yellow-eyed junco nestlings by twin-spotted rattlesnakes. Southwest. Natur. 35:367–368.

Guthrie, J. E. 1926. The snakes of Iowa. Iowa St. Col. Agric. Mech. Arts, Agric. Exp. St. Bull. (239):146–192.

Gutiérrez, J. M., and R. Bolaños. 1980. Karyotype of the yellow-bellied sea snake, *Pelamis platurus*. J. Herpetol. 14:161–165.

Haast, W. E., and R. Anderson. 1981. Complete guide to snakes of Florida. Phoenix Publ. Co., Inc., Miami, Florida. 139 pp.

Haller, R. 1971. The diamondback rattlesnakes. Herpetology 3(3):1–34.

Hallowell, E. 1854. Description of new reptiles from California. Proc. Acad. Natur. Sci. Philadelphia 7:91–97.

Hamilton, W. J., Jr. 1950. Food of the prairie rattlesnake (*Crotalus v. viridis* Rafinesque). Herpetologica 6:34.

Hamilton, W. J., Jr., and J. A. Pollack. 1955. The food of some crotalid snakes from Fort Benning, Georgia. Natur. Hist. Misc. (140):1–4.

Hammerson, G. A. 1981. Opportunistic scavenging by *Crotalus ruber* not field-proven. J. Herpetol. 15:125.

———. 1982. Amphibians and reptiles in Colorado. Colorado Div. Wildl. Dept. Natur. Res. 131 pp.

Hardy, D. L. 1983. Envenomation by the Mojave rattlesnake (*Crotalus scutulatus scutulatus*) in southern Arizona, U.S.A. Toxicon 21:111–118.

———. 1986. Fatal rattlesnake envenomation in Arizona: 1969–1984. Clinical Toxicol. 24:1–10.

Hardy, D. L., M. Jeter, and J. J. Corrigan, Jr. 1982. Envenomation by the northern blacktail rattlesnake (*Crotalus molossus molossus*): report of two cases and the *In Vitro* effects of the venom on fibrinolysis and platelet aggregation. Toxicon 20:487–492.

Harlan, R. 1827. Genera of North American Reptilia and a synopsis of the species. J. Acad. Natur. Sci. Philadelphia (1)5:317–372.

Harris, H. H., Jr. 1965. Case reports of two dusky pigmy rattlesnake bites (*Sistrurus miliarius barbouri*). Bull. Maryland Herpetol. Soc. 2:8–10.

Harris, H. S., Jr., and R. S. Simmons. 1972. An April birth record for *Crotalus lepidus* with a summary of annual broods in rattlesnakes. Bull. Maryland Herpetol. Soc. 8:54–56.

———. 1974. The New Mexican ridge-nosed rattlesnake. Natl. Parks Conserv. Mag. 48(3):22–24.

———. 1975. An endangered species, the New Mexican ridge-nosed rattlesnake. Bull. Maryland Herpetol. Soc. 11:1–7.

———. 1976. The paleogeography and evolution of *Crotalus willardi*, with a formal description of a new subspecies from New Mexico, United States. Bull. Maryland Herpetol. Soc. 12:1–22.

———. 1977. Additional notes concerning cannibalism in pit vipers. Bull. Maryland Herpetol. Soc. 13:121–122.

———. 1978. A preliminary account of the rattlesnakes with the descriptions of four new subspecies. Bull. Maryland Herpetol. Soc. 14:105–211.

Harrison, H. H. 1949–1950. Pennsylvania reptiles and amphibians. Pennsylvania Fish. Comm., Harrisburg. 23 pp.

———. 1971. The world of the snake. Lippincott, New York. 160 pp.

Hartline, P. H. 1974. Thermoreception in snakes. *In* Fessard, A. (ed.), Handbook of sensory physiology, vol. 3, pp. 297–312. Springer-Verlag, New York.

Hartline, P. H., and H. W. Campbell. 1969. Auditory and vibratory responses in the midbrains of snakes. Science 163:1221–1223.

Hartline, P. H., L. Kass, and M. S. Loop. 1978. Merging of modalities in the optic tectum: infrared and visual integration in rattlesnakes. Science 199:1225–1229.

Harwig, S. H. 1966. Rattlesnakes are where and when you find them. J. Ohio Herpetol. Soc. 5:163.

Hayes, W. K. 1986. Observations of courtship in the rattlesnake, *Crotalus viridis oreganus*. J. Herpetol. 20:246–249.

Hayes, W. K., and D. Duvall. 1991. A field study of prairie rattlesnake predatory strikes. Herpetologica 47:78–81.

Hayes, W. K., and J. G. Galusha. 1984. Effects of rattlesnake (*Crotalus viridis oreganus*) envenomation upon mobility of male wild and laboratory mice (*Mus musculus*). Bull. Maryland Herpetol. Soc. 54:171–175.

Hayes, W. K., I. I. Kaiser, and D. Duvall. 1991. The mass of venom expended by prairie rattlesnakes when feeding on rodent prey. *In* J. A. Campbell and E. D. Brodie, Jr. (eds.), Biology of pit vipers. Cornell Univ. Press, Ithaca, New York. In press.

Heath, W. G. 1961. A trailing device for small animals designed for field study of the Gila monster (*Heloderma suspectum*). Copeia 1961:491–492.

Heatwole, H. 1987. Sea snakes. New South Wales Univ. Press, Australia. 85 pp.

Heatwole, H., and E. P. Finnie. 1980. Seal predation on a sea snake. Herpetofauna 11:24.

Hecht, M. K., C. Kropach, and B. M. Hecht. 1974. Distribution of the yellow-bellied sea snake, *Pelamis platurus,* and its significance in relation to the fossil record. Herpetologica 30:387–396.

Hecht, M. M., and D. Marien. 1956. The coral snake mimic problem: a reinterpretation. J. Morphol. 98:335–366.

Hennessy, D. F., and D. H. Owings. 1988. Rattlesnakes create a context for localizing their search for potential prey. Ethology 77:317–329.

Hensley, M. M. 1949. Mammal diet of *Heloderma*. Herpetologica 5:152.

———. 1950. Notes on the natural history of *Heloderma suspectum*. Trans. Kansas Acad. Sci. 53:268–269.

Hermann, J. A. 1950. Mammals of the Stockton Plateau of northeastern Terrell County, Texas. Texas J. Sci. 2:368–393.

Herreid, C. F. II. 1961. Snakes as predators of bats. Herpetologica 17:271–272.

Heyrand, R. L., and A. Call. 1951. Growth and age in western striped racer and Great Basin rattlesnake. Herpetologica 7:28–40.

Hirth, H. F. 1966a. Weight changes and mortality of three species of snakes during hibernation. Herpetologica 22:8–12.

———. 1966b. The ability of two species of snakes to return to a hibernaculum after displacement. Southwest. Natur. 11:49–53.

Hirth, H. F., and A. C. King. 1968. Biomass densities of snakes in the cold desert of Utah. Herpetologica 24:333–335.

———. 1969. Body temperatures of snakes in different seasons. J. Herpetol. 3:101–102.

Hirth, H. F., R. C. Pendleton, A. C. King, and T. R. Downard. 1969. Dispersal of snakes from a hibernaculum in northwestern Utah. Ecology 50:332–339.

Hoessle, C. 1963. A breeding pair of western diamondback rattlesnakes, *Crotalus atrox*. Bull. Philadelphia Herpetol. Soc. 11:65–66.

Holbrook, J. E. 1836–1842. North American herpetology; or, a description of the reptiles inhabiting the United States. Vol. 1–3. J. Dobson, Philadelphia.

Holman, J. A. 1979. A review of North American tertiary snakes. Publ. Mus. Michigan St. Univ., Paleontol. Ser 1:201–260.

———. 1981. A review of North American Pleistocene snakes. Publ. Mus. Michigan St. Univ., Paleontol. Ser. 1:261–306.

———. 1982. The Pleistocene (Kansas) herpetofauna of Trout Cave, West Virginia. Ann. Carnegie Mus. Natur. Hist. 51:391–404.

Holman, J. A., G. Bell, and J. Lamb. 1990. A late Pleistocene herpetofauna from Bell Cave, Alabama. Herpetol. J. 1:521–529.

Holman, J. A., and F. Grady. 1989. The fossil herpetofauna (Pleistocene: Irvingtonian) of Hamilton Cave, Pendleton County, West Virginia. Natl. Speleol. Soc. Bull. 51:34–41.

Holman, J. A., J. H. Harding, M. M. Hensley, and G. R. Dudderar. 1989. Michigan snakes. Michigan St. Univ. Cooperat. Ext. Serv., East Lansing. 72 pp.

Howarth, B. 1974. Sperm storage: as a function of the female reprodutive system. *In* Johnson, A. D., and C. W. Foley (eds.), The oviduct and its functions, pp. 237–270. Academic Press, New York.

Howell, C. T., and S. F. Wood. 1957. The prairie rattlesnake at Gran Quivara National Monument, New Mexico. Bull. South. California Acad. Sci. 56:97–98.

Hudnall, J. A. 1979. Surface activity and horizontal movements in a marked population of *Sistrurus miliarius barbouri*. Bull. Maryland Herpetol. Soc. 15:134–138.

Hudson, G. E. 1942. The amphibians and reptiles of Nebraska. Univ. Nebraska, Nebraska Conserv. Bull. (24):1–146.

Huheey, J. E., and A. Stupka. 1967. Amphibians and reptiles of Great Smoky Mountains National Park. Univ. Tennessee Press, Knoxville. 98 pp.

Hulme, J. H. 1952. Observation of a snake bite by a cottonmouth moccasin. Herpetologica 8:51.

Hurter, J., Sr. 1911. Herpetology of Missouri. Trans. Acad. Sci. St. Louis 20:59–274.

Hutchison, R. H. 1929. On the incidence of snake bite poisoning in the United States and the results of the newer methods of treatment. Bull. Antivenin Inst. Amer. 3:43–57.

———. 1930 . Further notes on the incidence of snakebite poisoning in the United States. Bull. Antivenin Inst. Amer. 4:40–43.

Iverson, J. B. 1975. Notes on Nebraska reptiles. Trans. Kansas Acad. Sci. 78:51–62.

———. 1978. Reproductive notes on Florida snakes. Florida Sci. 41:201–207.

Jackson, D. L., and R. Franz. 1981. Ecology of the eastern coral snake (*Micrurus fulvius*) in northern peninsular Florida. Herpetologica 37:213–228.

Jackson, J. F., and D. L. Martin. 1980. Caudal luring in the dusky pygmy rattlesnake, *Sistrurus miliarius barbouri*. Copeia 1980:926–927.

Jackson, J. J. 1983. Snakes of the southeastern United States. Publ. Sect., Georgia Extension Serv. 112 pp.

Jacob, J. S. 1977. An evaluation of the possibility of hybridization between the rattlesnakes *Crotalus atrox* and *C. scutulatus* in the southwestern United States. Southwest. Natur. 22:469–485.

———. 1980. Heart rate–ventilatory response of seven terrestrial species of North American snakes. Herpetologica 36:326–335.

———. 1981. Population density and ecological requirements of the western pygmy rattlesnake in Tennessee. U.S. Fish Wildl. Serv., Denver. 45 pp.

Jacob, J. S., and J. S. Altenbach. 1977. Sexual color dimorphism in *Crotalus lepidus klauberi* Gloyd (Reptilia, Serpentes, Viperidae). J. Herpetol. 11:81–84.

Jacob, J. S., and S. L. Carroll. 1982. Effect of temperature on the heart rate–ventilatory response in the copperhead, *Agkistrodon contortrix* (Reptilia: Viperidae). J. Therm. Biol. 7:117–120.

Jacob, J. S., and C. W. Painter. 1980. Overwinter thermal ecology of *Crotalus viridis* in the north-central plains of New Mexico. Copeia 1980:799–805.

Jacob, J. S., S. R. Williams, and R. P. Reynolds. 1987. Reproductive activity of male *Crotalus atrox* and *C. scutulatus* (Reptilia: Viperidae) in northeastern Chihuahua, Mexico. Southwest. Natur. 32:273–276.

Jacobsen, N. 1977. The prairie rattler and its bite. North Dakota Outdoors 40(3):15–17.

Jaksic, F. M., and H. W. Greene. 1984. Emperical evidence of non-correlation between tail loss frequency and predation intensity on lizards. Oikos 42(3):407–411.

Jayne, B. C. 1986. Kinematics of terrestrial snake locomotion. Copeia 1986:915–927.

Jennings, M. R. 1984. Longevity records for lizards of the family Helodermatidae. Bull. Maryland Herpetol. Soc. 20:22–23.

John-Alder, H. B., C. H. Lowe, and A. F. Bennett. 1983. Thermal dependence of locomotory energetics and aerobic capacity of the Gila monster (*Heloderma suspectum*). J. Comp. Physiol. 151:119–126.

Johnson, B. D. 1968. Selected Crotalidae venom properties as a source of taxonomic criteria. Toxicon 6:5–10.

Johnson, B. D., J. Hoppe, R. Rogers, and H. L. Stahnke. 1968. Characteristics of venom from the rattlesnake *Crotalus horridus atricaudatus*. J. Herpetol. 2:107–112.

Johnson, B. D., H. L. Stahnke, and J. A. Hoppe. 1968. Variations of *Crotalus scutulatus* raw venom concentrations. J. Arizona Acad. Sci. 5:41–42.

Johnson, B. D., H. L. Stahnke, and R. Koonce. 1967. A method for estimating *Crotalus atrox* venom concentrations. Toxicon 5:35–38.

Johnson, D. H., M. D. Bryant, and A. H. Miller. 1948. Vertebrate animals of the Providence Mountains area of California. Univ. California Publ. Zool. 48:221–376.

Johnson, E. K. 1987. Stability of venoms from the northern Pacific rattlesnake (*Crotalus viridis oreganus*). Northwest Sci. 61:110–113.

Johnson, E. K., K. V. Kardong, and C. L. Ownby. 1987. Observations on white and yellow venoms from an individual southern Pacific rattlesnake (*Crotalus viridis helleri*). Toxicon 25:1169–1180.

Johnson, J. E., Jr. 1948. Copperhead in a tree. Herpetologica 4:214.

Johnson, L. F., J. S. Jacob, and P. Torrence. 1982. Annual testicular and androgenic cycles of the cottonmouth (*Agkistrodon piscivorus*) in Alabama. Herpetologica 38:16–25.

Johnson, R. G. 1955. The adaptive and phylogenetic significance of vertebral form in snakes. Evolution 9:367–388.

Johnson, T. B. 1983. Status Report: *Crotalus willardi willardi* (Meek, 1905). U.S. Fish Wildl. Serv. Contr. 14–16–0002–81–224. 70 pp.

———. 1987. Banded rock rattlesnake. Wildlife Views 30(5):4.

Johnson, T. B., and G. S. Mills. 1982. A preliminary report on the status of *Crotalus lepidus*, *C. pricei*, and *C. willardi* in southeastern Arizona. U.S. Fish Wildl. Serv. Contr. 14–16–0002–81–224. 24 pp.

Jones, J. M. 1976. Variations of venom proteins in *Agkistrodon* snakes from North America. Copeia 1976:558–562.

Jones, J. M., and P. M. Burchfield. 1971. Relationship of specimen size to venom extracted from the copperhead, *Agkistrodon contortrix*. Copeia 1971:162–163.

Jones, K. B. 1983. Movement patterns and foraging ecology of Gila monsters (*Heloderma suspectum* Cope) in northwestern Arizona. Herpetologica 39:247–253.

Julian, G. 1951. Sex ratios of the winter populations. Herpetologica 7:21–24.

Kaiser, E., and H. Michl. 1958. Die Biochemie der tierischen Gifte. Franz Deuticke, Wien. 258 pp.

Kardong, K. V. 1974. Kinesis of the jaw apparatus during the strike in the cottonmouth snake, *Agkistrodon piscivorus*. Forma Functio 7:327–354.

———. 1977. Kinesis of the jaw apparatus during swallowing in the cottonmouth snake, *Agkistrodon piscivorus*. Copeia 1977:338–348.

———. 1979. "Protovipers" and the evolution of snake fangs. Evolution 33:433–443.

———. 1980. Gopher snakes and rattlesnakes: presumptive Batesian mimicry. Northwest Sci. 54:1–4.

———. 1982a. The evolution of the venom apparatus in snakes from colubrids to viperids and elapids. Mem. Inst. Butantan 46:105–118.

———. 1982b. Comparative study of changes in prey capture behavior of the cottonmouth (*Agkistrodon piscivorus*) and Egyptian cobra (*Naja haje*). Copeia 1982:337–343.

———. 1986a. Predatory strike behavior of the rattlesnake, *Crotalus viridis oreganus*. J. Comp. Psychol. 100:304–314.

———. 1986b. The predatory strike of the rattlesnake: when things go amiss. Copeia 1986:816–820.

Kardong, K. V., and S. P. MacKessy. 1991. The strike behavior of a congenitally blind rattlesnake. J. Herpetol. 25:208–211.

Kauffeld, C. F. 1943a. Fieldnotes on some Arizona reptiles and amphibians. Amer. Midl. Natur. 29:342–358.

———. 1943b. Growth and feeding of new-born Price's and green rock rattlesnakes. Amer. Midl. Natur. 29:607–614.

———. 1957. Snakes and snake hunting. Hanover House, Garden City, New York. 266 pp.

———. 1961. Massasauga land. Bull. Philadelphia Herpetol. Soc. 9(3):7–13.

———. 1969. Snakes: the keeper and the kept. Doubleday, New York. 248 pp.

Keegan, H. L. 1944. Indigo snakes feeding upon poisonous snakes. Copeia 1944:59.

Keenlyne, K. D. 1972. Sexual differences in feeding habits of *Crotalus horridus horridus*. J. Herpetol. 6:234–237.

———. 1978. Reproductive cycles in two species of rattlesnakes. Amer. Midl. Natur. 100:368–375.

Keenlyne, K. D., and J. R. Beer. 1973. Food habits of *Sistrurus catenatus catenatus*. J. Herpetol. 7:382–384.

Keiser, E. D., Jr. 1971. The poisonous snakes of Louisiana and the emergency treatment of their bites. Louisiana Wildl. Fish. Comm. 16 pp.

Keiser, E. D., Jr., and L. D. Wilson. 1979. Checklist and key to the herpetofauna of Louisiana. 2d ed. Lafayette Natur. Hist. Mus. Tech. Bull. (1):1–49.

Keith, J., R. Lee, and D. Chiszar. 1985. Spatial orientation by cottonmouths (*Agkistrodon piscivorus*) after detecting prey. Bull. Maryland Herpetol. Soc. 21:145–149.

Kelly, H. A., A. W. Davis, and H. C. Robertson. 1936. Snakes of Maryland. Natur. Hist. Soc. Maryland, Baltimore. 103 pp.

Kennicott, R. 1860. Descriptions of new species of North American serpents in the museum of the Smithsonian Institution, Washington. Proc. Acad. Natur. Sci. Philadelphia 12:328–338.

———. 1861. On three new forms of rattlesnakes. Proc. Acad. Natur. Sci. Philadelphia. 13:206–208.

Kerfoot, W. C. 1969. Selection of an appropriate index for the study of variability of lizard and snake body scale counts. Systematic Zool. 18:53–62.

King, K. A. 1975. Unusual food item of the western diamondback rattle-snake (*Crotalus atrox*). Southwest. Natur. 20:416–417.

King, M., and D. Duvall. 1990. Prairie rattlesnake seasonal migrations: episodes of movement, vernal foraging and sex differences. Anim. Behav. 39:924–935.

King, M., D. McCarron, D. Duvall, G. Baxter, and W. Gern. 1983. Group avoidance of conspecific but not interspecific chemical cues by prairie rattlesnakes (*Crotalus viridis*). J. Herpetol. 17:196–198.

Kinghorn, J. R. 1956. The snakes of Australia. Angus and Robertson, London. 197 pp.

Kitchens, C. S., S. Hunter, and L. H. S. Van Mierop. 1987. Severe myonecrosis in a fatal case of envenomation by the canebrake rattle-snake (*Crotalus horridus atricaudatus*). Toxicon 25:455–458.

Kitchens, C. S., and L. H. S. Van Mierop. 1983. Mechanism of defibrination in humans after envenomation by the eastern dia-mondback rattlesnake. Amer. J. Hematol. 14:345–354.

Klauber, L. M. 1930. New and renamed subspecies of *Crotalus confluentus* Say, with remarks on related species. Trans. San Diego Soc. Natur. Hist. 6:95–144.

———. 1931. A statistical survey of the snakes of the southern border of California. Bull. Zool. Soc. San Diego 8:1–93.

———. 1935. A new subspecies of *Crotalus confluentus*, the prairie rattle-snake. Trans. San Diego Soc. Natur. Hist. 8:75–90.

———. 1936–40. A statistical study of the rattlesnakes. Parts I–VII. Trans. San Diego Soc. Natur. Hist. nos. 1–6.

———. 1943. Tail-length differences in snakes with notes on sexual dimorphism and the coefficient of divergence. Bull. Zool. Soc. San Diego 18:1–60.

———. 1944. The sidewinder, *Crotalus cerastes*, with description of a new subspecies. Trans. San Diego Soc. Natur. Hist. 10:91–126.

———. 1949a. Some new and revived subspecies of rattlesnakes. Trans. San Diego Soc. Natur. Hist. 11:61–116.

———. 1949b. The subspecies of the ridge-nosed rattlesnake, *Crotalus wil-lardi*. Trans. San Diego Soc. Natur. Hist. 11:121–140.

———. 1956. *Agkistrodon* or *Ancistrodon*? Copeia 1956:258–259.

———. 1972. Rattlesnakes: their habits, life histories, and influence on mankind. 2d ed. Univ. California Press, Berkeley. 1533 pp.

Klawe, W. L. 1964. Food of the black-and-yellow sea snake, *Pelamis platurus*, from Ecuadorian coastal waters. Copeia 1964:712–713.

Klimstra, W. D. 1959. Food habits of the cottonmouth in southern Illinois. Natur. Hist. Misc. (168):1–8.

Kochva, E., and C. Gans. 1966. Histology and histochemistry of venom glands of some crotaline snakes. Copeia 1966:506–515.

Kofron, C. P. 1978. Foods and habitats of aquatic snakes (Reptilia, Ser-pentes) in a Louisiana swamp. J. Herpetol. 12:543–554.

———. 1979. Reproduction of aquatic snakes in south-central Louisiana. Herpetologica 35:44–50.

Krempels, D. M. 1984. Near infrared reflectance by coral snakes: aposematic coloration? Progr. 6th Ann. Meet. Amer. Soc. Ichthyol. Herpetol.:142.

Kropach, C. 1971a. Sea snake (*Pelamis platurus*) aggregations on slicks in Panama. Herpetologica 27:131–135.

———. 1971b. Another color variety of the sea snake *Pelamis platurus* from Panama Bay. Herpetologica 27:326–327.

———. 1972. *Pelamis platurus* as a potential colonizer of the Caribbean Sea. Bull. Biol. Soc. Washington 2:267–269.

———. 1975. The yellow-bellied sea snake, *Pelamis*, in the eastern Pacific. *In* Dunson, W. A. (ed.), The biology of sea snakes, pp. 185–213. Univ. Park Press, Baltimore, Maryland.

Kropach, C., and J. D. Soule. 1973. An unusual association between an ectoproct and a sea snake. Herpetologica 29:17–19.

Lacépède, B. G. E. 1788–1789. Histoire naturelle des serpens. 2 vols. Paris.

Lamson, G. H. 1935. The reptiles of Connecticut. Connecticut St. Geol. Natur. Hist. Surv. Bull. (54):1–35.

Landreth, H. F. 1973. Orientation and behavior of the rattlesnake, *Crotalus atrox*. Copeia 1973:26–31.

LaPointe, J. 1953. Case report of a bite from the massasauga, *Sistrurus catenatus catenatus*. Copeia 1953:128–129.

Lardie, R. L. 1976. Large centipede eaten by a western massasauga. Bull. Oklahoma Herpetol. Soc. 1:40.

La Rivers, I., III. 1973. Effects of rattlesnake venom on rattlesnakes *Crotalus viridis lutosus*. HISS News-J. 1:161–162.

———. 1976. I—Some comments on snakebite treatment in the United States, and II—An account of a human envenomation by an adult western diamondback rattlesnake (*Crotalus atrox* Baird & Girard). Occ. Pap. Biol. Soc. Nevada (42):1–6.

Lasky, W. R. 1980. A case of spitting rattlesnake. Herpetology 10(3):14.

Laszlo, J. 1975. Probing as a practical method of sex recognition in snakes. Intern. Zoo Yearb. 15:178–179.

Latreille, P. A. 1801. *In* C. S. Sonnini and P. A. Latreille, Histoire naturelle des reptiles . . . Vol. 3. Deterville, Paris. 335 pp.

Laughlin, H. E. 1959. Stomach contents of some aquatic snakes from Lake McAlester, Pittsburgh County, Oklahoma. Texas J. Sci. 11:83–85.

Laughlin, H. E., and B. J. Wilks. 1962. The use of sodium pentobarbital in population studies of poisonous snakes. Texas J. Sci. 14:188–191.

Lee, D. S. 1968. Herpetofauna associated with Central Florida mammals. Herpetologica 24:83–84.

Lee. R. K. K., D. A. Chiszar, and H. M. Smith. 1988. Post-strike orientation of the prairie rattlesnake facilitates location of envenomated prey. J. Ethol. 6:129–134.

Leviton. A. E. 1972. Reptiles and amphibians of North America. Doubleday, New York. 250 pp.

Lewis, T. H. 1949. Dark coloration in the reptiles of the Tularosa Malpais, New Mexico. Copeia 1949:181–184.

———. 1951. Dark coloration in the reptiles of the Malpais of the Mexican Border. Copeia 1951:311–312.

Lillywhite, H. B. 1974. Activity of snakes in recently burned chaparral. Bull. Ecol. Soc. Amer. 55:43.

———. 1982. Cannibalistic carrion ingestion by the rattlesnake, *Crotalus viridis*. J. Herpetol. 16:95.

Lindner, D. 1962. Feeding observations of *Micruroides*. Bull. Philadelphia Herpetol. Soc. 10(2–3):31.

Lindsdale, J. M. 1940. Amphibians and reptiles in Nevada. Proc. Amer. Acad. Arts Sci. 73:197–257.

Lindsey, P. 1979. Combat behavior in the dusky pygmy rattlesnake, *Sistrurus miliarius barbouri*, in captivity. Herp. Review 10:93.

Linnaeus, C. 1758. Systema naturae . . . 10th ed. Vol. 1. Stockholm, Sweden. 826 pp.

———. 1766. Systema naturae . . . 12th ed. Stockholm, Sweden. 532 pp.

Linzey, D. W. 1979. Snakes of Alabama. Strode Publ., Inc., Huntsville, Alabama. 136 pp.

Linzey. D. W., and M. J. Clifford. 1981. Snakes of Virginia. Univ. Press Virginia, Charlottesville. 159 pp.

Liu, C.-S., C.-L. Wang, and R. Q. Blackwell. 1975. Isolation and partial characterization of pelamitoxin A from *Pelamis platurus* venom. Toxicon 13:31–36.

Logier, E. B. S. 1939. The reptiles of Ontario. Royal Ontario Mus. Handbk. (4):1–63.

———. 1958. The snakes of Ontario. Univ. Toronto Press, Toronto, Canada. 94 pp.

Logier, E. B. S., and G. C. Toner. 1961. Check list of amphibians and reptiles of Canada and Alaska. Contrib. Royal Ontario Mus. Zool. Palaeontol. (53):1–92.

Lohoefener, R., and R. Altig. 1983. Mississippi herpetology. Mississippi St. Univ. Res. Cent. Nat. Space Tech. Lab. Bull. (1):1–66.

Loomis, R. B. 1948. Notes on the herpetology of Adams County, Iowa. Herpetologica 4:121–122.

Loveridge, A. 1938. Food of *Micrurus fulvius fulvius*. Copeia 1938:201–202.

Lowe, C. H. 1942. Notes on the mating of desert rattlesnakes. Copeia 1942:261–262.

———. 1948a. Territorial behavior in snakes and the so-called courtship dance. Herpetologica 4:129–135.

———. 1948b. Effect of venom of *Micruroides* upon *Xantusia vigilis*. Herpetologica 4:136.

———(ed.). 1964. The vertebrates of Arizona. Univ. Arizona Press, Tucson. 270 pp.

Lowe, C. H., D. S. Hinds, P. J. Lardner, and K. E. Justice. 1967. Natural free-running period in vertebrate animal populations. Science 156:531–534.

Lowe, C. H., and K. S. Norris. 1950. Aggressive behavior in male sidewinders, *Crotalus cerastes*, with a discussion of aggressive behavior and territoriality in snakes. Natur. Hist. Misc. (66):1–13.

Lowe, C. H., C. R. Schwalbe, and T. B. Johnson. 1986. The venomous reptiles of Arizona. Arizona Game Fish Dept., Phoenix. 115 pp.

Lowell, J. A. 1957. A bite by a sidewinder rattlesnake. Herpetologica 13:135–136.

Ludlow, M. E. 1981. Observations on *Crotalus v. viridis* (Rafinesque) and the herpetofauna of the Ken-Caryl Ranch, Jefferson County, Colorado. Herp. Review 12:50–52.

Lynn, W. G. 1931. The structure and function of the facial pit of the pit vipers. Amer. J. Anat. 49:97–139.

Lyon, M. W., and C. Bishop. 1936. Bite of the prairie rattlesnake *Sistrurus catenatus* Raf. Proc. Indiana Acad. Sci. 45:253–256.

Lyons, W. J. 1971. Profound thrombocytopena associated with *Crotalus ruber* envenomation: a clinical case. Toxicon 9:237–240.

Macartney, J. M. 1985. The ecology of the northern Pacific rattlesnake, *Crotalus viridis oreganus*, in British Columbia. Unpublish. M.S. Thesis, Univ. Victoria, British Columbia.

———. 1989. Diet of the northern Pacific rattlesnake, *Crotalus viridis oreganus*, in British Columbia. Herpetologica 45:299–304.

Macartney J. M., and P. T. Gregory. 1988. Reproductive biology of female rattlesnakes (*Crotalus viridis*) in British Columbia. Copeia 1988:47–57.

Macartney, J. M., P. T. Gregory, and M. B. Charland. 1990. Growth and sexual maturity of the western rattlesnake, *Crotalus viridis*, in British Columbia. Copeia 1990:528–542.

Macartney, J. M., P. T. Gregory, and K. W. Larson. 1988. A tabular survey of data on movements and home ranges of snakes. J. Herpetol. 22:61–73.

Macartney, J. M., K. W. Larson, and P. T. Gregory. 1989. Body temperatures and movements of hibernating snakes (*Crotalus* and *Thamnophis*) and thermal gradients of natural hibernacula. Can. J. Zool. 67:108–114.

Macht, D. I. 1937. Comparative toxicity of sixteen specimens of *Crotalus* venom. Proc. Soc. Exp. Biol. Med. 36:499–501.

Mackessy, S. P. 1985. Fractionation of red diamond rattlesnake (*Crotalus ruber ruber*) venom: protease, phosphodiesterase, L-amino acid oxidase activities and effects of metal ions and inhibitors on protease activity. Toxicon 23:337–340.

———. 1988. Venom ontogeny in the Pacific rattlesnake *Crotalus viridis helleri* and *C. v. oreganus*. Copeia 1988:92–101.

MacMahon, J. A., and A. H. Hamer. 1975a. Hematology of the sidewinder (*Crotalus cerastes*). Comp. Biochem. Physiol. 51A:53–58.

———. 1975b. Effects of temperature and photoperiod on oxygenation and other blood parameters of the sidewinder (*Crotalus cerastes*): adaptive significance. Comp. Biochem. Physiol. 51A:59–69.

Manion, S. 1968. *Crotalus willardi*—the Arizona ridge-nosed rattlesnake. Herpetology 2(3):27–30.

Mao, S.-H., and B.-Y. Chen. 1980. Sea snakes of Taiwan. A natural history of sea snakes. Natl. Sci. Counc. Republic China, Taipei, Taiwan. 64 pp.

Mao, S.-H., B.-Y. Chen, F.-Yi Yin, and Y.-W. Guo. 1983. Immunotaxonomic relationships of sea snakes to terrestrial elapids. Comp. Biochem. Physiol. 74A:869–872.

Maple, W. T., and L. P. Orr. 1968. Overwintering adaptions of *Sistrurus catenatus* in northeastern Ohio. J. Herpetol. 2:179–180.

Marchisin, A. 1978. Observation on an audiovisual "warning" signal in the pigmy rattlesnake, *Sistrurus miliarius* (Reptilia, Serpentes, Crotalidae). Herp. Review 9:92–93.

Marion, K. R., and O. J. Sexton. 1984. Body temperatures and behavioral activities of hibernating prairie rattlesnakes, *Crotalus viridis*, in artificial dens. Prairie Natur. 16:111–116.

Markland, F. S., Jr. 1988. Fibrin(ogen)olytic enzymes from snake venoms. Hematology 7:149–172.

Markland, F. S., Jr., K. N. N. Reddy, and L. F. Guan. 1988. Purification and characterization of a direct-acting fibrinolytic enzyme from southern copperhead venom. Hematology 7:173–189.

Marr, J. C. 1944. Notes on amphibians and reptiles from the central United States. Amer. Midl. Natur. 32:478–490.

Martin, B. E. 1974. Distribution and habitat adaptations in rattlesnakes of Arizona. HERP: Bull. New York Herpetol. Soc. 10(3&4):3–12.

———. 1975a. Notes on a brood of the Arizona ridge-nosed rattlesnake *Crotalus willardi willardi*. Bull. Maryland Herpetol. Soc. 11:64.65.

———. 1975b. An occurrence of the Arizona ridge-nosed rattlesnake, *Crotalus willardi willardi*, observed feeding in nature. Bull. Maryland Herpetol. Soc. 11:66–67.

———. 1975c. A brood of Arizona ridge-nosed rattlesnakes (*Crotalus willardi willardi*) bred and born in captivity. Bull. Maryland Herpetol. Soc. 11:187–189.

———. 1976. A reproductive record for the New Mexican ridge-nosed rattlesnake (*Crotalus willardi obscurus*). Bull. Maryland Herpetol. Soc. 12:126–128.

Martin, D. L. 1984. An instance of sexual defense in the cottonmouth, *Agkistrodon piscivorus*. Copeia 1984:772–774.

Martin, J. H., and R. M. Bagby. 1972. Temperature-frequency relationship of the rattlesnake rattle. Copeia 1972:482–485.

———. 1973. Effects of fasting on the blood chemistry of the rattlesnake, *Crotalus atrox*. Comp. Biochem. Physiol. 44A:813–820.

Martin, J. R., and J. T. Wood. 1955. Notes on the poisonous snakes of the Dismal Swamp area. Herpetologica 11:237–238.

Martin, P. J. 1930. Snake hunt nets large catch. Bull. Antivenin Inst. Amer. 4:77–78.

Martin, W. H. 1981. The timber rattlesnake in the Northeast: its range. past and present. HERP: Bull. New York Herpetol. Soc. 17:15–20.

———. 1988. Life history of the timber rattlesnake. Catesbeiana 8(1):9–12.

———. 1990. The timber rattlesnake, *Crotalus horridus*, in the Appalachian Mountains of eastern North America. Catesbeiana 10:49.

Martinez, M., E. D. Rael, and N. L. Maddux. 1990. Isolation of a hemmorrhagic toxin from Mojave rattlesnake (*Crotalus scutulatus scutulatus*) venom. Toxicon 28:685–694.

Martof, B. S. 1956. Amphibians and reptiles of Georgia, a guide. Univ. Georgia Press, Athens. 94 pp.

Martof, B. S., W. M. Palmer, J. R. Bailey, J. R. Harrison III, and J. Dermid. 1980. Amphibians and reptiles of the Carolinas and Virginia. Univ. North Carolina Press, Chapel Hill. 264 pp.

Marx, N., and G. B. Rabb. 1972. Phyletic analysis of fifty characters of advanced snakes. Fieldiana: Zool. 63:1–321.

Matthey, R. 1931a. Chromosomes de sauriens: Helodermatidae, Varanidae, Xantusiidae, Anniellidae, Anguidae. Bull. Soc. Vaudoise Sci. Natur. 57:269.

———. 1931b. Chromosomes de Reptiles: Sauriens, ophidiens, cheloniens. L'evolution la formule chromosomiale chez les sauriens. Rev. Suisse Zool. 38:117–186.

Mattison, C. 1986. Snakes of the world. Facts on File Publ., New York. 190 pp.

McCauley, R. H., Jr. 1945. The reptiles of Maryland and the District of Columbia. Privately Publ., Hagerstown, Md. 194 pp.

McCoy, C. J., 1961. Birth season and young of *Crotalus scutulatus* and *Agkistrodon contortrix lacticinctus*. Herpetologica 17:140.

―――. 1980. Identification guide to Pennsylvania snakes. Carnegie Mus. Natur. Hist. Educ. Bull. (1):1–12.

―――. 1982. Amphibians and reptiles in Pennsylvania. Carnegie Mus. Natur. Hist. (6):1–91.

McCoy, C. J., and D. E. Hahn. 1979. The yellow-bellied seasnake, *Pelamis platurus* (Reptilia: Hydrophiidae), in the Philippines. Ann. Carnegie Mus. 48:231–234.

McCoy, C. J., and W. L. Minckley. 1969. *Sistrurus catenatus* (Reptilia: Crotalidae) from the Cuatro Ciénegas Basin, Coahuila, Mexico. Herpetologica 25:152–153.

McCranie, J. R. 1980a. *Crotalus adamanteus*. Catalog. Amer. Amphi. Rept. 252:1–2.

―――. 1980b. *Crotalus pricei*. Catalog. Amer. Amphib. Rept. 266:1–2.

―――. 1988. Description of the hemipenis of *Sistrurus ravus* (Serpentes: Viperidae). Herpetologica 44:123–126.

McCrystal, H. K., and R. J. Green. 1986. *Agkistrodon contortrix pictigaster* (Trans-Pecos Copperhead). Feeding. Herp. Review 17:61.

McDiarmid, R. W. (ed.). 1978. Rare and endangered biota of Florida. Vol. 3. Amphibians and reptiles. Univ. Press Florida, Gainesville. 74 pp.

McDowell, S. B. 1972. The genera of sea-snakes of the *Hydrophis* group (Serpentes: Elapidae). Trans. Zool. Soc. London 32:195–247.

―――. 1986. The architecture of the corner of the mouth of colubroid snakes. J. Herpetol. 20:353–407.

McDuffie, G. T. 1961. Notes on the ecology of the copperhead in Ohio. J. Ohio Herpetol. Soc. 3:26–27.

―――. 1963. Studies on the size, pattern and coloration of the northern copperhead (*Agkistrodon contortrix mokeson* Daudin) in Ohio. J. Herpetol. 4:15–22.

McKenna, M. G., and G. Allard. 1976. Rattlesnake research. North Dakota Outdoors 38(9):11–13.

Meachem, A., and C. W. Myers. 1961. An exceptional pattern varient of the coral snake, *Micrurus fulvius* (Linnaeus). Quart. J. Florida Acad. Sci. 24:56–58.

Mead, J. I., T. H. Heaton, and E. M. Mead. 1989. Late Quaternary reptiles from caves in the east-central Great Basin. J. Herpetol. 23:186–189.

Mead, J. I., E. L. Roth, T. R. Van Devender, and D. W. Steadman. 1984. The late Wisconsinan vertebrate fauna from Deadman Cave, southern Arizona. Trans. San Diego Soc. Natur. Hist. 20:247–276.

Mebs, D., and F. Kornalik. 1984. Intraspecific variation in content of a basic toxin in eastern diamondback rattlesnake (*Crotalus adamanteus*) venom. Toxicon 22:831–833.

Mebs, D., and H. W. Raudonat. 1966. Biochemical investigations on *Heloderma* venom. Mem. Inst. Butantan 33:907–912.

Meek, S. E. 1906. An annotated list of a collection of reptiles from southern California and northern Lower California. Field Mus. Publ., Zool. Ser. (1905) 7:3–19.

Megonigal, J. P. 1985. *Agkistrodon contortrix mokeson* (Northern Copperhead) and *Lampropeltis getulus getulus* (Eastern Kingsnake). Catesbeiana 5:16.

Mehrtens, J. M. 1987. Living snakes of the world in color. Sterling Publ. Co., Inc., New York. 480 pp.

Melcer, T., and D. Chiszar. 1989a. Strike-induced chemical preferences in prairie rattlesnakes (*Crotalus viridis*). Anim. Learn. Behav. 17:368–372.

———. 1989b. Striking prey creates a specific chemical search image in rattlesnakes. Anim. Behav. 37:477–486.

Melcer, T., D. Chiszar, and H. M. Smith. 1990. Strike-induced chemical preferences in rattlesnakes: role of chemical cues arising from the diet of prey. Bull. Maryland Herpetol. Soc. 26:1–4.

Melcer, T., K. Kandler, and D. Chiszar. 1988. Effects of novel chemical cues on predatory responses of rodent-specializing rattlesnakes. Bull. Psychonomic Soc. 26:580–582.

Menne, H. A. L. 1959. Lets over het voorkomen van ratelslangen in Canada. Lacerta 18:4–6.

Mertens, R. 1956a. ber reptilienbastarde, II. Senckenbergiana Biol. 37: 383–394.

———. 1956b. Das Problem der Mimikry bei Korallenschlangen. Zool. Jahrb. (System). 84:541–576.

———. 1957. Gibt es ein Mimikry bei Korallenschlangen? Natur u. Volk 87:56–66.

Meshaka, W. E., S. E. Trauth, B. P. Butterfield, and A. B. Bevill. 1989. Litter size and aberrant pattern in the southern copperhead, *Agkistrodon contortrix contortrix*, from northeastern Arkansas. Bull. Chicago Herpetol. Soc. 24:91–92.

Meszler, R. M., C. R. Auker, and D. O. Carpenter. 1981. Fine structure and organization of the infrared receptor relay, the lateral descending nucleus of the trigeminal nerve in pit vipers. J. Comp. Neurol. 196: 571–584.

Meszler, R. M., and D. B. Webster. 1968. Histochemistry of the rattlesnake facial pit. Copeia 1968:722–728.

Metter, D. E. 1963. A rattlesnake that encountered a porcupine. Copeia 1963:161.

Meylan, P. A. 1982. The squamate reptiles of the Ingles IA Fauna (Irvingtonian: Citrus County, Florida). Bull. Florida St. Mus. Biol. Sci. 27: 111–196.

Miller, A. H., and R. C. Stebbins. 1964. The lives of desert animals in Joshua Tree National Monument. Univ. California Press, Berkeley. 452 pp.

Miller, D. M., R. A. Young, T. W. Gatlin, and J. A. Richardson. 1982. Amphibians and reptiles of the Grand Canyon National Park. Grand Canyon Natur. Hist. Assoc. Monogr. (4):1–144.

Milstead, W. W., J. S. Mecham, and H. McClintock. 1950. The amphibians and reptiles of the Stockton Plateau in northern Terrell County, Texas. Texas J. Sci. 2:543–562.

Minckley, C. O., and W. E. Rinne. 1972. Another massasauga from Mexico. Texas J. Sci. 23:432–433.

Minton, J. E. 1949. Coral snake preyed upon by a bullfrog. Copeia 1949:288.

Minton, S. A., Jr. 1953. Variation in venom samples from copperheads (*Agkistrodon contortrix mokeson*) and timber rattlesnakes (*Crotalus horridus horridus*). Copeia 1953:212–215.

———. 1956. Some properties of North American pit viper venoms and their correlation with phylogeny. *In* Buckley, E., and N. Porges (eds.), Venoms, pp. 145–151. AAAS, Washington, D.C.

———. 1958. Observations on amphibians and reptiles of the Big Bend Region of Texas. Southwest. Natur. 3:28–54.

————. 1966. A contribution to the herpetology of West Pakistan. Bull. Amer. Mus. Natur. Hist. 134:27–184.

————. 1967. Observations on toxicity and antigenic makeup of venoms from juvenile snakes. Toxicon 4:294.

————. 1972. Amphibians and reptiles of Indiana. Indiana Acad. Sci. Mongr. (3):1–346.

————. 1974. Venom diseases. Charles C Thomas, Publ., Springfield, Illinois. 235 pp.

————. 1983. *Sistrurus catenatus*. Catalog. Amer. Amphib. Rept. 332:1–2.

Minton, S. A., Jr., H. G. Dowling, and F. E. Russell. 1968. Poisonous snakes of the world. U.S. Govt. Print. Office, Washington, D.C. 212 pp.

Minton, S. A., Jr., and M. R. Minton. 1969. Venomous reptiles. Scribner's, New York. 274 pp.

Minton, S. A., Jr., and S. A. Weinstein. 1984. Protease activity and lethal toxicity of venoms from some little known rattlesnakes. Toxicon 22:828–830.

Mitchell, J. C. 1974. The snakes of Virginia, part I. Poisonous snakes and their look-alikes. Virginia Wildlife 35(2):16–18, 28.

————. 1975. Notes on a cottonmouth from Petersburg, Virginia. Virginia Herpetol. Soc. Bull. 75:5.

————. 1977. An instance of cannibalism in *Agkistrodon contortrix* (Serpentes: Viperidae). Bull. Maryland Herpetol. Soc. 13:119.

————. 1980. Viper's brood. A guide to identifying some of Virginia's juvenile snakes. Virginia Wildlife 41(9):8–10.

————. 1981. Notes on male combat in two Virginia snakes, *Agkistrodon contortrix* and *Elaphe obsoleta*. Catesbeiana 1:7–9.

Mitchell, J. C., and W. H. Martin III. 1981. Where the snakes are. Virginia Wildlife 42(6):8–9.

Molenaar, G. J. 1974. An additional trigeminal system in certain snakes possessing infrared receptors. Brain Res. 78:340–344.

Monroe, J. E. 1962. Chromosomes of rattlesnakes. Herpetologica 17:217–220.

Montgomery, W. B., and G. W. Schuett. 1989. Autumnal mating with subsequent production of offspring in the rattlesnake *Sistrurus miliarius streckeri*. Bull. Chicago Herpetol. Soc. 24:205–207.

Moore, R. G. 1978. Seasonal and daily activity patterns and thermoregulation in the southwestern speckled rattlesnake (*Crotalus mitchelli pyrrhus*) and the Colorado Desert sidewinder (*Crotalus cerastes laterorepens*). Copeia 1978:439–442.

Mosauer, W. 1932. On the locomotion of snakes. Science 76:583–585.

————. 1935. How fast can snakes travel? Copeia 1935:6–9.

Mosauer, W., and E. L. Lazier. 1933. Death from insolation in desert snakes. Copeia 1933:149.

Mosimann, J. E., and G. B. Rabb. 1952. The herpetology of Tiber Reservoir area, Montana. Copeia 1952:23–27.

Mount, R. H. 1975. Reptiles and amphibians of Alabama. Auburn Univ., Agric. Exp. Stat. 347 pp.

————. 1981. The red imported fire ant, *Solenopsis invicta*, (Hymenoptera: Formicidae) as a possible serious predator on some native southeastern vertebrates: direct observations and subjective impressions. J. Alabama Acad. Sci. 52:71–78.

————(ed.). 1984. Vertebrate wildlife of Alabama. Alabama Agric. Exp. St., Auburn. 44 pp.

Mount, R. H., and J. Cecil. 1982. *Agkistrodon piscivorus* (cottonmouth). Hybridization. Herp. Review 13:95–96.

Muir, J. H. 1982. Notes on the climbing ability of a captive timber rattlesnake, *Crotalus horridus*. Bull. Chicago Herpetol. Soc. 17:22–23.

———. 1990. Three anatomically aberrant albino *Crotalus atrox* neonates. Bull. Chicago Herpetol. Soc. 25:41–42.

Munro, D. F. 1947. Effect of a bite by *Sistrurus* on *Crotalus*. Herpetologica 4:57.

Murphy, J. B., and B. L. Armstrong. 1978. Maintenance of rattlesnakes in captivity. Univ. Kansas Mus. Natur. Hist. Spec. Publ. (3):1–40.

Murphy, J. B., J. E. Rehg, P. F. A. Maderson, and W. B. McCrady. 1987. Scutellation and pigmentation defects in a laboratory colony of western diamondback rattlesnakes (*Crotalus atrox*): mode of inheritance. Herpetologica 43:292–300.

Murphy, J. B., and J. A. Shadduck. 1978. Reproduction in the eastern diamondback rattlesnake, *Crotalus adamanteus*, in captivity, with comments regarding taratoid birth anomaly. British J. Herpetol. 5:727–733.

Murphy, R. W. 1988. The problematic phylogenetic analysis of interlocus heteropolymer isozyme characters: a case study from sea snakes and cobras. Can. J. Zool. 66:2628–2633.

Murphy, R. W., and C. B. Crabtree. 1985. Evolutionary aspects of isozyme patterns, number of loci, and tissue-specific gene expression in the prairie rattlesnake, *Crotalus viridis viridis*. Herpetologica 41:451–470.

———. 1988. Genetic identification of a natural hybrid rattlesnake: *Crotalus scutulatus scutulatus* X *C. viridis viridis*. Herpetologica 44:119–123.

Murphy, T. D. 1964. Box turtle, *Terrapene carolina*, in stomach of copperhead, *Agkistrodon contortrix*. Copeia 1964:221.

Myers, C. W. 1956. An unrecorded food item of the timber rattlesnake. Herpetologica 12:326.

Myers, G. S. 1945. Nocturnal observations on sea snakes in Bahia Honda, Panama. Herpetologica 3:22–23.

Nauck, E. G. 1929. Untersuchungen über des Gift einer Seeschlange (*Hydrus platurus*) des Pazifischen Ozeans. Deutsche Tropenmed. Zeit. 33:167–170.

Neill, W. T. 1947. Size and habits of the cottonmouth moccasin. Herpetologica 3:203–205.

———. 1948. Hibernation of amphibians and reptiles in Richmond County, Georgia. Herpetologica 4:107–114.

———. 1951. Notes on the natural history of certain North American snakes. Publ. Res. Div. Ross Allen's Rept. Inst. 1:47–60.

———. 1957. Some misconceptions regarding the eastern coral snake, *Micrurus fulvius*. Herpetologica 13:111–118.

———. 1958. The occurrence of amphibians and reptiles in saltwater areas, and a bibliography. Bull. Mar. Sci. Gulf Caribbean 8:1–97.

———. 1960. The caudal lure of various juvenile snakes. Quart. J. Florida Acad. Sci. 23:173–200.

———. 1961. River frog swallows eastern diamondback rattlesnake. Bull. Philadelphia Herpetol. Soc. 9(1):19.

———. 1963. Polychromatism in snakes. Quart. J. Florida Acad. Sci. 26:194–216.

Neill, W. T., and E. R. Allen. 1956. Secondarily ingested food items in snakes. Herpetologica 12:172–174.

Newman, E. A., and P. H. Hartline. 1981. Integration of visual and infrared information in bimodal neurons of the rattlesnake optic tectum. Science 213:789–791.

Nicoletto, P. 1985. Some reptiles from Sinking Creek and Gap Mountains, Montgomery County, Virginia, April–June 1983. Catesbeiana 5(1):13–15.

Noble, G. K., and A. Schmidt. 1937. The structure and function of the facial and labial pits of snakes. Proc. Amer. Philos. Soc. 77:263–288.

Nussbaum, R. A., E. D. Brodie, Jr., and R. M. Storm. 1983. Amphibians and reptiles of the Pacific Northwest. Univ. Press Idaho, Moscow. 332 pp.

O'Connell, B., D. Chiszar, and H. M. Smith. 1981. Effect of poststrike disturbance on strike-induced chemosensory searching in the prairie rattlesnake (*Crotalus v. viridis*). Behav. Neurol. Biol. 32:343–349.

———. 1983. Strike-induced chemosensory searching in prairie rattlesnakes (*Crotalus viridis*) during daytime and at night. J. Herpetol. 17:193–196.

O'Connell, B., T. Poole, P. Nelson, H. M. Smith, and D. Chiszar. 1982. Strike-induced searching by prairie rattlesnakes (*Crotalus v. viridis*) after predatory and defensive strikes which made contact with mice. Bull. Maryland Herpetol. Soc. 18:152–160.

Odum, R. A. 1979. The distribution and status of the New Jersey timber rattlesnake including an analysis of Pine Barren populations. HERP: Bull. New York Herpetol. Soc. 15:27–35.

Oldfield, B. L., and D. E. Keyler. 1989. Survery of timber rattlesnake (*Crotalus horridus*) distribution along the Mississippi River in western Wisconsin. Trans. Wisconsin Acad. Sci. Arts Lett. 77:27–34.

Oliver. J. A. 1955. The natural history of North American amphibians and reptiles. Van Nostrand, Princeton, New Jersey. 359 pp.

———. 1958. Snakes in fact and fiction. MacMillan, New York. 199 pp.

Oliver, J. A., and J. R. Bailey. 1939. Amphibians and reptiles of New Hampshire. *In* Biological Survey of the Connecticut watershed, pp. 195–217. New Hampshire Fish and Game Dept., Concord.

Olson, R. E., B. Marx, and R. Rome. 1986. Descriptive dentition morphology of lizards of Middle and North America, I: Scincidae, Teiidae, and Helodermatidae. Bull. Maryland Herpetol. Soc. 22:97–124.

Ortenburger, A. I., and R. D. Ortenburger. 1926. Field observations on some amphibians and reptiles of Pima County, Arizona. Proc. Oklahoma Acad. Sci. 6:101–121.

Ortenburger, R. D. 1924. Notes on the Gila monster. Proc. Oklahoma Acad. Sci. 4:22.

Over, W. H. 1923. Amphibians and reptiles of South Dakota. South Dakota Geol. Natur. Hist. Surv. Bull. (12):1–31.

———. 1928. A personal experience with rattlesnake bite. Bull. Antivenin Inst. Amer. 2:8–10.

Ownby, C. L., and T. R. Colberg. 1990. Comparison of the immunogenicity and antigenic composition of several venoms of snakes in the family Crotalidae. Toxicon 28:189–199.

Palisot de Beauvois, A. M. F. J. 1799. Memoir on Amphibia. Serpents. Trans. Amer. Philos. Soc. 4:362–381.

Palmer, W. M. 1965. Intergradation among the copperheads (*Agkistrodon contortrix* Linnaeus) in the North Carolina coastal plain. Copeia 1965:246–247.

———. 1971. Distribution and variation of the Carolina pigmy rattlesnake, *Sistrurus miliarius miliarius* Linnaeus, in North Carolina. J. Herpetol. 5:39–44.

―――. 1974. Poisonous snakes of North Carolina. St. Mus. Natur. Hist. North Carolina, Raleigh. 22 pp.

―――. 1978. *Sistrurus miliarius*. Catalog. Amer. Amphib. Rept. 220:1–2.

Palmer, W. M., and G. M. Williamson. 1971. Observations on the natural history of the Carolina pigmy rattlesnake, *Sistrurus miliarius miliarius* Linnaeus. Elisha Mitchell Sci. Soc. J. 87:20–25.

Parker, H. W. 1977. Snakes of the world. Dover Publ., Inc., New York. 191 pp.

Parker, S. A., and D. Stotz. 1977. An observation on the foraging behavior of the Arizona ridge-nosed rattlesnake, *Crotalus willardi willardi* (Serpentes: Crotalidae). Bull. Maryland Herpetol. Soc. 13:123.

Parker, W. S., and W. S. Brown. 1973. Species composition and population changes in two complexes of snake hibernacula in northern Utah. Herpetologica 29:319–326.

―――. 1974. Mortality and weight changes of Great Basin rattlesnakes (*Crotalus viridis*) at a hibernaculum in northern Utah. Herpetologica 30:234–239.

Parmley, D. 1986. Herpetofauna of the Rancholabrean Schulze Cave local fauna of Texas. J. Herpetol. 20:1–10.

―――. 1988. Middle Holocene herpetofauna of Klein Cave, Kerr County, Texas. Southwest. Natur. 33:378–382.

Parrish, H. M., and M. S. Kahn. 1967. Bites by coral snakes: reports of 11 representative cases. Amer. J. Med. Sci. 81:561–568.

Parrish, H. M., and R. E. Thompson. 1958. Human envenomation from bites of recently milked rattlesnakes: a report of three cases. Copeia 1958:83–86.

Patten, R. B. 1981. Author's reply. J. Herpetol. 15:126.

Patten, R. B., and B. H. Banta. 1980. A rattlesnake, *Crotalus ruber*, feeds on a road-killed animal. J. Herpetol. 14:111–112.

Patterson, R. A. 1967a. Some physiological effects caused by venom from the Gila monster, *Heloderma suspectum*. Toxicon 5:5–10.

―――. 1967b. Smooth muscle stimulating action of venom from the Gila monster, *Heloderma suspectum*. Toxicon 5:11–15.

Patterson, R. A., and I. S. Lee. 1969. Effects of *Heloderma suspectum* venom on blood coagulation. Toxicon 7:321–324.

Pendlebury, G. B. 1977. Distribution and abundance of the prairie rattlesnake, *Crotalus viridis viridis*, in Canada. Can. Field-Natur. 91:122–129.

Penn, G. H., Jr. 1943. Herpetological notes from Cameron Parish, Louisiana. Copeia 1943:58–59.

Perkins, C. B. 1951. Hybrid rattlesnakes. Herpetologica 7:146.

Perry, J. 1978. An observation of "dance" behavior in the western cottonmouth, *Agkistrodon piscivorus leucostoma* (Reptilia, Serpentes, Viperidae). J. Herpetol. 12:428–429.

Peters. J. A. 1964. Dictionary of herpetology. Hafner Publ. Co., New York. 393 pp.

Petersen, R. C. 1970. Connecticut's venomous snakes. Connecticut St. Geol. Natur. Hist. Surv. Bull. (103):1–39.

Peterson, H. W., and H. M. Smith. 1973. Observations on sea snakes in the vicinity of Acapulco, Guerrero, Mexico. Bull. Chicago Herpetol. Soc. 8(3–4):29.

Peterson. K. H. 1982. Reproduction in captive *Heloderma suspectum*. Herp. Review 13:122–124.

―――. 1983. Reproduction of captive *Crotalus mitchelli mitchelli* and

216   BIBLIOGRAPHY

Crotalus durissus at the Houston Zoological Gardens. Proc. Rept. Symp. Capt. Prop. Husb. 6:323–327.

————. 1990. Conspecific and self-envenomation in snakes. Bull. Chicago Herpetol. Soc. 25:26–28.

Phelps, T. 1981. Poisonous snakes. Blandford Press, Poole, Dorset. 237 pp.

Pianka, E. R. 1967. Lizard species diversity. Ecology 48:333–351.

Pickwell, Gayle. 1972. Amphibians and reptiles of the Pacific states. Dover Publ., Inc. New York. 234 pp.

Pickwell, George V. 1971. Knotting and coiling behavior in the pelagic sea snake *Pelamis platurus* (L.). Copeia 1971:348–350.

————. 1972. The venomous sea snakes. Fauna 1(4):17–32.

Pickwell, George V., R. L. Bezy, and J. E. Fitch. 1983. Northern occurrences of the sea snake, *Pelamis platurus*, in the eastern Pacific, with a record of predation on the species. California Fish and Game 69:172–177.

Pickwell, George V., and W. A. Culotta. 1980. *Pelamis, Pelamis platurus*. Catalog. Amer. Amphib. Rept. 255:1–4.

Pickwell, George V., J. A. Vick, W. H. Shipman, and M. M. Grenan. 1972. Production, toxicity and preliminary pharmacology of venom from the sea snake, *Pelamis platurus*. *In* Worthen, L. R. (ed.), Proc. 3d Food-drugs from the sea conference, pp. 247–265. Marine Tech. Soc., Washington, D.C.

Pinney, R. 1981. The snake book. Doubleday, Garden City, New York. 248 pp.

Pisani, G. R., J. T. Collins, and S. R. Edwards. 1973. A re-evaluation of the subspecies of *Crotalus horridus*. Trans. Kansas Acad. Sci. 75:255–263.

Pollard, C. B., A. F. Novak, R. W. Harmon, and W. H. Runzler. 1952. A study of the toxicity and stability of dried moccasin (*Agkistrodon piscivorus*) venom. Quart. J. Florida Acad. Sci. 15:162–164.

Pope, C. H. 1937. Snakes alive and how they live. Viking Press, New York. 238 pp.

————. 1944. Amphibians and reptiles of the Chicago area. Chicago Natur Hist. Mus. 275 pp.

————. 1946. Snakes of the northeastern United States. New York Zool. Soc. 52 pp.

————. 1955. The reptile world. Alfred A. Knopf. New York. 325 pp.

Porter, C. A., M. J. Hamilton, J. W. Sites, Jr., and R. J. Baker. 1991. Location of ribosomal DNA in chromosomes of squamate reptiles: systematic and evolutionary implications. Herpetologica 47:271–280.

Porter, K. R. 1972. Herpetology. Saunders, Philadelphia. 524 pp.

Porter, T. 1983. Induced cannibalism in *Crotalus mitchelli*. Bull. Chicago Herpetol. Soc. 18:48.

Pough, F. H. 1966. Ecological relationships in southeastern Arizona with notes on other species. Copeia 1966:676–683.

————. 1988. Mimicry and related phenomena. *In* Gans, C., and R. B. Huey (eds.), Biology of the Reptilia, vol. 16, Defense and life history, pp. 153–234. Alan R. Liss, Inc. New York.

Pough, F. H., and H. B. Lillywhite. 1984. Blood volume and blood oxygen capacity of sea snakes. Physiol. Zool. 57:32–39.

Powell, R., M. Inboden, and D. D. Smith. 1990. Erstnachweis von Hybriden zwischen den Klapperschlangen *Crotalus cerastes laterorepens* Klauber, 1944 und *Crotalus scutulatus scutulatus* (Kennicott, 1861). Salamandra 26:319–329.

Powers, A. 1972. An instance of cannibalism in captive *Crotalus viridis helleri* with a brief review of cannibalism in rattlesnakes. Bull. Maryland Herpetol. Soc. 8:60–61.

———. 1973. A review of the purpose of the rattle in crotalids as a defensive diversionary mechanism. Bull. Maryland Herpetol. Soc. 9:30–32.

Prange, H. D., and S. P. Christman. 1976. The allometrics of rattlesnake skeletons. Copeia 1976:542–545.

Pregill, G. K., J. A. Gauthier, and H. W. Greene. 1986. The evolution of helodermatid squamates, with description of a new taxon and an overview of Varanoidea. Trans. San Diego Soc. Natur. Hist. 21:167–202.

Price. A. H. 1980. *Crotalus molossus*. Catalog. Amer. Amphib. Rept. 242: 1–2.

———. 1982. *Crotalus scutulatus*. Catalog. Amer. Amphib. Rept. 291:1–2.

———. 1988. Observations on maternal behavior and neonate aggregation in the western diamondback rattlesnake, *Crotalus atrox* (Crotalidae). Southwest. Natur. 33:370–374.

Price, R. 1989. A unified microdermoglyphic analysis of the genus *Agkistrodon*. The Snake 21:90–100.

Prieto, A. A., and E. R. Jacobson. 1968. A new locality for melanistic *Crotalus molossus molossus* in southern New Mexico. Herpetologica 24:339–340.

Quinn, H. R. 1977. Further notes on reproduction in *Crotalus willardi* (Reptilia, Serpentes, Crotalidae). Bull. Maryland Herpetol. Soc. 13:111.

———. 1979a. Sexual dimorphism in tail pattern of Oklahoma snakes. Texas J. Sci. 31:157–160.

———. 1979b. Reproduction and growth of the Texas coral snake *Micrurus fulvius tenere*. Copeia 1979:453–463.

Quinn, J. S. 1985. Caspian terns respond to rattlesnake predation in colony. Wilson Bull. 97:233–234.

Radcliffe, C. W., D. Chiszar, and B. O'Connell. 1980. Effects of prey size on poststrike behavior in rattlesnakes (*Crotalus durissus, C. enyo,* and *C. viridis*). Bull. Psychonomic Soc. 16:449–450.

Radcliffe, C. W., K. Estep, T. Boyer, and D. Chiszar. 1986. Stimulus control of predatory behavior in red spitting cobras (*Naja mossambica pallida*) and prairie rattlesnakes (*Crotalus viridis*). Anim. Behav. 34:804–814.

Rado, T. A., and P. G. Rowlands. 1981. A range extension and low elevational record for Arizona ridgenose rattlesnake (*Crotalus w. willardi*). Herp. Review 12:15.

Rael, E. D., R. A. Knight, and H. Zepeda. 1984. Electrophoretic variants of Mojave rattlesnake (*Crotalus scutulatus scutulatus*) venoms and migration differences of Mojave toxin. Toxicon 22:980–985.

Rafinesque, C. S. 1818. Further account of discoveries in natural history in the western states. Amer. Month. Mag. Crit. Rev. 4:39–42.

Rage, J.-C. 1984. Serpentes. Handbuch der Paläoherpetologie. Part II. Gustav Fischer, Stuttgart. 80 pp.

Rahn, H. 1942a. Effect of temperature on color change in the rattlesnake. Copeia 1942:178.

———. 1942b. The reproductive cycle of the prairie rattler. Copeia 1942:233–240.

Ramirez, G. A., P. L. Fletcher, Jr., and L. D. Possani. 1990. Characterization of the venom from *Crotalus molossus nigrescens* Gloyd (black tail rattlesnake): isolation of two proteases. Toxicon 28:285–297.

Ramsey, L. W. 1948. Combat dance and range extension of *Agkistrodon piscivorus leucostoma*. Herpetologica. 4:228.

Raun, G. G. 1965. A guide to Texas snakes. Texas Mem. Mus., Mus. Notes (9):1–85.

Raun, G. G., and F. R. Gehlbach. 1972. Amphibians and reptiles in Texas. Dallas Mus. Natur. Hist. Bull. (2):1–61.

Reese, A. M. 1947. The hemipenes of copperhead embryos. Herpetologica 3:206–208.

Reid, H. A. 1956. Sea-snake bites. British Med. J. (4984):73–78.

Reinert, H. K. 1978. The ecology and morphological variation of the massasauga rattlesnake (*Sistrurus catenatus*). M.S. thesis, Clarion St. Col., Clarion, Pennsylvania.

———. 1981. Reproduction by the massasauga (*Sistrurus catenatus catenatus*). Amer. Midl. Natur. 105:393–395.

———. 1984a. Habitat separation between sympatric snake populations. Ecology 65:478–486.

———. 1984b. Habitat variation within sympatric snake populations. Ecology 65:1673–1682.

———. 1991. A profile and impact assessment of organized rattlesnake hunts in Pennsylvania. J. Pennsylvania Acad. Sci., in press.

Reinert, H. K., D. Cundall, and L. M. Bushar. 1984. Foraging behavior of the timber rattlesnake, *Crotalus horridus*. Copeia 1984:976–981.

Reinert, H. K., and W. R. Kodrich. 1982. Movements and habitat utilization by the massasauga, *Sistrurus catenatus catenatus*. J. Herpetol. 16:162–171.

Reinert, H. K., and R. T. Zappalorti. 1988a. Timber rattlesnakes (*Crotalus horridus*) of the Pine Barrens: their movement patterns and habitat preference. Copeia 1988:964–978.

———. 1988b. Field observation of the association of adult and neonatal timber rattlesnakes, *Crotalus horridus*, with possible evidence for conspecific trailing. Copeia 1988:1057–1059.

Reynolds, R. P. 1982. Seasonal incidence of snakes in northeastern Chihuahua, Mexico. Southwest. Natur. 27:161–166.

Reynolds, R. P., and G. V. Pickwell. 1984. Records of the yellow-bellied sea snake, *Pelamis platurus*, from the Galápagos Islands. Copeia 1984:786–789.

Reynolds, R. P., and N. J. Scott, Jr. 1982. Use of a mammalian resource by a Chihuahuan snake community. *In* Scott, N. J., Jr. (ed.), Herpetological communities, pp. 99–118. U.S. Fish Wildl. Serv., Wildl. Res. Rep. 13.

Richards, R. L. 1990. Quaternary distribution of the timber rattlesnake (*Crotalus horridus*) in southern Indiana. Proc. Indiana Acad. Sci. 99:113–122.

Roberts, A. R. 1947. *Sistrurus* in Michigan. Herpetologica 4:6.

Robinson, B. G., and K. V. Kardong. 1991. Relocation of struck prey by venomoid (venom-less) rattlesnakes, *Crotalus viridis oreganus*. Bull. Maryland Herpetol. Soc. 27:23–30.

Roddy, H. J. 1928. Reptiles of Lancaster County and the state of Pennsylvania. Science Press, Lancaster, Pennsylvania. 53 pp.

Rogers, L. 1903. On the physiological action of the poison of the Hydrophidae. Proc. Royal Soc. London 71:481–496.

Romer, A. S. 1956. Osteology of the reptiles. Univ. Chicago Press, Chicago. 772 pp.

Rose, W. 1950. The reptiles and amphibians of southern Africa. Maskew Miller, Ltd., Cape Town. 378 pp.

Rossman, D. A. 1960. Herpetological survey of the Pine Hills area of southern Illinois. Quart. J. Florida Acad. Sci. 22:207–225.

Rowe, M. P., R. G. Coss, and D. H. Owings. 1986. Rattlesnake rattles and burrowing owl hisses: a case of acoustic Batesian mimicry. Ethology 72:53–71.

Roze, J. 1967. A check list of the New World venomous coral snakes (Elapidae), with descriptions of new forms. Amer. Mus. Novitates (2287):1–60.

———. 1974. *Micruroides, M. euryxanthus*. Catalog. Amer. Amphib. Rept. 163:1–4.

———. 1982. New World coral snakes (Elapidae): a taxonomic and biological summary. Mem. Inst. Butantan 46:305–338.

Roze, J. A., and G. M. Tilger. 1983. *Micrurus fulvius*. Catalog. Amer. Amphib. Rept. 316:1–4.

Ruben, J. A. 1983. Mineralized tissues and exercise physiology of snakes. Amer. Zool. 23:377–381.

Ruben. J. A., and C. Geddes. 1983. Some morphological correlates of striking in snakes. Copeia 1983:221–225.

Rubinoff, I., J. B. Graham, and J. Motta. 1986. Diving of the sea snake *Pelamis platurus* in the Gulf of Panama. I. Dive depth and duration. Marine Biol. 91:181–191.

Rubinoff, I., and C. Kropach. 1970. Differential reactions of Atlantic and Pacific predators to sea snake. Nature 228:1288–1290.

Ruiz, J. M. 1951. Sobre a distinção genérica dos Crotalidae (Ophidia: Crotaloidea) baseada em alguns caracteres osteológicos. (Nota preliminar). Mem. Inst. Butantan 23:109–114.

Russell, F. E. 1960. Snake venom poisoning in southern California. California Med. 93:347–350.

———. 1967a. Bites by the Sonoran coral snake, *Micruroides euryxanthus*. Toxicon 5:39–42.

———. 1967b. Gel diffusion study on human sera following rattlesnake venom poisoning. Toxicon 5:147–148.

———. 1983. Snake venom poisoning. Scholium Intern., Inc., Great Neck, New York. 562 pp.

Russell, F. E., and C. M. Bogert. 1981. Gila monster: its biology, venom and bite—a review. Toxicon 19:341–359.

Russell, F. E., J. A. Emery, and T. E. Long. 1960. Some properties of rattlesnake venom following 26 years storage. Proc. Soc. Exp. Biol. Med. 103:737–739.

Russell, F. E., and B. A. Michaelis. 1960. Zootoxicologic effects of *Crotalus* venoms. Physiologist 3:135.

Russell, F. E., N. Ružić, and H. Gonzalez. 1973. Effectiveness of antivenin (Crotalidae) polyvalent following injection of *Crotalus* venom. Toxicon 11:461–464.

Ruthven, A. G., C. Thompson, and H. T. Gaige. 1928. The herpetology of Michigan. Univ. Mus., Univ. Michigan Handb. Ser. (3):1–228.

Saint Girons, H. 1986. Les organes thermorécepteurs des serpents: fossettes loréales et labiales. J. Psychol. Norm. Pathol. 81(1986):357–367.

Sanders, J. S., and J. S. Jacob. 1981. Thermal ecology of the copperhead (*Agkistrodon contortrix*). Herpetologica 37:264–270.

Sanders, R. T. 1951. Effect of venom injections in rattlesnakes. Herpetologica 7:47–52.

Savage, J. M., and F. S. Cliff. 1953. A new subspecies of sidewinder, *Crotalus cerastes*, from Arizona. Natur. Hist. Misc. (119):1–7.

Savage, J. M., and J. B. Slowinski. 1990. A simple consistent terminology for the basic colour patterns of the venomous coral snakes and their mimics. Herpetol. J. 1:530–532.

Say, T. 1823. [Description of *Crotalus tergeminus*]. *In* Edwin James, compiler, Account of an expedition from Pittsburgh to the Rocky Mountains, performed in the years 1819 and '20 . . . , vol. 2, p. 499. Carey and Lea, Philadelphia.

Schaefer, N. 1976. The mechanism of venom duct to the fang in snakes. Herpetologica 32:71–76.

Schaefer, W. H. 1934. Diagnosis of sex in snakes. Copeia 1934:181.

Schaeffer, G. C. 1969. Sex independent ground color in the timber rattlesnake, *Crotalus horridus horridus*. Herpetologica 25:65–66.

Schaeffer, R. C., Jr., S. Bernick, T. H. Rosenquist, and F. E. Russell. 1972a. The histochemistry of the venom glands of the rattlesnake *Crotalus viridis helleri*—I. Lipid and non-specific esterase. Toxicon 10:183–186.

———. 1972b. The histochemistry of the venom glands of the rattlesnake *Crotalus viridis helleri*—II. Monoamine oxidase, acid and alkaline phosphotase. Toxicon 10:295–297.

Schaeffer, R. C., Jr., R. W. Carlson, H. Whigham. F. E. Russell, and M. H. Weil. 1973. Some hemodynamic effects of rattlesnake (*Crotalus viridis helleri*) venom. Proc. West. Pharmacol. Soc. 16:58–62.

Schaeffer, R. C., Jr., T. R. Pattabhiraman, R. W. Carlson, F. E. Russell, and M. H. Weil. 1979. Cardiovascular failure produced by a peptide from the venom of the southern Pacific rattlesnake, *Crotalus viridis helleri*. Toxicon 17:447–453.

Schmidt, K. P. 1928. Notes on American coral snakes. Bull. Antivenin Inst. Amer. 2:63–64.

———. 1932. Stomach contents of some American coral snakes, with the description of a new species of *Geophis*. Copeia 1932:6–9.

———. 1953. A check list of North American amphibians and reptiles. 6th ed. Amer. Soc. Ichthyol. Herpetol., Chicago. 280 pp.

Schmidt, K. P., and D. D. Davis. 1941. Field book of snakes of the United States and Canada. Putnam, New York. 322 pp.

Schmidt, K. P., and R. F. Inger. 1957. Living reptiles of the world. Hanover House, Garden City, New York. 287 pp.

Schmidt-Nielsen, K., and R. Fange. 1958. Salt glands in marine reptiles. Nature 182:783–785.

Schoener, T. W. 1977. Competition and the niche. *In* Gans, C., and D. W. Tinkle (eds.), Biology of the Reptilia, vol. 7, pp. 35–136. Academic Press, New York.

Schuett, G. W. 1982. A copperhead (*Agkistrodon contortrix*) brood produced from autumn copulations. Copeia 1982:700–702.

———. 1986. Selected topics on reproduction of the copperhead, *Agkistrodon contortrix* (Reptilia, Serpentes, Viperidae). Unpubl. Master's thesis, Central Michigan Univ., Mt. Pleasant.

Schuett, G. W., D. L. Clark, and F. Kraus. 1984. Feeding mimicry in the rattlesnake *Sistrurus catenatus*, with comments on the evolution of the rattle. Anim. Behav. 32:625–626.

Schuett, G. W., and J. C. Gillingham. 1986. Sperm storage and multiple paternity in the copperhead, *Agkistrodon contortrix*. Copeia 1986:807–811.

———. 1988. Courtship and mating of the copperhead, *Agkistrodon contortrix*. Copeia 1988:374–381.

Schultz, E., A. W. Clark, A. Susuki, and R. G. Cassens. 1980. Rattlesnake shaker muscle: 1. A light microscopic and histochemical study. Tissue Cell 12:323–334.

Schwab, D. 1988. Growth and rattle development in a captive timber rattlesnake, *Crotalus horridus*. Bull. Chicago Herpetol. Soc. 23:26–27.

Schwammer, H. 1983. Herpetologische Beobachtungen aus Colorado/ USA> Aasfressen bei *Sistrurus catenatus edwardsi/tergeminus* und Verhaltensmimikry bei *Pituophis melanoleucus*. Aquaria 30:90–93.

Scudder, K. M., and D. Chiszar. 1977. Effects of six visual stimulus conditions on defensive and exploratory behavior in two species of rattlesnakes. Psychol. Rec. 3:519–526.

Scudder, K. M., D. Chiszar, and H. M. Smith. 1983. Effects of environmental odors on strike-induced chemosensory searching by rattlesnakes. Copeia 1983:519–522.

Scudder, K. M., D. Chiszar, H. M. Smith, and T. Melcer. 1988. Response of neonatal prairie rattlesnakes (*Crotalus viridis*) to conspecific and heterospecific chemical cues. Psychol. Rec. 38:459–471.

Secor, S. M. 1991. A preliminary analysis of the movement and home range size of the sidewinder, *Crotalus cerastes*. In J. A. Campbell and E. D. Brodie, Jr. (eds.), Biology of pit vipers. Cornell Univ. Press, Ithaca, New York. In press.

Seigel, R. A. 1986. Ecology and conservation of an endangered rattlesnake, *Sistrurus catenatus*, in Missouri, U.S.A. Biol. Conserv. 35:333–346.

Seigel, R. A., J. T. Collins, and S. S. Novak (eds.). 1987. Snakes: ecology and evolutionary biology. MacMillan Publ. Co., New York. 529 pp.

Seigel, R. A., and H. S. Fitch. 1984. Ecological patterns of relative clutch mass in snakes. Oecologia (Berlin) 61:293–301.

———. 1985. Annual variation in reproduction in snakes in a fluctuating environment. J. Anim. Ecol. 54:497–505.

Seigel, R. A., H. S. Fitch, amd N. B. Ford. 1986. Variability in relative clutch mass in snakes among and within species. Herpetologica 42:179–185.

Seigel, R. A., L. E. Hunt, J. L. Knight, L. Malaret, and N. L. Zuschlag. 1984. Vertebrate ecology and systematics: a tribute to Henry S. Fitch. Univ. Kansas Mus. Natur. Hist. Spec. Publ. (10):1–278.

Sexton, O. J., and K. R. Marion. 1981. Experimental analysis of movements by prairie rattlesnakes, *Crotalus viridis*, during hibernation. Oecologia (Berlin) 51:37–41.

Seymour, R. S., R. G. Spragg, and M. T. Hartman. 1981. Distribution of ventilation and perfusion in the sea snake, *Pelamis platurus*. J. Comp. Physiol. 145:109–115.

Shannon, F. H. 1953. Case reports of two Gila monster bites. Herpetologica 9:125–127.

Shaw, C. E. 1948a. The male combat "dance" of some crotalid snakes. Herpetologica 4:137–145.

———. 1948b. A note on the food habits of *Heloderma suspectum* Cope. Herpetologica 4:145.

———. 1951. Male combat in American colubrid snakes with remarks on combat in other colubrid and elapid snakes. Herpetologica 7:149–168.

———. 1961. Snakes of the sea. Zoonooz 34(7):3–5.

———. 1962. Sea snakes at the San Diego Zoo. Intern. Zoo Yearb. 4:49–52.

———. 1964. Beaded lizards—dreaded, but seldom deadly. Zoonooz 3:10–15.

————. 1966. Southern Pacific rattlesnake. Zoonooz 39:19.

————. 1968. Reproduction of the Gila monster (*Heloderma suspectum*) at the San Diego Zoo. Zool. Gart. 35:1–6.

————. 1971. The coral snakes, genera *Micrurus* and *Micruroides* of the United States and northern Mexico. *In* W. Bücherl and E. E. Buckley (eds.), Venomous animals and their venoms, vol. 2, Venomous vertebrates, pp. 157–172. Academic Press, New York.

Shaw, C. E., and S. Campbell. 1974. Snakes of the American West. Alfred A. Knopf, New York. 332 pp.

Shine, R. 1978. Sexual size dimorphism and male combat in snakes. Oecologia (Berlin) 33:269–277.

Shipman, W. H., and G. V. Pickwell. 1973. Venom of the yellow-bellied sea snake (*Pelamis platurus*): some physical and chemical properties. Toxicon 11:375–377.

Shufeldt, R. W. 1890. Contributions to the study of *Heloderma suspectum*. Proc. Zool. Soc. London 1890:148–244.

————. 1891. The poison apparatus of the *Heloderma*. Nature (London) 43:514–515.

Sibley, H. 1951. Snakes are scared of you! Field and Stream 55(9):46–48.

Simons, L. H. 1986. *Crotalus atrox* (Western Diamondback Rattlesnake). Pattern. Herp. Review 17:20,22.

Sinclair, R., W. Hon, and R. B. Ferguson. 1965. Amphibians and reptiles of Tennessee. Tennessee Game Fish Comm., Nashville. 29 pp.

Smart, E. W. 1951. Color analysis in the Great Basin rattlesnake. Herpetologica 7:41–46.

Smith, H. M. 1946. Handbook of lizards of the United States and of Canada. Comstock Publ. Co., Ithaca, New York. 557 pp.

————. 1952. A revised arrangement of maxillary fangs of snakes. Turtox News 30:214–218.

————. 1956. Handbook of amphibians and reptiles of Kansas. Univ. Kansas Mus. Natur. Hist. Misc. Publ. 2d ed. (9):1–356.

————. 1990. Signs and symptoms following human envenomation by the Mojave rattlesnake, *Crotalus scutulatus*, treated without use of antivenom. Bull. Maryland Herpetol. Soc. 26:105–110.

Smith, H. M., and H. K. Gloyd. 1963. Nomenclatural notes on the snake names *Scytale*, *Boa scytale*, and *Agkistrodon mokasen*. Herpetologica 19:280–282.

Smith, H. M., and O. Sanders. 1952. Distributional data on Texan amphibians and reptiles. Texas J. Sci. 4:204–219.

Smith, M. A. 1926. Monograph of the sea snakes. British Museum (Natural History), London. 130 pp.

Smith, P. W. 1961. The amphibians and reptiles of Illinois. Illinois Natur. Hist. Surv. Bull. 28:1–298.

Smith, P. W., and M. M. Hensley. 1958. Notes on a small collection of amphibians and reptiles from the vicinity of the Pinacate Lava Cap in northwestern Sonora, Mexico. Trans. Kansas Acad. Sci. 61:64–76.

Smith, P. W., and L. M. Page. 1972. Repeated mating of a copperhead and timber rattlesnake. Herp. Review 4:196.

Smith, S. M. 1975. Innate recognition of coral snake pattern by a possible avian predator. Science 187:759–760.

Smith, S. M., and A. M. Mostrom. 1985. "Coral snake" rings: are they helpful in foraging? Copeia 1985:384–387.

Smits, A. W., and H. B. Lillywhite. 1985. Maintenance of blood volume in snakes: transcapillary shifts of extravascular fluids during acute hemorrhage. J. Comp. Physiol. B 155:305–310.

Snellings, E., Jr. 1986. The gentleman of snakes. Florida Natur. 59:6–8.

Snyder, D. H. 1972. Amphibians and reptiles of Land Between the Lakes. Tennessee Valley Auth. 90 pp.

Snyder, D. H., D. F. Burchfield, and R. W. Nall. 1967. First records of the pigmy rattlesnake in Kentucky. Herpetologica 23:240–241.

Soto, J. G., J. C. Perez, M. M. Lopez, M. Martinez, T. B. Quintanilla-Hernandez, M. S. Santa-Hernandez, K. Turner, J. L. Glenn, R. C. Straight, and S. A. Minton. 1989. Comparative enzymatic study of HPLC-fractionated *Crotalus* venoms. Comp. Biochem. Physiol. 93B:847–855.

Speake, D. W., and R. H. Mount. 1973. Some possible ecological effects of "Rattlesnake Roundups" in the southeastern coastal plain. Proc. 27th Ann. Conf. S. E. Assoc. Game Fish Comm. 1973:267–277.

Stabler, R. M. 1939. Frequency of skin shedding in snakes. Copeia 1939:227–229.

Stadelman, R. E. 1929a. Some venom extraction records. Bull. Antivenin Inst. Amer. 3:29.

———. 1929b. Further notes on the venom of newborn copperheads. Bull. Antivenin Inst. Amer. 3:81.

Stahnke, H. L. 1950. The food of the Gila monster. Herpetologica 6:103–106.

———. 1952. A note on the food of the Gila monster, *Heloderma suspectum* Cope. Herpetologica 8:64–65.

Stark, M. 1984. A prairie rattlesnake drinking water! Blue Jay 42:195–196.

———. 1986. Overwintering by an ambystomid salamander in a prairie rattlesnake hibernaculum. Herp. Review 17:7.

———. 1987. An active prairie rattlesnake den taken over by foxes. Blue Jay 45:53–54.

Stebbins, R. C. 1954. Amphibians and reptiles of western North America. McGraw-Hill Book Co., Inc., New York. 536 pp.

———. 1985. A field guide to western reptiles and amphibians, 2d ed. Houghton Mifflin Co., Boston. 336 pp.

Stechert, R. 1980. Observations on northern snake dens. HERP: Bull. New York Herpetol. Soc. 15:7–14.

———. 1981. Historical depletion of timber rattlesnake colonies in New York state. HERP: Bull. New York Herpetol. Soc. 17:23–24.

Steehouder, T. 1988. *Agkistrodon piscivorus,* the cottonmouth. Litteratura Serpentium 8:173–181.

Stejneger, L. 1898. The poisonous snakes of North America. Smithsonian Inst. Report 1898:338–487.

———. 1903. The reptiles of the Huachuca Mountains, Arizona. Proc. U.S. Natl. Mus. 25:149–158.

Stewart, M.M., G. E. Larson, and T. H. Mathews. 1960. Morphological variation in a litter of timber rattlesnakes. Copeia 1960:366–367.

Stickel, W. H. 1952. Venomous snakes of the United States and treatment of their bites. U.S. Dept. Int., Wildl. Leafl. (339):1–29.

Stille, B. 1987. Dorsal scale microdermatoglyphics and rattlesnake (*Crotalus* and *Sistrurus*) phylogeny (Reptilia: Viperidae: Crotalinae). Herpetologica 43:98–104.

Stimson, A. C., and H. T. Engelhardt. 1960. The treatment of snakebite. J. Occupational Med. 2:163–168.

St. John, A. D. 1980. Knowing Oregon reptiles. Salem Audubon Soc.

Stoddard, H. L. 1942. The bobwhite quail: its habits, preservation and increase. Charles Scribner's Sons, New York. 559 pp.

Storer, D. H. 1839. Reptiles of Massachusetts. Rept. Comm. Zool. Surv. Massachusetts:203–253.

Storer, T. I. 1931. *Heloderma* poisoning in man. Bull. Antivenin Inst. Amer. 5:12–15.

Storment, D. 1990. Field observations of sexual dimorphism in head pattern (markings) in timber rattlesnakes (*Crotalus horridus*). Bull. Chicago Herpetol. Soc. 25:160–162.

Stringer, J. M., R. A. Kainer, and A. T. Tu. 1972. Myonecrosis induced by rattlesnake venom. An electron microscope study. Amer. J. Pathol. 67:127–140.

Stubbs, T. H. 1979. Moccasin. Florida Natur. 52(4):2–4.

Studenroth, K. R. 1991. *Agkistrodon piscivorus conanti* (Florida Cottonmouth). Foraging. Herp. Review 22:60.

Stürzebecher, J., U. Neumann, and J. Meier. 1991. Inhibition of the protein C activator Protac[R], a serine proteinase from the venom of the southern copperhead snake *Agkistrodon contortrix contortrix*. Toxicon 29:151–155.

Surface, H. A. 1906. The serpents of Pennsylvania. Bull. Pennsylvania State Dept. Agric. Div. Zool. (4):133–208.

Sutherland, I. D. W. 1958. The "combat dance" of the timber rattlesnake. Herpetologica 14:23–24.

Swanson, P. L. 1930. Notes on the massasauga. Bull. Antivenin Inst. Amer. 4:70–71.

———. 1933. The size of *Sistrurus catenatus catenatus* at birth. Copeia 1933:37.

———. 1952. The reptiles of Venango County, Pennsylvania. Amer. Midl. Natur. 47:161–182.

Sweet, S. S. 1985. Geographic variation, convergent crypsis and mimicry in gopher snakes (*Pituophis melanoleucus*) and western rattlesnakes (*Crotalus viridis*). J. Herpetol. 19:55–67.

Tan, N.-H., and G. Ponnudurai. 1990. A comparative study of the biological activities of venoms from snakes of the genus *Agkistrodon* (moccasins and copperheads). Comp. Biochem. Physiol. 95B:577–582.

Tanner, W. W. 1960. *Crotalus mitchelli pyrrhus* Cope in Utah. Herpetologica 16:140.

———. 1975. Checklist of Utah amphibians and reptiles. Proc. Utah Acad. Sci., Arts, Lett. 52:4–8.

Taub, A. M., and W. A. Dunson. 1967. The salt gland in a sea snake (*Laticauda*). Nature (London) 215:995–996.

Taylor, E. H. 1953. Early records of the seasnake *Pelamis platurus* in Latin America. Copeia 1953:124.

Taylor, W. P. 1935. Notes on *Crotalus atrox* near Tucson, Arizona, with a special reference to its breeding habits. Copeia 1935:154–155.

Telford, S. R., Jr. 1952. A herpetological survey of Lake Shipp, Polk County, Florida. Quart. J. Florida Acad. Sci. 15:175–185.

———. 1955. A description of the eggs of the coral snake, *Micrurus f. fulvius*. Copeia 1955:258.

Tennant, A. 1985. A field guide to Texas snakes. Texas Monthly Press, Austin. 260 pp.

Tevis, L. 1943. Field notes on a red rattlesnake in Lower California. Copeia 1943:241–245.

Thayer, F. D., Jr. 1988. *Crotalus atrox* (Western Diamondback Rattlesnake). Hunting behavior. Herp. Review 19:35.

Thireau, M. 1991. Types and historically important specimens of rattlesnakes in the Muséum National d'Histoire Naturelle (Paris). Smithsonian Herpetol. Inform. Serv. 87:1–10.

Thomas, R. G., and F. H. Pough. 1979. The effects of rattlesnake venom on digestion of prey. Toxicon 17:221–228.

Thompson, S. W. 1982. Snakes of South Dakota. South Dakota Conserv. Digest 49(4):12–18.

Thorne, E. T. 1977. Sybille Creek snake dance. Wyoming Wild Life 41(6):14.

Tinkham, E. R. 1971a. The biology of the Gila monster. *In* Bücherl, W., and E. E. Buckley (eds.), Venomous animals and their venoms, vol. 2, Venomous vertebrates, pp. 381–413. Academic Press, New York.

———. 1971b. The venom of the Gila monster. *In* Bücherl, W., and E. E. Buckley (eds.), Venomous animals and their venoms, vol. 2, Venomous vertebrates, pp. 415–422. Academic Press, New York.

Tinkle, D. W. 1959. Observations of reptiles and amphibians in a Louisiana swamp. Amer. Midl. Natur. 62:189–205.

———. 1962. Reproductive potential and cycles in female *Crotalus atrox* from northwestern Texas. Copeia 1962:306–313.

Todd, R. E., Jr. 1973. The cottonmouth in Kentucky. Kentucky Herpetol. 4(2–4):5–9.

Tomko, D. S. 1975. The reptiles and amphibians of the Grand Canyon. Plateau 47:161–166.

Toweill, D. E. 1982. Winter foods of eastern Oregon bobcats. Northwest Sci. 56:310–315.

Trapido, H. 1937. The snakes of New Jersey: a guide. Newark Mus. 60 pp.

———. 1939. Parturition in the timber rattlesnake, *Crotalus horridus horridus* Linné. Copeia 1939:230.

Troost, G. 1836. On a new genus of serpents, and new species of the genus *Heterodon*, inhabiting Tennessee. Ann. Lyc. Natur. Hist. New York 3:174–190.

Tryon, B. W. 1978. Reproduction in a pair of captive Arizona ridge-nosed rattlesnakes, *Crotalus willardi willardi* (Reptilia, Serpentes, Crotalidae). Bull. Maryland Herpetol. Soc. 14:83–88.

———. 1985. Snake hibernation and breeding: in and out of the zoo. *In* Townson, S., and K. Lawrence (eds.), Reptiles: breeding, behaviour, and veterinary aspects, pp. 19–31. British Herpetol. Soc., London, England.

Tryon, B. W., and H. K. McCrystal. 1982. *Micrurus fulvius tenere* reproduction. Herp. Review 13:47–48.

Tu, A. T. 1976. Investigation of the sea snake, *Pelamis platurus* (Reptilia, Serpentes, Hydrophiidae), on the Pacific Coast of Costa Rica, Central America. J. Herpetol. 10:13–18.

———. 1977. Venoms: chemistry and molecular biology. John Wiley & Sons, New York. 560 pp.

———(ed.). 1982. Rattlesnake venoms: their actions and treatment. Mercel Dekker, Inc., New York. 393 pp.

Tu, A. T., and B. L. Adams. 1968. Phylogenetic relationships among venomous snakes of the genus *Agkistrodon* from Asia and the North American continent. Nature (London) 217:760–762.

Tu, A. T., T. S. Lin, and A. L. Bieber. 1975. Purification and chemical characterization of the major neurotoxin from the venom of *Pelamis platurus*. Biochemistry 14:3408–3413.

Tu, A. T., and D. S. Murdock. 1967. Protein nature and some enzymatic properties of the lizard *Heloderma suspectum suspectum* (Gila monster) venom. Comp. Biochem. Physiol. 22:389–396.

Tyning, T. 1987. In the path of progress. Sanctuary (Lincoln) 26(9):3–5.

Uhler, F. M., C. Cottam, and T. E. Clarke. 1939. Food of snakes of the George Washington National Forest, Virginia. Trans. 4th North American Wildl. Conf., pp. 605–622.

Underwood, G. 1967. A contribution to the classification of snakes. Publ. British Mus. Natur. Hist. (653):1–179.

Vaeth, R. H. 1984. A note on courtship and copulatory behavior in *Micrurus fulvius*. Bull. Chicago Herpetol. Soc. 18:86–88.

Van Bourgondien, T. M., and R. C. Bothner. 1969. A comparative study of the arterial systems of some New World Crotalinae (Reptilia: Ophidia). Amer. Midl. Natur. 81:107–147.

Van Denburgh, J. 1895. Description of a new rattlesnake (*Crotalus pricei*) from Arizona. Proc. California Acad. Sci. Ser. 2. 5:856–857.

Van Devender, T. R., and J. I. Mead. 1978. Early Holocene and late Pleistocene amphibians and reptiles in Sonoran Desert packrat middens. Copeia 1978:464–475.

Van Devender, T. R., A. M. Phillips III, and J. I. Mead. 1977. Late Pleistocene reptiles and small mammals from the Lower Grand Canyon of Arizona. Southwest. Natur. 22:49–66.

Van Devender, T. R., A. M. Rea, and W. E. Hall. 1991. Faunal analysis of late Quaternary vertebrates from Organ Pipe Cactus National Monument, southwestern Arizona. Southwest. Natur. 36:94–106.

Van Devender, T. R., A. M. Rea, and M. L. Smith. 1985. The Sangamon interglacial vertebrate fauna from Rancho la Brisca, Sonora, Mexico. Trans. San Diego Soc. Natur. Hist. 21:23–55.

Vandeventer, T. L. 1977. Report of a double pre-molt period in a mottled rock rattlesnake. Bull. Chicago Herpetol. Soc. 12:60.

Van Riper, W. 1954. Measuring the speed of a rattlesnake's strike. Anim. Kingdom 57:50–53.

———. 1955. How a rattlesnake strikes. Natur. Hist. 64:308–311.

Verkerk, J. W. 1986. Verzorging en kweek van de dwergratelslang, *Sistrurus miliarius barbouri*. Lacerta 45:15–20.

———. 1987. Enkele aanvullende opmerkingen over het gedrag van de dwergratelslang (*Sistrurus miliarius barbouri*). Lacerta 45:142–143.

Vermersch, T. G., and R. E. Kuntz. 1986. Snakes of south central Texas. Eakin Press, Austin. 137 pp.

Vetas, B. 1951. Temperatures of entrance and emergence. Herpetologica 7:15–20.

Vial, J. L., T. L. Berger, and W. T. McWilliams, Jr. 1977. Quantitative demography of copperheads, *Agkistrodon contortrix* (Serpentes: Viperidae). Res. Popul. Ecol. Kyoto Univ. 18:223–234.

Vigle, G. O., and H. Heatwole. 1978. A bibliography of the Hydrophiidae. Smithson. Herpetol. Inform. Serv. (41):1–20.

Villeneuve, M., and D. Rivard. 1986. Massasauga rattlesnake management at Georgian Bay Islands National Park, Ontario, Canada. Newslett. Bull. Can. Soc. Environ. Biol. 42(4):27–29.

Vincent, J. W. 1982a. Color pattern variation in *Crotalus lepidus lepidus* (Viperidae) in southwestern Texas. Southwest. Natur. 27:263–272.

———. 1982b. Phenotypic variation in *Crotalus lepidus lepidus* (Kennicott). J. Herpetol. 16:189–191.

Visser, J. 1967. Color varieties, brood size, and food of South African *Pelamis platurus* (Ophidia: Hydrophiidae). Copeia 1967:219.

Vitt, L. J. 1974. Body temperatures of high latitude reptiles. Copeia 1974:255–256.

Vitt, L. J., and A. C. Hulse. 1973. Observations on feeding habits and tail display of the Sonoran coral snake, *Micruroides euryxanthus*. Herpetologica 29:302–304.

Vitt, L. J., and R. D. Ohmart. 1978. Herpetofauna of the lower Colorado River: Davis Dam to the Mexican border. Proc. West. Found. Vert. Zool. 2:33–72.

Vogler, J. 1973. Teach respect—not fear: getting along with rattlesnakes. Wyoming Wild Life 35(11):22–25.

Vogt, R. C. 1981. Natural history of amphibians and reptiles of Wisconsin. Milwaukee Public Mus., Milwaukee, Wisconsin. 205 pp.

Vorhies, C. T. 1929. Feeding of *Micrurus euryxanthus*, the Sonoran coral snake. Bull. Antivenin Inst. Amer. 2:98.

———. 1948. Food items of rattlesnakes. Copeia 1948:302–303.

Vorhies, C. T., and W. P. Taylor. 1940. Life history and ecology of the white-throated wood rat, *Neotoma albigula albigula*, in relation to grazing in Arizona. Univ. Arizona Coll. Agric. Tech. Bull. (86):453–529.

Voris, H. K. 1977. A phylogeny of the sea snakes (Hydrophiidae). Fieldiana: Zool. 70:79–166.

———. 1983. *Pelamis platurus* (Culebra del Mar, Pelagic sea snake). *In* Janzen, D. H. (ed.), Costa Rican Natural History, pp. 411–412. Univ. Chicago Press, Chicago, Illinois.

Voris, H. K., H. Voris, and W. B. Jeffries. 1983. Sea snakes: mark, release, recapture. Field Mus. Natur. Hist. Bull. 54(9):5–10.

Wagner, E., R. Smith, and F. Slavens. 1976. Breeding the Gila monster *Heloderma suspectum* in captivity. Intern. Zoo Yearb. 17:74–78.

Wagner, F. W., and J. M. Prescott. 1966. A comparative study of proteolytic activities in the venoms of some North American snakes. Comp. Biochem. Physiol. 17:191–201.

Wagner, R. T. 1962. Notes on the combat dance in *Crotalus adamanteus*. Bull. Philadelphia Herpetol. Soc. 10(1):7–8.

Walker, J. M. 1963. Amphibians and reptiles of Jackson Parish, Louisiana. Proc. Louisiana Acad. Sci. 26:91–101.

Wallace, R. L., and L. V. Diller. 1990. Feeding ecology of the rattlesnake, *Crotalus viridis oreganus*, in northern Idaho. J. Herpetol. 24:246–253.

Wallach, V. 1990. A record brood for the southern copperhead, *Agkistrodon c. contortrix* Linnaeus. Bull. Maryland Herpetol. Soc. 26:17–20.

Walley, H. D. 1963. The rattlesnake, *Crotalus horridus horridus*, in north-central Illinois. Herpetologica 19:216.

Webb, R. G. 1970. Reptiles of Oklahoma. Univ. Oklahoma Press, Norman. 370 pp.

Weinstein, S. A., P. J. Lafaye, and L. A. Smith. 1991. Observations on a vemon neutralizing fraction isolated from serum of the northern copperhead, *Agkistrodon contortrix mokasen*. Copeia 1991:777–786.

Weinstein, S. A., S. A. Minton, and C. E. Wilde. 1985. The distribution among ophidian venoms of a toxin isolated from the venom of the Mojave rattlesnake (*Crotalus scutulatus scutulatus*). Toxicon 23:825–844.

Weinstein, S. A., and L. A. Smith. 1990. Preliminary fractionation of tiger rattlesnake (*Crotalus tigris*) venom. Toxicon 28:1447–1455.

Weldon, P. J. 1988. Feeding responses of Pacific snappers (genus *Lutjanus*) to the yellow-bellied sea snake (*Pelamis platurus*). Zool. Sci. (Tokyo) 5:443–448.

Weldon, P. J., and D. B. Fagre. 1989. Responses by canids to scent gland secretions of the western diamondback rattlesnake (*Crotalus atrox*). J. Chem. Ecol. 15:1589–1604.

Weldon, P. J., H. W. Sampson, L. Wong, and H. A. Lloyd. 1991. Histology and biochemistry of the scent glands of the yellow-bellied sea snake (*Pelamis platurus:* Hydrophiidae). J. Herpetol. 25:367–370.

Werler, J. E. 1950. The poisonous snakes of Texas and the first aid treatment of their bites. Texas Game Fish Bull. 31:1–40.

———. 1951. Miscellaneous notes on the eggs and young of Texan and Mexican reptiles. Zoologica 36:37–48.

Werler, J. E., and D. M. Darling. 1950. A case of poisoning from the bite of a coral snake, *Micrurus f. tenere* Baird and Girard. Herpetologica 6:197–199.

Wetmore, A. 1965. The birds of the Republic of Panama, Part 1. Tinamidae (Timamous) to Rynchopidae (Skimmers). Smithsonian Misc. Coll. 150:1–483.

Wharton, C. H. 1960. Birth and behavior of a brood of cottonmouths, *Agkistrodon piscivorus piscivorus*, with notes on tail-luring. Herpetologica 16:125–129.

———. 1966. Reproduction and growth in the cottonmouths, *Agkistrodon piscivorus* Lacépède, of Cedar Keys, Florida. Copeia 1966:149–161.

———. 1969. The cottonmouth moccasin on Sea Horse Key, Florida. Bull. Florida St. Mus. Biol. Sci. 14:227–272.

Wheeler, G. C. 1947. The amphibians and reptiles of North Dakota. Amer. Midl. Natur. 38:162–190.

Wheeler, G. C., and J. Wheeler. 1966. The amphibians and reptiles of North Dakota. Univ. North Dakota Press, Grand Forks. 104 pp.

White, A. M. 1979. An unusually large brood of northern copperheads (*Agkistrodon contortrix mokeson*) from Ohio. Ohio J. Sci. 79:78.

White, F. N., and R. C. Lasiewski. 1971. Rattlesnake denning: theoretical considerations on winter temperatures. J. Theor. Biol. 30:553–557.

Whitt, A. L., Jr. 1970. Some mechanisms with which *Crotalus horridus horridus* responds to stimuli. Trans. Kentucky Acad. Sci. 31:45–48.

Wickler, W. 1968. Mimicry in plants and animals. McGraw-Hill Book Co., Inc., New York. 255 pp.

Wiley, G. O. 1929. Notes on the Texas rattlesnake in captivity with special reference to the birth of a litter of young. Bull. Antivenin Inst. Amer. 3:8–14.

Wilkinson, J. A., J. L. Glenn, R. C. Straight, and J. W. Sites, Jr. 1991. Distribution and genetic variation in vemon A and B populations of the Mojave rattlesnake (*Crotalus scutulatus scutulatus*) in Arizona. Herpetologica 47:54–68.

Williamson, M. A. 1971. An instance of cannibalism in *Crotalus lepidus* (Serpentes: Crotalidae). Herp. Review 3:18.

Wilson, A. B., and S. A. Minton. 1983. *Agkistrodon piscivorus leucostoma* (Western Cottonmouth). USA: Indiana. Herp. Review 14:84.

Wilson, S. C. 1954. Snake fight. Texas Game Fish 12(5):16–17.

Wingert, W. A., T. R. Pattabiraman, D. Powers, and F. E. Russell. 1981. Effect of a rattlesnake venom (*Crotalus viridis helleri*) on bone marrow. Toxicon 19:181–183.

Wittner, D. 1978. A discussion of venomous snakes of North America. HERP: Bull. New York Herpetol. Soc. 14:12–17.

Wolff, N. O., and T. S. Githens. 1939a. Record venom extraction from water moccasin. Copeia 1939:52.

———. 1939b. Yield and toxicity of venom from snakes extracted over a period of two years. Copeia 1939:234.

Wood, J. T. 1954. The distribution of poisonous snakes in Virginia. Virginia J. Sci. 5:152–167.

Woodbury, A. M. 1929. A new rattlesnake from Utah. Bull. Univ. Utah (Biol. Ser.) 20(6):104.

———. 1942. Status of the name *Crotalus concolor*. Copeia 1942:258.

———. 1951. Symposium: a snake den in Toole County, Utah. Introduction—a ten year study. Herpetologica 7:4–14.

Woodbury, A. M., and R. M. Hansen. 1950. A snake den in Tintic Mountains, Utah. Herpetologica 6:66–69.

Woodbury, A. M., and R. Hardy. 1947. The Mojave rattlesnake in Utah. Copeia 1947:66.

Woodin, W. W. 1953. Notes on some reptiles from the Huachuca area of southeastern Arizona. Bull. Chicago Acad. Sci. 9:285–296.

Woodson, W. D. 1947. Toxicity of *Heloderma* venom. Herpetologica 4:31–33.

———. 1949. Summary of *Heloderma*'s food habits. Herpetologica 5:91–92.

Wright, A. H., and A. A. Wright. 1957. Handbook of snakes of the United States and Canada. Comstock Publ. Assoc., Ithaca, New York, Vols. I, II. 1105 pp.

———. 1962. Handbook of snakes of the United States and Canada. Vol. III. Bibliography. Edwards Brothers, Inc. Ann Arbor, Michigan. 179 pp.

Wright, B. A. 1941. Habit and habitat studies of the massasauga rattlesnake (*Sistrurus catenatus catenatus* Raf.) in northeastern Illinois. Amer. Midl. Natur. 25:659–672.

Wright, R. A. S. 1987. Natural history observations on venomous snakes near the Peaks of Otter, Bedford County, Virginia. Catesbeiana 7(2):2–9.

Yatkola, D. A. 1976. Fossil *Heloderma* (Reptilia, Helodermatidae). Occ. Pap. Mus. Natur. Hist. Univ. Kansas (51):1–14.

Yerger, R. W. 1953. Yellow bullhead preyed upon by cottonmouth moccasin. Copeia 1953:115.

Young, N. 1940. Snakebite: treatment and nursing care. Amer. J. Nursing 40:657–660.

Young, R. A., and D. M. Miller. 1980. Notes on the natural history of the Grand Canyon rattlesnake, *Crotalus viridis abyssus* Klauber. Bull. Chicago Herpetol. Soc. 15:1–5.

Young, R. A., D. M. Miller, and D. C. Ochsner. 1980. The Grand Canyon rattlesnake (*Crotalus viridis abyssus*): comparison of venom protein profiles with other *viridis* subspecies. Comp. Biochem. Physiol. 66B:601–603.

Zann, L. P., R. J. Cuffey, and C. Kropach. 1975. Fouling organisms and parasites associated with the skin of sea snakes. *In* Dunson, W. A. (ed.), The biology of sea snakes, pp. 251–265. Univ. Park Press, Baltimore, Meryland.

Zegel, J. C. 1975. Notes on collecting and breeding the eastern coral snake, *Micrurus fulvius fulvius*. Bull. Southwest. Herpetol. Soc. 1:9–10.

Zimmermann, A. A., and C. H. Pope. 1948. Development and growth of the rattle of rattlesnakes. Fieldiana: Zool. 32:355–413.

Zimmerman, E. G., and C. W. Kilpatrick. 1973. Karyology of North American crotaline snakes (family Viperidae) of the genera *Agkistrodon*, *Sistrurus*, and *Crotalus*. Can. J. Genet. Cytol. 15:389–395.

Glossary of scientific names

The phonetic spellings below are not to be taken as gospel, but represent a consensus of a poll of herpetologists from various parts of the country. Phonetic notation generally follows Webster III and includes the following sounds: \a\ as in cap; \ā\ as in rate; \ä\ as in calm, box; \ai\ as in air; \au̇\ as in plow; \e\ as in get; \ē\ as in me; \i\ as in big; \ī\ as in bite; \ō\ as in hope; \ȯ\ as in saw, ought, fall; \ȯi\ as in boy, coin; \u̇\ as in pull, book; \ü\ as in rule, too; and the schwa \ə\ which sounds like the a in alone, the e in system, the i in easily, the o in gallop, and the u in circus. The reader is reminded of E. B. White's advice: If you don't know how to pronounce a word, say it loud.

abyssus (ə-bis'əs): bottomless (Grand Canyon)
adamanteus (ad"ə-man'tē-əs): diamondlike (pattern)
Agkistrodon (ag-kis'trō-dän"): hooked tooth
atricaudatus (a"tri-cȯ-dā'təs): black-tailed)
atrox (a'träks): savage, fierce, cruel
barbouri (bär'bər-ī): named for Thomas Barbour
catenatus (kat"ə-nā'təs): chainlike (pattern)
cerastes (sə-ras'tēz): horned
cerberus (sər'bər-əs, sər-bair'əs): black watchdog

cercobombus (sər"cō-bäm'bəs): tail buzzer

cinctum (sink'təm): banded, girdled

conanti (kō'nənt-i): named for Roger Conant

concolor (kän'kəl"ər): similarly colored, uniform

contortrix (kän-tōr'triks): twister

Crotalus (krō'tə-ləs): rattle (tail)

edwardsi (ed'wərds-ī): named for L. A. Edwards

euryxanthus (yər"i-zan'thəs): broad yellow (bands)

fulvius (fül'vē-əs): reddish-yellow, orange (bands)

helleri (hel'ər-ī): named for Edmond Heller

Heloderma (hē"lō-dər'mə): nail (studded) skin

horridus (hòr'ə-dəs): horrid, dreadful

klauberi (klaù'bər-ī): named for Laurence M. Klauber

laterorepens (lat"ə-rō-rē'pens, lat"ə-ro-re'pens,): side creeping

laticinctus (lat-i-sink'təs): broad band

lepidus (lep'i-dəs): pretty, attractive; scaly

leucostoma (lü"cō-stō'mə): white mouth

lutosus (lü-tō'səs): muddy

Micruroides (mī"krür-öi'des): like *Micrurus*

Micrurus (mī-krür'əs): small tail

miliarius (mil"ē-air'ē-əs): millet-like (pattern)

mitchelli (mich'əl-ī): named for S. Weir Mitchell

mokasen (mäk'ə-sən): mocassin

molossus (mə-läs'əs, mō-ləs'əs): after the Molossian wolfdog of
   antiquity

nuntius (nùn'tē-əs): messenger

obscurus (äb-skyùr'əs): hidden, faded

oreganus (òr"ə-gän'əs): from Oregon

Pelamis (pe-lam'əs, pe-lām'is, pel'ə-məs): fishlike

phaeogaster (fā'ō-gas"tər, fē'ō-gas"tər): dark belly

pictigaster (pik'ti-gas"tər): painted belly

piscivorus (pə-siv'ə-rəs, pis"ki–vòr'əs, pī"si–vòr'əs) fish eating

platurus (plə-tú'rəs): flat tail

prīceī (pris'i): named for W. W. Price

pyrrhus (pir'əs): flame-colored, reddish

ruber (rü'bər): red

scutulatus (skyü"tyü-lā'təs, skü"chü-lā'təs): small shield, diamondlike pattern

Sistrurus (sis-trúr'əs): rattle tail

stephensi (stē'vənz-ī, ste'fənz-ī): named for Frank Stephens

streckeri (strek'ər-ī): named for John K. Strecker, Jr.

suspectum (səs-pek'təm): suspected, distrusted

tener (ten'ər): tender, delicate (appearance)

tergeminus (tər-jem'i-nəs): triple, threefold (spotted pattern)

tigris (tī'grəs): tiger (pattern)

viridis (vir'ə-dəs): green

willardi (wil'ərd-i): named for Frank C. Willard

Index

Page numbers in bold indicate descriptions of species or subspecies.